Methods for Preparation of Media, Supplements, and Substrata for Serum-Free Animal Cell Culture

Cell Culture Methods for Molecular and Cell Biology

David W. Barnes, David A. Sirbasku, and Gordon H. Sato, *Editors*

Volume 1: Methods for Preparation of Media, Supplements, and Substrata for Serum-Free Animal Cell Culture

Volume 2: Methods for Serum-Free Culture of Cells of the Endocrine System

Volume 3: Methods for Serum-Free Culture of Epithelial and Fibroblastic Cells

Volume 4: Methods for Serum-Free Culture of Neuronal and Lymphoid Cells

Methods for Preparation of Media, Supplements, and Substrata for Serum-Free Animal Cell Culture

Editors

David W. Barnes

Department of Biological Sciences
University of Pittsburgh
Pittsburgh, Pennsylvania

David A. Sirbasku

Department of Biochemistry
and Molecular Biology
University of Texas Medical School
Houston, Texas

Gordon H. Sato

W. Alton Jones Cell Science Center
Lake Placid, New York

Alan R. Liss, Inc., New York

Library of Congress Cataloging in Publication Data

Main entry under title:

Methods for preparation of media, supplements, and
    substrata for serum-free animal cell culture.
    (Cell culture methods for molecular and cell biology;
v. 1)
    Includes index.
    1. Culture media (Biology) 2. Cell culture.
I. Barnes, David W. (David William), 1949–
II. Sirbasku, David A. (David Andrew), 1941–
III. Sato, Gordon. IV. Title: Serum-free animal cell
culture. V. Series.
QH585.M46   1984   591'.07'24   84-7203
ISBN 0-8451-3800-6

To Emily, Donna, and Miyo

Contents

METHODS FOR PREPARATION OF BASAL NUTRIENT MEDIA

METHODS FOR PREPARATION OF MITOGENIC PEPTIDES

METHODS FOR PREPARATION OF SUBSTRATA

Contributors

Steven K. Akiyama, Membrane Biochemistry Section, Laboratory of Molecular Biology, National Cancer Institute, National Institutes of Health, Bethesda, MD 20205 [215]

Mario A. Anzano, Laboratory of Chemoprevention, National Cancer Institute, National Institutes of Health, Bethesda, MD 20205 [181]

Richard K. Assoian, Laboratory of Chemoprevention, National Cancer Institute, National Institutes of Health, Bethesda, MD 20205 [181]

David W. Barnes, Department of Biological Sciences, University of Pittsburgh, Pittsburgh, PA 15260 [xxi,245]

Ralph A. Bradshaw, Department of Biological Chemistry, California College of Medicine, University of California at Irvine, Irvine, CA 92717 [139]

Albert E. Chung, Department of Biological Sciences, University of Pittsburgh, Pittsburgh, PA 15260 [321]

Paula Fehnel, Laboratory of Tumor Immunology and Biology, National Cancer Institute, National Institutes of Health, Bethesda, MD 20205 [295]

Charles A. Frolik, Laboratory of Chemoprevention, National Cancer Institute, National Institutes of Health, Bethesda, MD 20205 [181]

Denis Gospodarowicz, Cancer Research Institute and the Departments of Medicine and Ophthalmology, University of California Medical Center, San Francisco, CA 94143 [69,275]

Lawrence A. Greenstein, Endocrinology Section, Metabolism Branch, National Cancer Institute, National Institutes of Health, Bethesda, MD 20205 [111]

Richard G. Ham, Department of Molecular, Cellular, and Developmental Biology, University of Colorado, Boulder, CO 80309 [3]

Robert A. Harper, Department of Dermatology, Temple University School of Medicine, Philadelphia, PA 19140 [147]

John R. Hassell, Laboratory of Developmental Biology and Anomalies, National Institute of Dental Research, National Institutes of Health, Bethesda, MD 20205 [231]

A. Tyl Hewitt, Laboratory of Developmental Biology and Anomalies, National Institute of Dental Research, National Institutes of Health, Bethesda, MD 20205 [239]

Michael Klagsbrun, Department of Surgical Research, Children's Hospital Medical Center, Boston, MA 02115 [159]

Hynda K. Kleinman, Laboratory of Developmental Biology and Anomalies, National Institute of Dental Research, National Institutes of Health, Bethesda, MD 20205 [231]

Steven R. Ledbetter, Laboratory of Developmental Biology and Anomalies, National Institute of Dental Research, National Institutes of Health, Bethesda, MD 20205 [231]

Lilly Lee, Endocrinology Section, Metabolism Branch, National Cancer Institute, National Institutes of Health, Bethesda, MD 20205 [111]

Lance A. Liotta, Laboratory of Pathology, National Cancer Institute, National Institutes of Health, Bethesda, MD 20205 [295]

Thomas Maciag, Department of Cell Biology, Revlon Biotechnology Research Center, Springfield, VA 22151 [195]

Shing Mai, Department of Biological Sciences, University of Pittsburgh, Pittsburgh, PA 15260 [321]

George R. Martin, Laboratory of Developmental Biology and Anomalies, National Institute of Dental Research, National Institutes of Health, Bethesda, MD 20205 [231,239]

Wallace L. McKeehan, W. Alton Jones Cell Science Center, Lake Placid, NY 12946 [209]

Alan C. Moses, Endocrinology Diabetes Unit, Beth Israel Hospital, Boston, MA 02215 [111]

S. Peter Nissley, Endocrinology Section, Metabolism Branch, National Cancer Institute, National Institutes of Health, Bethesda, MD 20205 [111]

Mounanandham Panneerselvam, Laboratory of Tumor Immunology and Biology, National Cancer Institute, National Institutes of Health, Bethesda, MD 20205 [295]

Elaine W. Raines, Department of Pathology, University of Washington, Seattle, WA 98146 [89]

Matthew M. Rechler, Section on Biochemistry of Cell Regulation, Laboratory of Biochemical Pharmacology, NIADDK, National Institutes of Health, Bethesda, MD 20205 [111]

Anita B. Roberts, Laboratory of Chemoprevention, National Cancer Institute, National Institutes of Health, Bethesda, MD 20205 [181]

Russell Ross, Departments of Pathology and Biochemistry, University of Washington, Seattle, WA 98146 [89]

Jeffrey S. Rubin, Department of Biological Chemistry, Washington University School of Medicine, St. Louis, MO 63110 [139]

Atul Sahai, Laboratory of Tumor Immunology and Biology, National Cancer Institute, National Institutes of Health, Bethesda, MD 20205 [295]

David S. Salomon, Laboratory of Tumor Immunology and Biology, National Cancer Institute, National Institutes of Health, Bethesda, MD 20205 [295]

Gordon H. Sato, W. Alton Jones Cell Science Center, Lake Placid, NY 12946 [xxi]

C. Richard Savage, Jr., Department of Biochemistry, Temple University School of Medicine, Philadelphia, PA 19140 **[147]**

Yuen W. Shing, Department of Surgical Research, Children's Hospital Medical Center and Harvard Medical School, Boston, MA 02115 **[159]**

Patricia A. Short, Endocrinology Section, Metabolism Branch, National Cancer Institute, National Institutes of Health, Bethesda, MD 20205 **[111]**

Janet Silnutzer, Department of Biological Sciences, University of Pittsburgh, Pittsburgh, PA 15260 **[245]**

David A. Sirbasku, Department of Biochemistry and Molecular Biology, University of Texas Medical School, Houston, TX 77225 **[xxi]**

Michael B. Sporn, Laboratory of Chemoprevention, National Cancer Institute, National Institutes of Health, Bethesda, MD 20205 **[181]**

K.S. Stenn, Department of Dermatology, Yale University School of Medicine, New Haven, CT 06510 **[269]**

Victor P. Terranova, Laboratory of Developmental Biology, National Institute of Dental Research, National Institutes of Health, Bethesda, MD 20205 **[295]**

Hugh H. Varner, Laboratory of Developmental Biology and Anomalies, National Institute of Dental Research, National Institutes of Health, Bethesda, MD 20205 **[239]**

Charity Waymouth, The Jackson Laboratory, Bar Harbor, ME 04609 **[23]**

Robert Weinstein, Departments of Pathology and Medicine, Charles A. Dana Research Institute, Harvard-Thorndike Laboratory, Harvard Medical School and Beth Israel Hospital, Boston, MA 02215 **[195]**

Kenneth M. Yamada, Membrane Biochemistry Section, Laboratory of Molecular Biology, National Cancer Institute, National Institutes of Health, Bethesda, MD 20205 **[215]**

Yvonne W.-H. Yang, Section on Biochemistry of Cell Regulation, Laboratory of Biochemical Pharmacology, NIADDK, National Institutes of Health, Bethesda, MD 20205 **[111]**

Contents of Volumes 2, 3, and 4

Volume 3:   Methods for Serum-Free Culture of Epithelial and Fibroblastic Cells

SERUM-FREE CULTURE OF EPITHELIAL CELLS

SERUM-FREE CULTURE OF FIBROBLASTIC CELLS

SERUM-FREE CULTURE OF LYMPHOID CELLS

Preface

For several of the past decades, experimental approaches to the study of cellular physiology were limited to either studies in the whole animal, or short-term studies with isolated cells or slices from various tissues. On the one hand, animal studies have provided our basic understanding of organ system physiology, but on the other hand this approach was limited because the in vivo measurements represented the net effect of many homeostatic changes that occurred during the experiments. In short, what was needed was a new experimental approach to cellular physiology in which isolated, functionally differentiated cells could be maintained in culture under conditions that allowed direct manipulations of the environment and measurement of the resulting changes in the function of a single cell type. The maintenance of cells in vitro has proven to be a major scientific undertaking of great importance, and it is now universally accepted that in vitro methods are indispensible to the study of cellular physiology.

Cell culture methods have been evolving rapidly. The first researchers in the field recognized that to establish even short-term cultures of animal cells in vitro, a basic requirement was the design of chemically defined synthetic media that contained the necessary mixtures of nutrients, vitamins, ions, and other essential metabolites, all of which were then to be maintained at the proper pH for life functions to be expressed. Pioneers such as Ted Puck, Charity Waymouth, Harry Eagle, Richard Ham, and others, led the efforts to provide all the workers of this field with formulations useful for specific cell types or purposes, and other formulations of synthetic media that were useful with a wide variety of cell types.

As these types of studies progressed, attention began to turn toward the problem of the most undefined component used in tissue culture media, namely the animal serum or plasma supplement. For a variety of cells in culture, it was widely accepted that the addition of this fluid was essential for growth. Serum obviously was growth-stimulatory for some cell types; what was less obvious was that the serum supplement was toxic to other cell types or inhibitory to some physiological responses of interest. In nearly all cases, serum is the source of both growth-promoting and growth-inhibiting substances and, taken as a whole, serum-supplemented media greatly favor proliferation of mesenchymal-origin cells (i.e., fibroblasts) over growth of other cell types. Problems of apparent cellular "dedifferentiation" or fibroblast overgrowth were encountered by many researchers, using conventional culture techniques, and this led to a widely held view that cells which expressed differentiated functions could not be maintained in culture. This position was in part proven incorrect by workers demonstrating that functionally differentiated tumor cell types could be maintained in culture and that these cells continued expression of tissue-specific functions. But even with these initial successes, the problems remained of how to establish highly differentiated normal cells in culture selectively and how to support their proliferation in the absence of the undefined serum supplement.

In these four volumes, a different and unifying approach to cell culture is described. The concepts basic to this approach evolved from ideas contributed by investigators in areas of physiology and endocrinology, as well as those in fields more traditionally associated with cell culture methodology, and spring from attempts to replace serum in culture medium with purified components chosen to serve for cells the role usually provided by serum components. The methods are based on the experimentally established concepts that the function of the serum supplement is to supply hormones, growth factors, cell attachment proteins, extracellular enzymes and enzyme inhibitors, metal ion transport proteins and other binding proteins, lipoproteins, essential fatty acids, and other specific nutritional requirements for growth and tissue-specific expression by cells in culture. By supplementing media with appropriate mixtures of these defined substances, various differentiated cell types can be maintained in culture and be shown to express tissue-specific functions not previously identified in vitro.

In an attempt to satisfy the needs of both the novice and seasoned investigators wishing to utilize serum-free culture methods, we have organized the series beginning with this volume, which contains a description of the general methods of preparation of many of the common media components, followed by three other volumes, each dealing with detailed formulations for specific cell types. In this first volume, methods are presented for establishing the basal nutrient requirements of various cells and preparation of serum-free media, as well as complete methods of preparation of several of the most common polypeptide growth factors used in defined media. Also, in this first volume, several important methods are described for the preparation of cell attachment proteins and extracellular matrix, which have proven to be essential components of many formulations of defined media. Careful evaluation of the properties and functions of the components described in Volume 1 will greatly aid the reader in the process of selecting which are likely to be important additions to the serum-free media desired. Volume 1 is organized to act as a companion to any of the other three, allowing the researcher to select one or more additional specialized volume of interest.

Beyond this general volume, the other three incorporate useful examples of how to apply these methods to individual problems. For example, in the second volume, entitled "Methods For Serum-Free Culture of Cells of the Endocrine System," methods are described for specific formulations effective with adrenal, pituitary, testicular, ovary, prostate, and mammary cells. The section on mammary cell culture is especially useful, since it is one of the most complete available on serum-free methods for cells from this organ.

Volume 3 of this series, "Methods for Serum-Free Culture of Epithelial and Fibroblastic Cells," describes much of what is currently known about the growth of these important cell types in a defined environment. This volume addresses the issues of culturing both normal and malignant epithelial cells, and the conditions that must be met to maximize expression of tissue-specific differentiated functions. Methods also described in Volume 3 deal directly with the uses of defined media to characterize the phenotypic and genotypic changes in untransformed and transformed fibroblasts. This field of research has received new interest with the discovery of oncogenes using in vitro transformation systems, and formulations of serum-free defined media for normal and transformed cells may become even more important with applications in these new directions.

Volume 4, the final of this series, is entitled "Methods for Serum-Free Culture of Neuronal and Lymphoid Cells." The first part of the volume focuses on applications to neuronal-origin cells that have been difficult to maintain in conventional serum-containing media. In addition, the preparation of neurotrophic factors is discussed in detail. The second part of Volume 4 addresses the methods of preparation of several of the important lymphokines, details growth conditions for lymphocytes, and also describes approaches to the growth of hybridomas and production of monoclonal antibody in serum-free, defined media. The latter method promises wide applications and is a broadly useful technical advance.

It has been our goal to provide a collection of culture methods that are detailed and useful at the laboratory bench. For this reason, less emphasis has been placed in this series on the conceptual importance and implications of the various aspects of the serum-free systems described. More thorough treatments of these aspects can be found in several recent publications: "Growth of Cells in Hormonally Defined Media" (Cold Spring Harbor Press, 1982) and "The Use of Serum-Free and Hormonally Defined Media" (J. Mather, ed., Plenum Press, 1984). The editors thank the contributors for sharing their detailed methods and for their helpful comments which have greatly improved these volumes. We hope that the information provided here will serve as a catalyst for future advances and that these volumes will inspire new applications moving well beyond our present-day technology.

David W. Barnes
David A. Sirbasku
Gordon H. Sato

Methods for Preparation of Basal Nutrient Media

Methods for Preparation of Media, Supplements, and Substrata
for Serum-Free Animal Cell Culture, pages 3-21
© 1984 Alan R. Liss, Inc., 150 Fifth Avenue, New York, NY 10011

1
Formulation of Basal Nutrient Media

Richard G. Ham

The importance of using a basal nutrient medium that has been optimized for the specific cell type that is being cultured has been greatly increased by two recent trends: 1) the replacement of serum and other grossly undefined supplements with defined or semidefined supplements; and 2) increased emphasis on growth of epithelial and other cell types that do not grow well in traditional media with any amount of supplementation. Since the term "basal nutrient medium" has been used in a number of different ways, it is desirable to begin by examining what is meant by it.

A complete cell culture medium can be considered, at least in theory, to be composed of two distinct parts: 1) a basal nutrient medium that satisfies all cellular requirements for nutrients; and 2) a set of supplements that satisfy other types of cellular growth requirements and make possible growth of the cells in the basal nutrient medium. From a cell culture perspective, a nutrient is defined as a chemical substance that enters a cell and is used as a structural component, as a substrate for biosynthesis or energy metabolism, or in a catalytic (cofactor) role in such metabolism [Bettger and Ham, 1982]. Anything else needed for cellular proliferation is normally classified as a supplement, including all undefined additives such as serum and pituitary extract.

The term "nutrient" is usually assumed to exclude substances whose roles are primarily regulatory, such as hormones and growth factors, despite the fact that such substances often enter cells and undergo catabolism. Attachment factors, extracellular matrix, and carrier proteins are also generally not classified as nutrients, although the carrier proteins often promote cellular multipli-

Department of Molecular, Cellular, and Developmental Biology, University of Colorado, Boulder, Colorado 80309

cation by making nutrients such as iron and lipids more readily available to the cells. On the other hand, essential inorganic ions are generally classified as nutrients, although they also play major roles that fall outside the strict definition of a nutrient, such as maintenance of membrane potential and osmotic balance. The specific nutrient requirements of cultured cells have been reviewed in detail elsewhere [Rizzino et al., 1979; Ham, 1981; Bettger and Ham, 1982] and are considered in this chapter only in the context of developing optimized basal nutrient media for specific cell types.

One of the major problems involved in trying to arrive at an exact definition of the term "basal nutrient medium" stems from the fact that the supplements that were used in classical culture systems also satisfied nutrient requirements in many cases. Thus, for example, cells grown in the minimal essential medium (MEM) developed by Eagle [1959] were dependent on dialyzed serum (or contaminants in the chemicals used to prepare the medium) for biotin, vitamin B_{12}, iron, selenium, and other trace nutrients [Ham, 1981]. Biotin and vitamin B_{12} have since been incorporated into many basal nutrient media. Nevertheless, the cellular requirement for biotin continues to be rediscovered with distressing regularity as cell types previously grown in MEM plus dialyzed serum (which contains bound biotin) are switched to serum-free systems. Iron is functionally absent from many basal nutrient media because of the extreme insolubility of ferric hydroxide, and it is frequently supplied as a supplement in the form of transferrin [Barnes and Sato, 1980a,b; Walthall and Ham, 1981; Bettger et al., 1981]. Selenium was found to be an essential cellular nutrient only a few years ago [McKeehan et al., 1976; Guilbert and Iscove, 1976] and is still frequently classified as a supplement, since many widely used basal nutrient media do not contain it. Certain other recently discovered cellular growth requirements whose roles appear to be primarily as nutrients, such as phosphoethanolamine [Kano-Sueoka et al., 1979] and ethanolamine [Kano-Sueoka and Errick, 1981], also continue to be added separately as supplements, rather than being incorporated into basal nutrient media in most laboratories, including that of the author [Tsao et al., 1982; Hammond et al., 1983].

In theory, the term "basal nutrient medium" should refer to an aqueous solution of nutrients and buffering agents that satisfies all of the nutrient and physiologic requirements for growth of the cell type in question in the presence of appropriate hormones, growth factors, attachment factors, etc. However, the practical reality is that nutrients whose role in promoting cellular multiplication became evident only when defined supplements began to be used, and nutrients that are difficult to work with in aqueous solution often continue to be added separately and regarded as supplements, rather than as part of the basal nutrient medium. Thus, in pragmatic terms, the basal nutrient medium must be regarded simply as the nutrient solution that is used as a starting point for the addition of supplements (including supplemental nutrients) to generate a complete growth medium.

The sections that follow discuss first the history and advantages of optimized basal nutrient media, and then practical procedures for developing such media.

HISTORY AND ADVANTAGES OF OPTIMIZED BASAL NUTRIENT MEDIA
Early Optimized Media

The recent attention that has been given to optimized basal nutrient media makes it easy to assume that optimization is a new concept. However, that is not the case. What is new is the realization of how greatly qualitative and quantitative growth requirements differ from one cell type to another. Optimization for the cell type being studied has always been an important aspect of the development of synthetic culture media (those composed of mixtures of purified chemicals rather than of complex natural materials such as serum and embryo extract). Thus, for example, the early history of medium 199, which is the oldest synthetic medium still in widespread use, is one of repeated qualitative and quantitative adjustments seeking to increase the survival time of chicken embryo heart fibroblasts in the complete absence of undefined supplements [Morgan et al., 1950]. Similarly, the history of Eagle's MEM (and of its predecessor, Eagle's basal medium) is one of precise optimization in order to obtain the best possible growth of mouse L and HeLa cells with a small amount of dialyzed serum as the only undefined supplement [Eagle, 1955, 1959]. The resulting medium, MEM, contained only those components whose requirements could be demonstrated, with each adjusted to its experimentally determined optimum concentration. This process of careful optimization is undoubtedly one of the major reasons why MEM has continued to be so widely used for so many years.

Unfortunately, the period of creative medium development during the 1950s degenerated in the 1960s to a process of circular reasoning that led most investigators to abandon medium development studies. The cell lines used in the pioneering studies of the 1950s had nearly always undergone extensive adaptation to grow in "natural" media (typically composed of mixtures of serum, embryo extract, and saline, with few, if any, added nutrients). Development of a synthetic medium was generally a two-step process, consisting first of an attempt to imitate the nutrient composition of a natural medium, followed by optimization of the nutrient composition of the preliminary synthetic medium to obtain the best growth of an already adapted cell line.

The new media developed in this manner were then used to establish new cell lines, again by a process of adaptation. When the nutrient requirements of these newly adapted cell lines were found to be similar to those of the original adapted lines, many investigators assumed that all cell types had rather similar nutrient requirements and turned their attention to other areas (see Ham [1974] and Ham [1981] for a more extended discussion of this problem).

Defined and Undefined Supplements

It was evident by the early 1960s that not all cellular growth requirements were the same. However, it was usually assumed that the differences were in requirements for supplements, rather than components of the basal nutrient medium. Supplementation with serum, and sometimes also with other complex natural materials such as embryo extract, was needed to establish new cultures. Normal cells typically required substantially larger amounts of undefined supplements than established cell lines. A variety of permanent cell lines could be grown in basal nutrient media without protein or lipid supplementation (see reviews by Higuchi [1973] and Katsuta and Takaoka [1973]). However, this had been achieved by an extensive process of cellular adaptation, which was generally assumed to involve "weaning" of the cells from dependence on supplements, with little or no direct involvement of the basal nutrient medium.

Sato [1975] proposed that the role of serum in promoting cellular multiplication consisted of supplying hormones, growth factors, attachment factors, etc., that were needed by the cells in relatively small amounts, and that serum could be replaced with appropriate mixtures of such factors. This approach proved to be highly successful for a wide variety of permanent cell lines whose initial serum requirements in conventional basal nutrient media were reasonably low [Hayashi and Sato, 1976; Bottenstein et al., 1979; Barnes and Sato 1980a,b]. However, with a few notable exceptions, such as kidney epithelial cells [Taub and Sato, 1980; Chung et al., 1982], direct replacement of serum with defined supplements has generally not been very successful for normal cells in conventional basal nutrient media.

Advantages of Optimized Basal Nutrient Media

One of the early indications of the importance of optimizing the basal nutrient medium specifically for the cell type to be grown in it came from studies of the nutrient requirements of Chinese hamster lines that required large amounts of serum for growth in existing basal nutrient media. Qualitative and quantitative optimization of the basal nutrient medium, together with gentle cell dispersal procedures to minimize cellular damage and trypsin carryover, progressively reduced the requirement for serum protein [Ham, 1962, 1963] and ultimately resulted in clonal growth in the total absence of serum or any added proteins, first in medium F12 [Ham, 1965], and more recently in an improved formulation, MCDB 302 [Hamilton and Ham, 1977]. Adaptive pressures were deliberately avoided in these studies by always maintaining stock cultures in media with optimal levels of serum supplementation. The final compositions of the protein-free defined media for these cells (F12 and MCDB 302) were quite different from those of earlier protein-free media for highly adapted cell lines (see Ham and McKeehan [1979] for a detailed comparison).

A similar optimization process has more recently resulted in monolayer growth of mouse neuroblastoma C1300 cells in basal nutrient medium MCDB 411 with no protein supplementation, and clonal growth with insulin as the only macromolecular supplement [Agy et al., 1981]. However, complete elimination of the requirement for supplements by optimization appears to be the exception rather than the rule, and has thus far not been achieved for any type of normal cell or "nontransformed" permanent cell line. What has been achieved for such cells by optimization, however, is a major reduction in the amount of undefined supplementation needed for good growth, which in turn has opened the way for the remaining requirements to be replaced with defined supplements, as described below.

Serum, even when extensively dialyzed, possesses a substantial ability to correct for quantitative problems (both deficiencies and excesses) in the composition of a basal nutrient medium. The mechanisms responsible for these effects have not been explored precisely, but they probably involve stimulation of transport to permit efficient utilization of nutrients that are present in suboptimal amounts and reversible binding to reduce the effective concentrations of nutrients that are inhibitory when present at excessive levels (see Ham and McKeehan [1978b] for an extended discussion of other possible roles of serum).

When the concentration of serum (or other undefined supplement) in a culture medium is reduced to a level where growth becomes suboptimal, it is frequently possible to improve growth by quantitative adjustment to bring all of the compoments of the basal nutrient medium within the optimum concentration ranges for growth with the reduced amount of supplement. In other cases, it may be necessary to add a nutrient (e.g., selenium or biotin) that has been previously supplied in adequate amounts in a bound form by the dialyzed supplement.

The cumulative effect of many such adjustments is to reduce the requirement for dialyzed supplements very substantially as the basal nutrient medium
is optimized. Examples of qualitatively and quantitatively optimized media that
have greatly reduced the requirement for serum or other supplements include
MCDB 104 for human diploid fibroblasts [McKeehan et al., 1977], MCDB
202 for chicken embryo fibroblasts [McKeehan and Ham, 1977], MCDB 402
for Swiss 3T3 cells [Shipley and Ham, 1981], MCDB 151 and MCDB 152 for
human epidermal keratinocytes [Peehl and Ham, 1980; Tsao et al., 1982],
and MCDB 131 for human microvascular endothelial cells [Knedler and Ham,
1983].

Neither optimization of the basal nutrient medium nor replacement of
serum with defined supplements works very well for growth of normal cells in
defined media when tried alone. However, a combination of these two
approaches has been highly successful. Optimization of the basal nutrient
medium for growth with minimal amounts of undefined supplements followed
by replacement with defined supplements has resulted in good growth in
defined media of normal human fibroblasts [Bettger et al., 1981], chicken
embryo fibroblasts [Graves and Ham, 1982], human epidermal keratinocytes
[Tsao et al., 1982], and human mammary epithelial cells [Hammond et al.,
1983]. With slightly different sets of defined supplements, the optimized basal
nutrient media developed for these cells have also supported defined medium
growth of a number of other cell types that have similar nutrient requirements,
including sheep preadipocyte fibroblasts [Broad and Ham, 1983], rabbit chondrocytes [Jennings and Ham, 1983a], human chondrocytes [Jennings and
Ham, 1983b], human bronchial epithelial cells [Lechner et al., 1982], and
chicken embryo myoblasts [Dollenmeier et al., 1982].

Another major advantage of using optimized basal nutrient media is that
they make possible good growth of cell types that are very difficult to grow in
conventional media. During the process of optimization for diverse cell types,
it became evident that different types of cells have quite different quantitative
nutrient requirements for growth with minimal amounts of undefined (or
defined) supplements, and that a medium optimized for a particular type of
cell could be distinctly selective against growth of certain other cell types
[Peehl and Ham, 1980; Tsao et al., 1982]. In retrospect, it is now clear that
most classical culture media (as well as modern media optimized for fibroblast-
like cells) selectively favor growth of fibroblasts and closely related cell types
and inhibit growth of most types of epithelial cells [Ham, 1984].

When a basal nutrient medium is employed that has been optimized for
the cell type in question (or for another cell type with similar nutrient requirements), many cell types that have previously been considered very difficult to

grow can be grown readily. Examples include human epidermal keratinocytes [Peehl and Ham, 1980; Tsao et al., 1982; Boyce and Ham, 1983], human bronchial epithelial cells [Lechner et al., 1982], rat prostate epithelial cells [McKeehan et al., 1982], human bladder epithelial cells [Kaighn et al., 1983], human mammary epithelial cells [Hammond et al., 1983], and human microvascular endothelial cells [Knedler and Ham, 1983].

In summary, the major advantages of using an optimized basal nutrient medium are: 1) reduction in the amount of supplementation (defined or undefined) needed for growth of the cell type in question; 2) improved growth of cell types that cannot be grown well in conventional media; and 3) enhanced selectivity of growth by minimizing the buffering effects of serum and other undefined supplements. The advantages of employing an optimized basal nutrient medium have been discussed in greater detail in Ham [1982]. The concept of selective growth and the possibility of further enhancing selectivity by optimizing media for maximum selective advantage are explored in Ham [1984]. Practical procedures for developing optimizing basal nutrient media are summarized in the following section.

PROCEDURES FOR DEVELOPING OPTIMIZED BASAL NUTRIENT MEDIA
Overview

The key to developing an optimized basal nutrient medium is to make as few assumptions as possible and to allow the cells for which the medium is being developed to guide every possible step. It is not safe, for example, to assume that all types of cells grow optimally with the same concentrations of cysteine [Ham et al., 1977; Hammond et al., 1983], histidine [Shipley and Ham, 1981], calcium [Peehl and Ham, 1980; Hennings et al., 1980; Boyce and Ham, 1983], or magnesium [Knedler and Ham, 1983], or even the same osmolarity [Iscove and Melchers, 1978] or pH [Ceccarini and Eagle, 1971].

The procedures used in the author's laboratory for developing an optimized basal nutrient medium for a particular cell type have been described in detail as a part of an overall 21-step procedure for the development of defined media [Ham, 1981:29-33]. The overall procedure will be summarized here only briefly, with primary emphasis on the optimization process and on new observations since the 1981 paper was written. It is recommended that the entire 21-step procedure in Ham [1981] be read carefully by anyone planning to undertake optimization for the first time. In the discussion that follows, the numbering of the 21 steps is the same as in the 1981 paper.

The first major phase (step 1) consists of obtaining growth of the cell type in question by any means necessary. Growth-response assays cannot be performed if there is no growth. Therefore, as will be discussed below, a "no holds barred" approach is fully justified at this stage if it is needed. The second major phase (steps 2-7) is to refine the assay system so that the small-molecule composition of test media can be controlled and manipulated precisely. This is usually achieved through the use of dialyzed supplements, although in some cases it may be possible to proceed directly to defined supplements prior to optimizing the basal nutrient medium. The third phase (steps 8-15) is the optimization process itself, in which the qualitative and quantitative composition of the basal nutrient medium is modified to yield the best growth with the least amount of supplementation. The fourth phase (steps 16-18) consists of replacing the remaining undefined supplements with defined supplements to generate a defined medium. The final phase (steps 19-21) consists of identification of any contaminants that may be contributing to the cellular growth in the defined medium and a final round of optimization of the basal nutrient medium for the best growth with defined supplements.

Initiating Growth of the Desired Cell Type

Many investigators have invested large amounts of time seeking to grow a wide variety of cell types in culture. Thus, it is usually possible to find a procedure in the literature that will yield a viable inoculum and at least some proliferation in vitro for virtually any cell type that is capable of multiplication in vivo. It is generally best to begin with such procedures and then try to improve them as needed for adequate growth response assays. If a satisfactory starting point is not available in the literature, a wide variety of basal nutrient media and complex supplements should be tested, starting with those that have worked well for related cell types. Careful attention should also be given to finding an isolation procedure that will yield a viable inoculum that is enriched in the desired cell type. For some kinds of epithelial cells, a supplement consisting of epidermal growth factor (EGF), insulin, hydrocortisone, and bovine pituitary extract may be superior to serum, which is sometimes inhibitory to epithelial cells [Ham, 1982, 1983, 1984; Boyce and Ham, 1983; Hammond et al., 1983; Kaighn et al., 1983; Lechner et al., 1982]. Conditioned media [Stampfer et al., 1980] or feeder layers [Puck and Marcus, 1955; Rheinwald and Green, 1975] may prove to be helpful. It is also sometimes helpful to use tumor promoters and/or agents such as cholera toxin that elevate intracellular cyclic AMP levels to obtain adequate initial growth of difficult cell types [Green, 1978; Stampfer, 1982; Eisinger and Marko, 1982]. In extreme cases, it may prove useful to begin with naturally malignant strains or with cells transformed with chemicals, viruses, or radiation, and then work backward to the normal cells after growth of the transformed cells has been improved.

Refining the Assay System

The next phase (steps 2-7) consists of refining the assay sytem to the point where it can be used effectively for the optimization process. It is generally desirable to develop a clonal growth assay (step 2), since colony formation provides the most sensitive means for determining the nutritional adequacy and freedom from toxicity of a culture medium [Ham, 1972]. However, it is also possible to optimize basal nutrient media for other purposes, such as growth of monolayer cultures to maximum density or expression of differentiated properties, by appropriate modification of the assay system. In particular, it should be noted that a medium that is optimized for clonal growth may have nutrient concentrations that are too low to support growth of dense cellular populations without rapid depletion.

The beginner is likely to be tempted to launch directly into optimization as soon as an adequate assay system is available. However, optimization is a slow, grueling, and tedious process that deserves to be bypassed as much as possible. It is therefore well worth the effort to survey a wide variety of existing basal nutrient media and supplements (step 3) to determine which combination provides the best starting point for optimization. It should also be emphasized that we do not yet know enough about cellular growth requirements to make accurate predictions. For example, by doing a broad survey we found that medium MCDB 202, which had originally been developed for chicken embryo fibroblasts [McKeehan and Ham, 1977], worked so well for human mammary epithelial cells that we were able to proceed directly to defined supplementation in that medium without a preliminary round of optimization [Ham, 1983]. Subsequent optimization under defined conditions has yielded an even better basal nutrient medium for the mammary cells, MCDB 170 [Hammond et al., 1983]. However, many months of preliminary optimization studies were saved by testing a medium that would not have been predicted to be particularly good for an epithelial cell type.

Surveying existing media will not be equally beneficial for all cell types, however. In many cases, the optimization process will uncover unusual quantitative requirements that have not been observed previously and that could not easily have been predicted. An example of this is the requirement by normal human microvascular endothelial cells of 10.0 mM magnesium, a level far above that in other culture media [Knedler and Ham, 1983]. Nevertheless, even in this case, the preliminary survey revealed that an unlikely basal nutrient medium, MCDB 402, which had been optimized for Swiss mouse 3T3 cells [Shipley and Ham, 1981], was the best starting point for optimization.

In most cases, after completion of the survey of basal nutrient media and supplements (step 3 in Ham [1981]), it is possible to proceed rather quickly through the remaining preliminary steps if they are needed at all. The survey will usually identify some combination of medium and supplements that will support at least minimal growth without a feeder layer (step 4) or conditioning (step 5). All that is needed initially is enough growth to perform the growth response assays, since the optimization process itself can be expected to result in major improvement in growth in cases where growth is initially poor.

In cases where dialyzed supplements fail completely to support growth (step 6), it may be necessary to isolate and characterize a new small-molecule growth factor (step 7). However, before a major effort is invested in such an undertaking, it is desirable to test all suspected nutrient requirements that may not be satisfied by the basal nutrient medium (e.g., biotin, ethanolamine, selenium, etc.). In addition, it may be desirable to reduce the undefined small-molecule supplement to a distinctly suboptimal level and perform a preliminary survey of the effects of quantitative adjustment of nutrient concentrations in the basal nutrient medium (step 9). A few simple experiments of this sort can sometimes avoid the frustration of investing a large effort in the isolation and purification of a known nutrient that is either missing from the basal nutrient medium or present at a concentration that will not support growth of the cell type in question.

Theoretical Basis for Optimization Procedure

The procedure that is employed for optimization of the basal nutrient medium is based on two theories concerning cellular growth requirements, both of which have been thoroughly substantiated by experimental data. The first theory is that complex interactions occur among the components that meet the individual growth requirements of any given type of cell. We have found it useful to adopt a holistic view in which all of the environmental variables that influence cellular multiplication are regarded as a single interacting set whose organization is such that a change in any member of the set can potentially alter the cellular growth response to any other member of the set [Ham and McKeehan, 1978a; Ham, 1981]. In a simple case, increasing the concentration of one amino acid might require a compensatory increase in the concentrations of other amino acids to maintain an appropriate balance [Ham, 1974]. In a more complex case, reducing the amount of dialyzed serum might increase the concentration of an essential amino acid that is needed to support satisfactory growth, perhaps due to changes in efficiency of active transport of the amino acid into the cells [Ham and McKeehan, 1978b].

We know that the need for serum can be greatly reduced by optimization of the basal nutrient medium and that serum can be replaced by appropriate mixtures of hormones and growth factors. The nature of the serum factors being replaced by optimization remains totally unknown, and it is likely in many cases that the serum factors that are replaced by hormones and growth factors are not chemically identical to the defined supplements that replace them. Thus, although not rigorously proven, it appears likely that the same requirements can, in some cases, be satisfied in three distinctly different ways, by qualitative and quantitative optimization of the basal nutrient medium, by supplying undefined serum factors, and by supplying mixtures of hormones and growth factors.

The practical implications of this pattern of holistic interaction are: 1) that optimization of the basal nutrient medium can have profound effects on other cellular growth requirements; 2) that a medium optimized for growth with a particular type and amount of supplementation may no longer be optimal if the supplementation is altered; and 3) that changes in the composition of the basal nutrient medium can themselves create the need for additional changes.

The second major theory is that under conditions of suboptimal growth the rate of cellular multiplication is determined by a first-limiting factor and is essentially unaffected by other components of the culture system. Thus, although a cell has many different growth requirements and a number of them may not be optimally satisfied, one (or a very few) condition(s) will be more restrictive to growth than all others and will determine the actual rate of multiplication.

This theory leads to two predictions that are of major importance to optimization. The first is that, for any given nutrient (as well as for other components of the complete growth medium), there will be a "plateau" region in the growth response curve, where the substance has ceased to be the rate-limiting factor for multiplication due to deficiency, and where its concentration has not yet reached a level that is rate-limiting for multiplication because of inhibitory or toxic effects [Ham and McKeehan, 1978a]. Thus, in theory (and in most cases, in practice), there is a reasonably wide range over which the concentration of the nutrient in question has little effect on growth rate. One of the objectives of optimization is to adjust all components of the basal nutrient medium to concentrations that are near the midpoints (on a log scale) of their plateaus of optimum growth, such that minor perturbations or holistic interactions are less likely to render them first-limiting for multiplication.

The second prediction is that it is necessary to identify and correct whatever is first-limiting for multiplication before the effects of other conditions that are not optimal can be detected [Ham and McKeehan, 1978b]. Thus, for example, if nutrient A is at a concentration that will support 40% of optimal growth in a complete medium and nutrient B is at a concentration that will support

60%, growth cannot be improved by adjusting the concentration of nutrient B until the concentration of nutrient A has been raised to a level capable of supporting more than 60% of optimal growth (Fig. 1).

One practical implication of the first-limiting factor theory is that optimization must be done sequentially. Thus, each time that an improvement is made in the basal nutrient medium, it becomes necessary to test again all of the variables that were previously found not to be first-limiting to see if one of them has become the new first-limiting factor.

A second practical implication that somewhat relieves the burden of the first is that it is usually possible to predict from partial growth response curves which of the nutrients in a medium are likely to become first-limiting as optimization proceeds. This is done by determining whether the concentration of the nutrient is close to the edge of the observed plateau of maximum growth in the current basal nutrient medium. In Figure 1, the effect of a growth-limiting concentration of nutrient A was to truncate the growth response curve for nutrient B, making the plateau of maximum growth lower and wider (dashed line) than it would have been with an optimum amount of

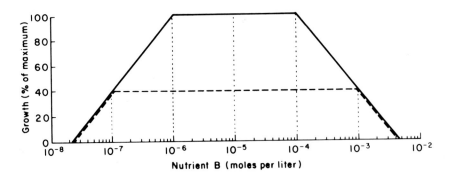

Fig. 1. Idealized growth response curves. Solid line: Response to nutrient B in an otherwise complete medium. Nutrient B is rate-limiting for growth below 10^{-6} M and inhibitory to growth above 10^{-4} M. Between 10^{-6} M and 10^{-4} M, growth is limited by inherent properties of the cells and is not affected by the concentration of nutrient B. Dashed line: Response to nutrient B when growth is limited to 40% of maximum by a deficiency in nutrient A and there is no interaction between nutrients A and B. The plateau of maximum response to nutrient B now extends from 10^{-7} M to 10^{-3} M. Under these conditions, growth is the same with 2×10^{-7} M nutrient B (which would support about 60% of maximum growth in a complete medium) as with 1×10^{-6} M nutrient B (which would support 100% of maximum growth in a complete medium). Thus, as long as growth is limited to 40% by nutrient A, growth cannot be improved above that level by adjusting nutrient B. However, if nutrient B is below 10^{-7} M or above 10^{-3} M, it will become rate-limiting and will reduce growth below the 40% level supported by nutrient A.

nutrient A (solid line). As long as nutrient A is first-limiting, growth will remain the same whether nutrient B is at a level that would support 60% of maximum growth in a complete medium (2×10^{-7} M) or a level that would support 100% (1×10^{-6} M). Thus, growth cannot be improved by adjusting the concentration of nutrient B. However, growth is further impaired when nutrient B is reduced below 10^{-7} M or increased above 10^{-3} M.

Such impairment makes it possible to perform a simple test to determine if nutrient B (or any other nutrient) is close to being rate-limiting. The procedure is to eliminate the nutrient from the basal nutrient medium and add it back at $0.1 \times$, $1.0 \times$, and $10.0 \times$ its normal concentration. If the nutrient is near the center of a broad optimum plateau in the truncated growth response curve (e.g., 10^{-5} M nutrient B in Fig. 1), it is unlikely to become rate-limiting even when the truncation is removed. In such cases, a 10-fold increase or decrease in the concentration of the nutrient will usually have no effect on growth. However, if the concentration is near the edge of the plateau (e.g., 2×10^{-7} M or 5×10^{-4} M nutrient B), either a 10-fold increase or a 10-fold decrease is likely to reduce growth substantially below the truncated maximum. A 10-fold increase or decrease is also likely to reduce growth in cases where the optimum plateau is very narrow. Thus, a preliminary survey in which each nutrient is tested at three concentrations, $0.1 \times$, $1.0 \times$, and $10.0 \times$ the current amount, will usually identify those nutrients that are the most likely to become rate-limiting during early phases of sequential optimization.

Optimization Procedures

The optimization procedures described in steps 8–15 of Ham [1981] are relatively detailed and have not been changed significantly since that article was written. If good growth was obtained with the dialyzed supplement (step 6), it is desirable to begin by reducing the supplementation until growth is distinctly suboptimal (step 8). This assures that growth is limited by the overall composition of the culture medium, which in turn makes it feasible to attempt to improve growth by altering the composition of the basal nutrient medium in a manner that renders growth less dependent on supplementation. If growth is already poor with the dialyzed supplement, qualitative and quantitative optimization can be started without making any further reduction in the level of supplementation. In all cases, however, each time that growth is improved it is necessary to reduce the supplementation further before continuing the optimization process (step 14). This allows the next most limiting aspect of the basal nutrient medium, which had previously been compensated for by the supplementation, to be detected and corrected.

The question of whether to begin with quantitative changes (step 9) or qualitative changes (step 10) depends on the nature of the basal nutrient medium that is being used as a starting point. If it has obvious qualitative deficiencies in vitamins, trace elements, or other nutrients that are known to be essential for some types of cells (or that are suspected to be of special significance for the cell type that is being studied), such nutrients should be tested at the beginning of the optimization process. If not, it is usually best to begin with a preliminary quantitative survey in which each component of the basal nutrient medium is tested at $0.1 \times$, $1.0 \times$, and $10.0 \times$ its initial concentration, as described in the previous section.

For each nutrient whose growth response is altered by a 10-fold increase or decrease in concentration, a complete growth response titration encompassing deficiency, positive growth response, the optimum plateau, and the inhibitory range should be performed. The growth response should then be plotted on semilogarithmic paper, and the midpoint on a log scale of the plateau of maximum growth selected as the new concentration of that nutrient in the basal nutrient medium [Ham and McKeehan, 1978a]. This adjustment should be made whenever a nutrient concentration is not already near the midpoint, whether or not it is currently rate-limiting for growth. The cumulative effect of such adjustments is to establish nutrient concentrations that are less likely to become rate-limiting as optimization progresses. This, in turn, greatly reduces the total number of titrations and adjustments that must be done during optimization. If a clonal assay is being used and growth of monolayer cultures is also a major concern, it may be desirable to set nutrient concentrations somewhat above the middle of the plateau, but still well away from the inhibitory range.

It is generally worthwhile to continue the optimization process as long as it continues to reduce the amount of supplementation (undefined or defined) that is needed. After the first rate-limiting aspects of the original medium have been identified and corrected and the amount of supplementation reduced substantially, it will probably be useful to repeat the survey of all components at $0.1 \times$, $1.0 \times$, and $10.0 \times$ their current concentrations to determine which are most likely to yield further improvement in growth by quantitative adjustment. Often the overall effect of many small adjustments that barely alter growth rate can be quite large.

One class of nutrients that is often overlooked in optimization studies is the lipids. As long as there is a reasonable amount of dialyzed serum present, lipids are supplied as bound lipoprotein complexes. However, as optimization proceeds and the serum concentration is reduced, lipids may become rate-limiting. It is therefore desirable to test lipid supplements, such as the liposome preparation of Bettger et al. [1981], for their ability to support a further reduction in serum supplementation. Trace elements are also often carried in bound form on serum macromolecules. Particular attention needs to be given to selenium (generally supplied as a selenite salt) and iron (generally supplied as a freshly prepared ferrous sulfate solution or as a transferrin complex), both of which frequently become rate-limiting with minimal levels of undefined or defined supplementation [McKeehan et al., 1976; Barnes and Sato, 1980a,b; Ham, 1981, 1984].

When the point is reached where quantitative adjustments are no longer beneficial and all of the obvious qualitative changes have been tried, it is sometimes useful to test complex mixtures of low-molecular-weight substances (hydrolysates, extracts, ultrafiltrates) for their ability to reduce the amount of dialyzed supplement required (step 11). In addition, low-molecular-weight growth-promoting substances can sometimes be separated from higher-molecular-weight carriers by fractionation under dissociating conditions (step 12). It is also desirable to attempt to reduce the amount of supplement needed by modification of the culture surface, and by employing less damaging techniques, such as low-temperature trypsinization [McKeehan, 1977], to dissociate the cells used an inoculum (step 13).

The question of how far to carry optimization with an undefined supplement (steps 8-15) before attempting to obtain growth with defined supplements (steps 16-18) has no absolute answer. Experience in the author's laboratory suggests that replacement of undefined supplements with defined is generally feasible when dialyzed serum is reduced to 2% (1.0 mg/ml dialyzed serum solids) or dialyzed bovine pituitary extract is reduced to a Lowry protein concentration of 70 μg/ml. Although only very limited studies have been done on the subject, optimization also clearly affects the kinds and amounts of defined supplementation needed for cellular proliferation [Wu and Sato, 1978; Ham, 1982]. Optimization studies can (and ideally should) be continued after defined medium growth has been achieved. This is done by reducing the macromolecular supplementation to rate-limiting levels and attempting further improvement in growth by qualitative and quantitative adjustment of the basal nutrient medium.

CONCLUSIONS

To the person who has never attempted it, optimization is often viewed as a process that is so complex and tedious that it should be avoided at any cost. Unfortunately, that cost is likely to be continued use of large amounts of undefined supplementation that mask and confuse experimental results, or even worse, inability to grow a particular cell type well with any amount of supplementation.

There is no denying that optimization is an unpleasant task. However, with careful attention to the procedures and shortcuts outlined in this chapter and in Ham [1981], optimization can be approached in a systematic manner that greatly reduces the number of separate experiments that must be performed and makes it much more likely that significant improvement will be achieved early in the process.

In addition, for those who are willing to endure the process, the rewards can be very great. A number of cell types that were previously considered extremely difficult to grow in culture and many others that previously required large amounts of undefined supplementation can now be grown quite easily in defined media, or with minimal undefined supplementation. These benefits have been achieved as a result of optimization of the basal nutrient medium specifically for each cell type.

Another important benefit from current optimization studies is the increased number of optimized media for diverse cell types that are available for testing during the preliminary screening phase. It is becoming increasingly common to find an already developed medium that will support good growth with a small amount of undefined supplementation. This, in turn, makes it possible to proceed directly to defined supplements without a preliminary round of optimization.

However, for those cell types that are still difficult to grow or require large amounts of supplementation in currently available basal nutrient media, there appears to be no reasonable alternative to optimization. Since the number of such cell types remains quite large, it is desirable for far more laboratories to undertake optimization studies than is now the case.

REFERENCES

Agy PC, Shipley GD, Ham RG (1981): Protein-free medium for C-1300 mouse neuroblastoma cells. In Vitro 17:671-680.

Barnes D, Sato G (1980a): Methods for growth of cultured cells in serum free medium. Anal Biochem 102:255-270.

Barnes D, Sato G (1980b): Serum-free cell culture: A unifying approach. Cell 22:649-655.

Bettger WJ, Ham RG (1982): The nutrient requirements of mammalian cells grown in culture. Adv Nutr Res 4:249-286.

Bettger WJ, Boyce ST, Walthall BJ, Ham RG (1981): Rapid clonal growth and serial passage of human diploid fibroblasts in a lipid-enriched synthetic medium supplemented with EGF, insulin, and dexamethasone. Proc Natl Acad Sci USA 78:5588-5592.

Bottenstein J, Hayashi I, Hutchings S, Masui H, Mather J, McClure DB, Ohasa S, Rizzino A, Sato G, Serrero G, Wolfe R, Wu R (1979): The growth of cells in serum-free hormone-supplemented media. Methods Enzymol 58:94-109.

Boyce ST, Ham RG (1983): Calcium-regulated differentiation of normal human epidermal keratinocytes in chemically defined clonal culture and serum-free serial culture. J Invest Dermatol 81:33s-40s.

Broad TE, Ham RG (1983): Growth and adipose differentiation of sheep preadipocyte fibroblasts in serum free medium. Eur J Biochem 135:33-39.

Ceccarini C, Eagle H (1971): pH as a determinant of cellular growth and contact inhibition. Proc Natl Acad Sci USA 68:229-233.

Chung SD, Alavi N, Livingston D, Hiller S, Taub M (1982): Characterization of primary rabbit kidney cultures that express proximal tubule functions in a hormonally defined medium. J Cell Biol 95:118-126.

Dollenmeier P, Turner DC, Eppenberger HM (1982): Proliferation and differentiation of chick skeletal muscle cells cultured in a chemically defined medium. Exp Cell Res 135:47-61.

Eagle H (1955): Nutrition needs of mammalian cells in tissue culture. Science 122:501-504.

Eagle H (1959): Amino acid metabolism in mammalian cell cultures. Science 130:432-437.

Eisinger M, Marko O (1982): Selective proliferation of normal human melanocytes in vitro in the presence of phorbol ester and cholera toxin. Proc Natl Acad Sci USA 79:2018-2022.

Graves DC, Ham RG (1982): Serum-free clonal growth of chicken embryo fibroblasts. In Vitro 18:305 (Abstract No. 127).

Green H (1978): Cyclic AMP in relation to proliferation of the epidermal cells: A new view. Cell 15:801-811.

Guilbert LJ, Iscove NN (1976): Partial replacement of serum by selenite, transferrin, albumin, and lecithin in haemopoietic cell cultures. Nature 263:594-595.

Ham RG (1962): Clonal growth of diploid Chinese hamster cells in a synthetic medium supplemented with purified protein fractions. Exp Cell Res 28:489-500.

Ham RG (1963): An improved nutrient solution for diploid Chinese hamster and human cell lines. Exp Cell Res 29:515-526.

Ham RG (1965): Clonal growth of mammalian cells in a chemically defined, synthetic medium. Proc Natl Acad Sci USA 53:288-293.

Ham RG (1972): Cloning of mammalian cells. Methods Cell Physiol 5:37-74.

Ham RG (1974): Nutritional requirements of primary cultures. A neglected problem of modern biology. In Vitro 10:119-129.

Ham RG (1981): Survival and growth requirements of nontransformed cells. Handbook Exp Pharmacol 57:13-88.

Ham RG (1982): Importance of the basal nutrient medium in the design of hormonally defined media. Cold Spring Harbor Conf Cell Prolif 9:39-60.

Ham RG (1983): Growth of normal human cells in defined media. In Fischer G, Wieser RJ (eds): "Hormonally Defined Media: A Tool in Cell Biology." Berlin: Springer Verlag, pp 16-30.

Ham RG (1984): Selective media. In Pretlow TG II, Pretlow T (eds): "Cell Separation: Methods and Selected Applications." New York: Academic, Vol. 3 (in press).

Ham RG, McKeehan WL (1978a): Development of improved media and culture conditions for clonal growth of normal diploid cells. In Vitro 14:11-22.

Ham RG, McKeehan WL (1978b): Nutritional requirements for clonal growth of nontransformed cells. In Katsuta H (ed): "Nutritional Requirements of Cultured Cells." Tokyo: Japan Scientific Societies Press, pp 63-115.

Ham RG, McKeehan WL (1979): Media and growth requirements. Methods Enzymol 58:44-93.

Ham RG, Hammond SL, Miller LL (1977): Critical adjustment of cysteine and glutamine concentrations for improved clonal growth of WI-38 cells. In Vitro 13:1-10.

Hamilton WG, Ham RG (1977): Clonal growth of Chinese hamster cell lines in protein-free media. In Vitro 13:537-547.

Hammond SL, Ham RG, Stampfer MR (1983): Defined medium for normal human mammary epithelial cells. In Vitro 19:252 (Abstract No. 72).

Hayashi I, Sato GH (1976): Replacement of serum by hormones permits growth of cells in a defined medium. Nature 259:132-134.

Hennings H, Michael D, Cheng C, Steinert P, Holbrook K, Yuspa SH (1980): Calcium regulation of growth and differentiation of mouse epidermal cells in culture. Cell 19:245-254.

Higuchi K (1973): Cultivation of animal cells in chemically defined media, a review. Adv Appl Microbiol 16:111-136.

Iscove NN, Melchers F (1978): Complete replacement of serum by albumin, transferrin, and soybean lipid in cultures of lipopolysaccharide-reactive B lymphocytes. J Exp Med 147:923-931.

Jennings SD, Ham RG (1983a): Clonal growth of primary cultures of rabbit ear chondrocytes in a lipid-supplemented defined medium. Exp Cell Res 145:415-423.

Jennings SD, Ham RG (1983b): Clonal growth of primary cultures of human hyaline chondrocytes in a defined medium. Cell Biol Int Rep 7:149-159.

Kaighn ME, Kirk D, Kigawa S, Vener G, Narayan SK (1983): Replicative cultures of normal human bladder epithelial cells in serum-free medium. In Vitro 19:252 (Abstract No. 74).

Kano-Sueoka T, Errick JE (1981): Effects of phosphoethanolamine and ethanolamine on growth of mammary carcinoma cells in culture. Exp Cell Res 136:137-145.

Kano-Sueoka T, Cohen DM, Yamaizumi Z, Nishimura S, Mori M, Fujiki H (1979): Phosphoethanolamine as a growth factor of a mammary carcinoma cell line of rat. Proc Natl Acad Sci USA 76:5741-5744.

Katsuta H, Takaoka T (1973): Cultivation of cells in protein- and lipid-free synthetic media. Methods Cell Biol 6:1-42.

Knedler A, Ham RG (1983): Optimized medium for clonal growth of human microvascular endothelial cells. In Vitro 19:254 (Abstract No. 81).

Lechner JF, Haugen A, McClendon IA, Pettis EW (1982): Clonal growth of normal adult human bronchial epithelial cells in a serum-free medium. In Vitro 18:633-642.

McKeehan WL (1977): The effect of temperature during trypsin treatment on viability and multiplication potential of single normal human and chicken fibroblasts. Cell Biol Int Rep 1:335-343.

McKeehan WL, Ham RG (1977): Methods for reducing the serum requirement for growth in vitro of nontransformed diploid fibroblasts. Dev Biol Standard 37:97-108.

McKeehan WL, Hamilton WG, Ham RG (1976): Selenium is an essential trace nutrient for growth of WI-38 diploid human fibroblasts. Proc Natl Acad Sci USA 73:2023-2027.

McKeehan WL, McKeehan KA, Hammond SL, Ham RG (1977): Improved medium for clonal growth of human diploid fibroblasts at low concentrations of serum protein. In Vitro 13:399-416.

McKeehan WL, Adams PS, Rosser MP (1982): Modified nutrient medium MCDB 151, defined growth factors, cholera toxin, pituitary factors, and horse serum support epithelial cell and suppress fibroblast proliferation in primary cultures of rat ventral prostate cells. In Vitro 18:87-91.

Morgan JF, Morton HJ, Parker RC (1950): Nutrition of animal cells in tissue culture. I. Initial studies on a synthetic medium. Proc Soc Exp Biol Med 73:1-8.

Peehl DM, Ham RG (1980): Clonal growth of human keratinocytes with small amounts of dialyzed serum. In Vitro 16:526-540.

Puck TT, Marcus PI (1955): A rapid method for viable cell titration and clone production with HeLa cells in tissue culture: The use of X-irradiated cells to supply conditioning factors. Proc Natl Acad Sci USA 41:432-437.

Rheinwald JG, Green H (1975): Serial cultivation of strains of human epidermal keratinocytes: The formation of keratinizing colonies from single cells. Cell 6:331-344.

Rizzino A, Rizzino H, Sato G (1979): Defined media and the determination of nutritional and hormonal requirements of mammalian cells in culture. Nutr Rev 37:369-378.

Sato GH (1975): The role of serum in cell culture. In Litwack G (ed): "Biochemical Actions of Hormones." New York: Academic, Vol 3, pp 391-396.

Shipley GD, Ham RG (1981): Improved medium and culture conditions for clonal growth with minimal serum protein and for enhanced serum-free survival of Swiss 3T3 cells. In Vitro 17:656-670.

Stampfer MR (1982): Cholera toxin stimulation of human mammary epithelial cells in culture. In Vitro 18:531-537.

Stampfer M, Hallowes RC, Hackett AJ (1980): Growth of normal human mammary cells in culture. In Vitro 16:415-425.

Taub M, Sato G (1980): Growth of functional primary cultures of kidney epithelial cells in defined medium. J Cell Physiol 105:369-378.

Tsao MC, Walthall BJ, Ham RG (1982): Clonal growth of normal human epidermal keratinocytes in a defined medium. J Cell Physiol 110:219-229.

Walthall BJ, Ham RG (1981): Multiplication of human diploid fibroblasts in a synthetic medium supplemented with EGF, insulin, and dexamethasone. Exp Cell Res 134:303-311.

Wu R, Sato G (1978): Replacement of serum in cell culture by hormones: A study of hormonal regulation of cell growth and specific gene expression. J Toxicol Environ Health 4:427-448.

Methods for Preparation of Media, Supplements, and Substrata
for Serum-Free Animal Cell Culture, pages 23–68
© 1984 Alan R. Liss, Inc., 150 Fifth Avenue, New York, NY 10011

2
Preparation and Use of Serum-Free Culture Media

Charity Waymouth

WHY USE SERUM-FREE MEDIA?

The media used in the earliest cell, tissue, and organ cultures were biological fluids. Lymph was used by Harrison [1908] and plasma or serum and embryonic extracts by Burrows [1910, 1911], on the assumption that these materials resembled the "natural" environment and sources of nutrition of cells and tissues in vivo. The first attempts to identify nutritional factors, especially in plasma or serum, were made by Lewis and Lewis [1911a, 1912], who tried protein-free preparations, such as chicken bouillon [Lewis and Lewis, 1911b]. By the 1930s, media in which many necessary low-molecular-weight components were identified (salts, glucose, some amino acids, and vitamins) began to be designed, e.g., by Vogelaar and Erlichman [1933] and by Baker and Ebeling [1939]; but these still required supplementation with serum, serum fractions, or protein digests to permit survival and growth. The early literature has been reviewed [Waymouth, 1954, 1965a, 1972].

Over the past 40 years, a number of much more complete defined media, containing salts (including trace elements), sugars, amino acids, vitamins, and sometimes other components (e.g., purines and pyrimidines, polyamines, lipids, hormones, and growth factors), have been prepared. Until recently, it was commonly assumed that a general formulation, perhaps with one or two supplements, would serve for a variety of cell types. Increasingly, this has been demonstrated to be invalid, and media carefully designed to meet the requirements, either for proliferation, survival, differentiation, or function—or

The Jackson Laboratory, Bar Harbor, Maine 04609

some combination of these functions—have become available. There are basic qualitative similarities in the nutritional requirements, but usually a constellation of additional factors, such as trace elements, hormones, and growth factors, more specific to the needs of individual cell types, are required. Ham [1981] has proposed some useful definitions of *nutrient, growth requirement, growth factor, mitogen,* and *hormone,* as these apply to components of media for cell culture. It may be noted that "growth factors" include some, such as nerve growth factor (NGF), that promote differentiated function, not proliferation; and the pituitary-derived fibroblast growth factor (FGF), which acts not only on fibroblasts [Kamely and Rudland, 1976] but, for example, on adrenocortical cells [Hornsby et al., 1979]. The roles of growth factors have been reviewed by Rudland and Jiménez de Asúa [1979], by Barnes and Sato [1980a,b], and by Vasiliev and Gelfand [1981].

It has remained a common practice, however, to use an incomplete culture medium and to supplement it with serum, usually in the amount of 2–20% by volume or more. This is becoming indefensible for studies that require well-controlled and well-defined nutritional conditions. Moreover, it is evident that the nutrition required for the growth (cell multiplication), or differentiation, or function, of each cell type is quite specific, and that optimal conditions cannot be attained by use of a "general" medium for cells with distinctive metabolic patterns and therefore distinctive needs. There are differences between the requirements, not only of cells of different embryological origin (ectodermal, mesodermal, endodermal)—or (to use the functional classification introduced by Willmer [1970]), between epitheliocytes, mechanocytes, amoebocytes, and neural cells—but also between cells from organisms of different species and ages, and between cells that are "normal" and diploid, and those that have undergone transformation, whether functional or neoplastic. Whereas it seems to be the rule that different kinds of epithelial cells have in common many requirements (as well as some distinctions), Willmer [1970] has pointed out that "epitheliocytes can emerge from ectoderm (skin), mesoderm (e.g., kidney), and endoderm (e.g., intestine), and so far as tissue-culture tests go, there is no general character by which to differentiate between them according to their origin." As the identification for specific cells of stimulants and their receptors proceeds, we may expect to be able to make these distinctions. For the foregoing reasons, one must regard with caution the commercialization of single "growth supplements" claiming to replace serum and containing many nutrients, hormones, and growth factors, because a single mixture not designed for any particular cell type may be adequate, but can no more be expected to be optimal for most cells than the serum it is supposed to replace.

For the cultivation of each cell type, then, it must be the objective to achieve a particular combination of factors for optimal effect in vitro, whether that effect is maximal proliferation or promotion of differentiation, or expression of function. A number of models now exist, but for many cell types definitive conditions remain to be worked out. This means, in large part, identifying factors that have been provided in the past by serum or other biological materials. The inclusion of serum has provided many nutrients and factors, some required in minute amounts. Many of these have been identified, although the precise roles of each are not always understood. These include peptide and steroid hormones, growth factors, trace elements, and lipids. Other serum factors, less thoroughly studied, include enzymes that may act upon medium components or cellular products.

It is sometimes overlooked, in replacing serum by serum-free media, that fetal sera (and the most common supplement has been fetal bovine serum) contain much higher concentrations of some components than maternal or other adult serum. Examples are arginase, which is present at concentrations at least 10 times those in newborn calf serum, or in adult horse or human serum [Kihara and de la Flor, 1968]; the folic acid complex [Grosswicz et al., 1960]; and most of the amino acids (except glutamine and arginine) [Corfield and Hay, 1978]. Some amino acids may be present in fetal sera at very high levels [Hopkins et al., 1971; Hill and Young, 1973; Young and McFadyen, 1973].

Sera carry beneficial factors and nutrients but also some constituents that prove detrimental. Inhibitors of growth [De Luca et al., 1966a,b] and of cellular functions [Erickson et al., 1982] have been attributed to serum. For example, the lifespan of tracheal epithelial cells was reduced in the presence of serum [Wu et al., 1982], and effects on lifespan have been shown to vary greatly with the batch of serum [Poiley et al., 1978; Schneider et al., 1978]. Effects may also be species-linked, so that the differential attachment of pancreatic endocrine cells was found to be less in calf serum than in serum-free, or human serum-containing, media [Nielsen and Brunstedt, 1982]. Among the causes of serum inhibition is the presence, in the serum of ungulates, of amine oxidases [Blaschko and Hawes, 1959; Blaschko and Bonney, 1962; Waymouth et al., 1982], which can convert polyamines to the highly cytotoxic product acrolein [Alarcon, 1964; Jensen and Therkelsen, 1982]. This toxicity may vary with cell type, source, and degree of differentiation, and this fact has been used as the basis for a method of cell selection [Webber and Chaproniere-Rickenberg, 1980]. In some systems, polyamines are, however, beneficial [Ham, 1964; Stoner et al., 1980].

Variations in the composition of sera from species to species, and within single or pooled batches from one species [Olmsted, 1967; Honn et al., 1975; Boone et al., 1972; Waymouth, 1977, 1978], including sometimes great variations in hormones [Esber et al., 1973], contribute to the importance for reproducible studies of providing the right components in the right proportions in every medium for cell or organ culture.

PURITY AND STABILITY OF CHEMICALS
Water

Water of the highest quality is of the greatest importance for preparing culture media [Pumper, 1973]. It should be recognized that ion exchange resins, which by definition remove ions, may contribute unknown amounts of nonionic materials to the effluent. Distillation, at least twice from an efficient still system, is a procedure that should *follow* passage through an ion exchange column, if one is used. A simple control on the quality of water is conductivity (or its reciprocal, electrical resistance), which measures free ions. Water acceptable for culture media should have a resistance of 1.5-2.0 mohms when freshly distilled. Some water stills have a built-in conductivity meter and a system for rejecting, from collection for storage, water with resistance below acceptable levels (e.g., 0.6-1.0 mohms). Water should preferably be used freshly distilled, because materials may be dissolved from storage vessels. These may be organic materials from plastics, or principally ions from glass. The acquisition of ions from glass storage vessels was documented many years ago by Healy et al. [1952]. A rise in conductivity upon storage caused by solution of atmospheric CO_2 is acceptable, but ions or molecules (or microorganisms) from a storage tank or bottle are not, if a well-defined medium is the final objective.

Purity of Chemicals

Media should be prepared from chemicals of the highest available purity. When a reliable source has been located, it is generally wise to continue to purchase from that source. Even the best commercial chemicals inevitably contain traces of contaminants [Aronowitz, 1981], and for some of these, maximum permissible concentrations are listed on the labels. Other trace substances—some of which may be toxic (e.g., Hg), but others of which may be found among those that have beneficial effects for some cells (e.g., Se, Cr, V, As, Sn) [McKeehan et al., 1976; Nielsen, 1981; Bettger and Ham, 1982a]—may also be present (but unlisted) adventitiously in components such as the major salts or the sugars. This caution applies with equal or greater force to biologicals, such as hormones, where other peptides or steroids, as well as trace elements and unrelated biological contaminants, may be present in small amounts.

Stability and Shelf Life

Inorganic chemicals that go into culture media are in general indefinitely stable, although some, such as $MgCl_2 \cdot 6H_2O$, are deliquescent, making them hard to weigh accurately after they have taken up water. Others may be subject to oxidation. Less stable components, such as some of the vitamins, hormones, and growth factors, should be stored in a refrigerator or frozen, as their properties demand.

Several compounds included in many media are quite unstable in solution. These include L-glutamine (also one of the most costly components) and ascorbic acid, which is readily oxidized, especially in neutral or basic solutions. The stability of some components is highly pH-dependent, e.g., some of the B vitamins [Frost and McIntire, 1944], a property that is taken advantage of in the preparation of autoclavable media (see pp. 41 and 47, below).

Shelf life is dependent not only on the natural decay rate of unstable compounds, pH, temperature of storage, and access to oxygen, but also to exposure to near-ultraviolet or "daylight" fluorescent light (see p. 30, below). Media should usually be stored at $+5°C$, because frozen storage may result in loss of some poorly soluble components (e.g., tyrosine), which may not redissolve completely when the medium is thawed. It is rarely desirable to store a complete medium for more than 3 weeks, although this time limit may depend upon the amounts and kinds of unstable constituents. For example, media containing insulin, which attaches to glass surfaces, may lose hormone on storage from this cause, if stored in glass. Insulin does not attach so readily to polycarbonate.

NUTRIENTS AND SPECIFIC FACTORS

Although Ham [1981], who reviewed this subject exhaustively, properly distinguishes nutrients and growth requirements from mitogens, growth factors, and hormones, we shall here distinguish more broadly between basic components generally contained in culture media (which include some items with identifiable specific functions, e.g., K^+, Ca^{2+}, or Zn^{2+} ions) from those added to produce or stimulate specific functions. The distinction is arbitrary, since several components fall into both categories, and some of the differences are quantitative rather than qualitative.

Basic Components

Inorganic salts. The major ions—Na^+, K^+, Ca^{2+}, Mg^{2+}, Cl^-, and HPO_4^{2-}, and usually, unless a means of generating endogenous CO_2 (e.g., pyruvate) is provided, HCO_3^-—are principally involved in maintaining electrolyte balance and contributing to the osmotic equilibrium of the system. Trace elements have been included in many media since their importance was established by Shooter and Gey [1952], e.g., the media of Kitos et al. [1962], Thomas and Johnson [1967], and Higuchi [1970], and most recently, in those of Ham and McKeehan (see review by Ham [1981]). Significant amounts of Fe^{2+}, as well as some Zn^{2+} and Cu^{2+}, may be contributed to a medium that contains methylcellulose (Methocel), which is often included as an "inert" protective agent, especially in spinner cultures. Interrelationships exist between the Fe^{2+}, Zn^{2+}, and Cu^{2+} ions, which are needed for many cells, and Mn^{2+}, which may be inhibitory, but the effects of these interactions can be overcome by increasing the Fe^{2+} [Thomas and Johnson, 1967]. Most media being devised today include at least Fe^{2+}, Cu^{2+}, Zn^{2+}, Co^{2+}, and $HSeO_3^-$. Major as well as minor ions may act as regulators of growth or function. For example, the level of K^+ is important for heart cell function [Pappano and Sperelakis, 1969; De Haan, 1967, 1970; Vahouny et al., 1970]; for gonad primordia [Fisher and Liversage, 1973]; for neurons [Scott and Fisher, 1970, 1971; Scott, 1971; Lasher and Zagon, 1972]; and for glia [Hertz et al., 1973], and it is noteworthy that fetal kidney requires high K^+ for normal development in vitro, in view of the fact that fetal serum (mouse) contains levels of K^+ about twice as great as maternal serum [Crocker and Vernier, 1970; Waymouth, 1972]. Ionized calcium plays an important role in the control of mitosis [Perris and Whitfield, 1967; Perris et al., 1968; Boynton et al., 1974; Boynton and Whitfield, 1976; Whitfield et al., 1971, 1981]. Calcium, and the ratio of Ca^{2+} to Mg^{2+}, are proving to be important in controlling cell proliferation and transformation in vitro [Whitfield et al., 1969; Balk et al., 1973; Swierenga et al., 1976; Tupper and Zorgniotti, 1977; Powers et al., 1982]. Polycations such as poly-L-lysine, which are often used to promote attachment of cells to culture dishes, may release Ca^{2+} from its binding sites [Whitfield et al., 1968]. This is important, because calcium also affects functions in vitro that relate to differentiation. For example, Peehl and Ham [1980b] noted that, in medium MCDB 151 supplemented with 1 mg/ml of dialyzed fetal bovine serum protein (FBSP), epidermal keratinocytes grew as monolayers when the Ca^{2+} was low (0.03 mM), but stratified when the concentration was raised to 1 mM. In a similar medium (MCDB 152, containing growth factors, hormones, and some trace elements not in MCDB 151, and no FBSP), a similar effect was observed [Tsao et al., 1982].

As the main components of culture media are more completely purified, the roles of trace elements are being recognized. Serum-free media may contain some of these important elements, e.g., significant amounts of Fe, Cu, and Mn were found in (not added to) a serum-free medium by Takaoka and Katsuta [1971]. The elements Fe, Cu, Zn, Co, Mo, Mn have been included in some media for a number of years. More recently, others have been identified. The recent work on requirements for these elements by cells in vitro has been reviewed by Nielsen [1981] and by Ham [1981]. Among the newer candidates for activities in vitro are Se, Sn, V, Cr, Al, and As [Bettger and Ham, 1982a]. The effect of selenium upon clonal growth of cells was identified by McKeehan et al. [1976] and Hamilton and Ham [1977], and this element is proving to be important for many cell types [Bettger and Ham, 1982a], usually at 1 or 3×10^{-5} mM. A much higher concentration of selenium was used by Simms et al. [1980], who demonstrated that human oat cell carcinoma cells can *survive* in a hormone-supplemented RPMI 1640 medium but, for continuous *growth* after passage, require 3×10^{-1} mM selenium. The interaction of numerous ions with important metabolites in vivo has been documented by Levander and Cheng [1980] and by McKeehan [1982b], and many of these are likely to have counterparts in vitro.

Amino acids. All media contain amino acids, but the number and proportions vary widely. As building blocks for protein synthesis, and also as energy sources, amino acids are highly important, but the classification into "essential" and "nonessential," on the basis of their requirement by whole organisms, appears to be invalid when applied to cells in vitro. Also no longer appropriate is the hypothesis upon which the mixture of 19 amino acids that was included in medium 199 [Morgan et al., 1950] was based, namely, that the amino acids should be provided in the proportion in which they are found by analysis to occur in tissues [Block and Bolling, 1945]. Not only are the requirements of different cells different [Christensen, 1982], but the abilities of different cells to act upon media components vary, so that some amino acids exert sparing effects on others, by active interconversion, transamination, and amination of keto acids [Patterson, 1972]. Media with high levels of amino acids provide insurance against depletion (although too high concentrations may also be toxic) and also an energy source. The different abilities of cells to metabolize amino acids is the basis of a useful method for selecting epithelial cells in mixed populations. Fibroblasts, at least from human and rodent sources [Gilbert and Migeon, 1975], and human lymphoid cells [Benke and Dittmar, 1976], do not contain the enzyme D-amino acid oxidase and therefore cannot, as epithelial cells can, convert D-valine to L-valine. Media in which D-valine is substituted for L-valine can be used in the early stages of long-term culture, to select for epithelial cells, and be replaced by an L-valine-containing medium after the fibroblasts have been eliminated. An analogous procedure, described by Kao and Prockop [1977], uses the proline analog cis-hydroxy-L-proline to remove fibroblasts. Hepatocytes are unique in containing the enzyme transcarbamylase and can therefore convert L-ornithine to L-arginine. An arginine-free, ornithine-containing medium has therefore been used as a basis for selecting hepatocytes in liver cultures [Leffert and Paul, 1973; Salas-Prato, 1982].

It was for long rather generally believed that L-glutamine was "essential," in the sense that it must be added to every medium. It does play several important roles in cell metabolism [McKeehan, 1982a], but in many systems it can be dispensed with by substitution of sufficient amounts of other amino acids (e.g., by Pasieka et al. [1958]; Griffiths and Pirt [1967]; Nagle and Brown [1971]; Waymouth [1979, 1981a]). Advantage is taken of this in the preparation of glutamine-free autoclavable media (see p. 41, below), where the heat-lability of glutamine prohibits its use.

Vitamins. Members of the B vitamin group, most of which act as cofactors in enzyme reactions, are requisite medium components, often (e.g., biotin and vitamin B_{12}) in very low concentrations. Choline and i-inositol are usually grouped with the B vitamins. Inositol has been shown to be required for some but not all cell lines [Jackson and Shin, 1982]. The requirements for the B vitamins have been reviewed by Waymouth [1954, 1972], Ham [1981], Morton [1981], and Bettger and Ham [1982a].

Wang has drawn attention to a phenomenon with important practical consequences, namely the formation of cytotoxic products upon exposure of solutions containing riboflavin and the aromatic amino acids tyrosine and tryptophan to sources of "daylight" fluorescent or near-UV light of wavelengths 300–420 nm [Wang et al., 1974; Stoien and Wang, 1974; Wang, 1975, 1976; Nixon and Wang, 1977]. The principal phototoxic product has been identified as hydrogen peroxide [Wang and Nixon, 1978]. The cobalamins (B_{12}) are also decomposed by exposure to light in the presence of oxygen [Hogenkamp, 1966]. To avoid the problem of loss of activity of these components, media should be prepared and used under light of greater than 500 nm wavelength, and storage of media and solutions containing riboflavin and vitamin B_{12} should be in the dark. Visible and near-UV light can also damage cells [Sanford et al., 1978].

Vitamin C, although quite unstable except in acid solutions, has been included in several media, primarily for its contribution to the redox potential, e.g., in conjunction with glutathione and cysteine [Waymouth, 1959; Fell and Rinaldini, 1965; Morton, 1981]. Specific effects identified with ascorbic acid include a role in proline hydroxylation [Peterkofsky, 1972; Feng et al., 1977; Faris et al., 1978].

Although some earlier serum-free media (e.g., medium 199 of Morgan et al. [1950]), and NCTC 107 [Evans et al., 1956] and NCTC 135 [Evans et al., 1964], contain several fat-soluble vitamins (as well as many coenzymes, fatty acids, and nucleic acid precursors), the fat-soluble vitamins have not generally been included as basic components of the more recent media. Vitamin E (α-tocopherol) has been used principally as an antioxidant [Bannai et al., 1977]. It has proven to be essential for the survival and for optimal functioning, in a serum-free medium containing several growth factors and hormones, of porcine Leydig cells [Mather et al., 1982]. Vitamin A and its analogs, especially retinoic acid, have important effects on cell differentiation and function [Fell and Rinaldini, 1965; Sporn et al., 1975; Jones-Villeneuve et al., 1982] and on membrane stability [Smith, 1981].

Lipids. The roles of lipids and fat-soluble components generally have recently come under scrutiny [King and Spector, 1981]. In serum-containing media fatty acids, sterols, phospholipids, and many other metabolites are carried on macromolecules, principally proteins [Fillerup et al., 1958; Yamane et al., 1975; Yamane, 1978a,b]. Fatty acids may bind to serum proteins in high proportions (e.g., bovine serum albumin [BSA] may be present in a BSA:fatty acids ratio of 9:1), so that, when proteins are present, they may both release beneficial fatty acids and bind those that may be inhibitory [Steele and Jenkin, 1977a,b]. In serum-free media, fatty acids have also been pro- vided in bound form, either to BSA as a carrier for oleic acid [Jenkin and Anderson, 1970; Morrison and Jenkin, 1972; Makino and Jenkin, 1975; Guskey and Jenkin, 1976], or to dialyzed fetal bovine serum protein for linoleic acid, in medium F-12 [Ham, 1965], and the subsequent MCDB series of Ham and his colleagues [McKeehan et al., 1977; Peehl and Ham, 1980b]. Linoleic acid is a precursor of the prostaglandins [Hinman, 1972], a group of compounds that are now included in several media, especially those for kidney cells [Taub and Livingston, 1981]. Many cells, particularly nontrans- formed cells, require fatty acids. Cholesterol [Chen and Kandutsch, 1981] and phospholipids are likely to become regular components of basic serum- free media. Ethanolamine or phosphoethanolamine already appear in several formulations [Murakami et al., 1982; Kano-Sueoka and Errick, 1981]. It is no longer necessary to add lipids in solvents such as ethanol, because various kinds and combinations of lipids (e.g., lecithin, cholesterol, sphingomyelin, vitamin E, and phospholipids) can be effectively introduced into serum-free media in the form of phospholipid-enclosed vesicles or liposomes [Poste et al., 1976; Bettger et al., 1981; Bettger and Ham, 1982b].

Energy sources. Glucose, the predominant sugar in adult sera, has been the principal energy source in most serum-free media. Fructose, the most abundant carbohydrate in mammalian fetal blood, has also been used, and it as well as other sugars (including galactose, mannose, maltose) is found to be utilized by many cell types in vitro [Harris and Kutsky, 1953; Eagle et al., 1958; Morgan and Morgan, 1960; Baugh et al., 1967]. A possible advantage of galactose as a substitute for glucose is the reported inability of some mycoplasmas to utilize it [Hu et al., 1975]. It was included by Leibovitz [1963] in his L-15 medium, on the basis that it produced less lactic acid than glucose, and therefore assisted in maintaining a stable pH.

The concentrations of sugars used in serum-free media have usually exceeded those in circulating blood, which are about 1,000 μg/ml for mammals and 2,000 μg/ml for birds. Concentrations of 5,000 μg/ml (27.8 mM) of D-glucose are not uncommon and allow for an adequate reserve between medium changes. High glucose levels appear to be important for cultures of nerve tissues (see Shapiro and Schrier [1973], who used 26 mM). The use of a high glucose concentration must be one of the first recorded methods for cell selection by medium manipulation. Latta and Buchholz [1939] reported that fibroblast growth ceased, but embryonic muscle migration continued, in 5,000 μg/ml glucose.

Contributions to the energy needs of cells are made by amino acids and α-keto acids (e.g., pyruvate, α-ketoglutarate, or oxaloacetate) [McKeehan and McKeehan, 1979; Groelke et al., 1979], but they cannot substitute for a carbohydrate. The complexities of the energy metabolism of cells in culture are discussed and reviewed by Gregg [1972].

Purines and pyrimidines. The requirements of cells in culture for nucleotide precursors were reviewed extensively by Kelley [1972]. Most cells appear to be able to synthesize nucleotides de novo, although providing purine or pyrimidine bases or nucleosides may increase efficiency. The bases and nucleosides have been included in many serum-free media, but with little consistency, qualitative or quantitative. Kelley [1972] speculated that "in general, the free bases would be more effective than the nucleoside derivatives as a source of preformed purines, whereas the nucleoside derivatives would be more effective than the free bases as a source of preformed pyrimidines." The bases most commonly found in serum-free media are adenine, hypoxanthine, and 5-methylcytosine. Nucleosides found are mainly thymidine, deoxyadenosine, deoxycytidine, and deoxyguanosine. Many of the early media of the NCTC series [Evans and Sanford, 1978] also contained a group of six nucleotide coenzymes, but they were not included in the more recent members of that series.

An unusual effect of adenine was observed by Peehl and Ham [1980a], who found that, in medium MCDB 151, epidermal keratinocytes required a high concentration (0.18 mM) of adenine, and also that this high level of adenine contributed to the retardation of fibroblast growth.

Hormones and growth factors. All cells have receptors for at least several, and probably many, hormones and growth factors [Cuatrecasas and Hollenberg, 1976]. These substances mediate many essential functions, including events in the cell cycle, synthesis of new cellular components, expression and repression of differentiation, and the production of specific products [McCarty and McCarty, 1977; Rudland and Jiménez de Asúa, 1979; Vasiliev and Gelfand, 1981; Mather and Sato, 1982]. Many peptide hormones are members of families [Blundell and Humbel, 1980] and may compete for common receptors.

The distinction between hormones (and other factors) generally required by many cell types, and therefore considered "basic," and those that determine specific pathways and functions of only a few cells cannot be rigidly made. Some may, at different concentrations, exert both metabolic and special stimulatory functions. Thus, most hormones and growth factors will be discussed below, under the heading Special Factors. Hormones now generally included as basic components, at least for epithelial cells, are insulin, which is multifunctional [Jiménez de Asúa et al., 1973; Steiner et al., 1978; Pawelek et al., 1982], and cortisol (hydrocortisone), or its synthetic analog dexamethasone [Walthall and Ham, 1981]. Cortisol inhibits fibroblasts [Noyes, 1973], enhances the activity of fibroblast growth factor [Gospodarowicz, 1974], binds insulin and epidermal growth factor (EGF) [Baker et al., 1978], and stimulates the production of fibronectin, a cell product that is important in the attachment of cells to surfaces [Marceau et al., 1980]. Dexamethasone has been found to elicit different responses from different clones of melanoma cells [Buzard and Horn, 1982].

Special Factors

Serum-free media for many cell types now include a variety of hormones and growth factors, tailored to the needs of the particular cells [Gospodarowicz and Moran, 1976; Barnes and Sato, 1980a,b]. The first "growth factor" so named was the nerve growth factor [Levi-Montalcini, 1952, 1966; Levi-Montalcini and Angeletti, 1968]. This has been followed by a long and still growing list [Pledger et al., 1982]. Some affect a relatively narrow spectrum of cells, e.g., the quite specific NGF [Bradshaw and Young, 1976; Server and Shooter, 1977]; factors specific to glial cells [Monard et al., 1975], namely the glial factor (GLF) that affects glial differentiation [Pfeiffer et al., 1977], and the glial maturation factor (GMF) [Lim et al., 1977]; and the connective tissue-activating peptides (CTAPs) [Castor and Cobel-Geard, 1980; Castor and Fremuth, 1982; Castor et al., 1981]. Others are less specific, e.g., epidermal growth factor [Cohen and Taylor, 1974; Carpenter and Cohen,

1979; Carpenter, 1980; Carpenter et al., 1982; Das, 1982], which affects not only epithelial cell proliferation and maturation but also the growth of fibroblasts [Hollenberg and Cuatrecasas, 1973; McKeehan, 1982b; McKeehan et al., 1982], and the lifespan of epidermal keratinocytes [Rheinwald and Green, 1977]; fibroblast growth factor [Gospodarowicz, 1974, 1975; Gospodarowicz et al., 1982]; platelet-derived growth factor (PDGF) [Ross et al., 1974; Kohler and Lipton, 1974; Ross et al., 1982]; somatomedins [Van Wyk et al., 1974; Van Wyk and Furlanetto, 1982; Westermark et al., 1981; Rothstein, 1982]; the prostaglandins [Taub, 1982; Mather et al., 1982]; and transformation factors [Roberts et al., 1982; Twardzik et al., 1982].

Growth factors that act specifically as proliferation-controlling signals are listed and discussed by Leffert and Koch [1977]. These range from ions, nutrients, cyclic nucleotides, hormones, and specific factors to cell density, pH, and gas tension. Mitogens of plant origin, such as phytohemagglutinin and pokeweed mitogen, are usually not considered as medium components. Barnes and Sato [1980b] show 20 growth factors and hormones, together with the iron-binding protein transferrin and the fatty acid-binding BSA, that are now regularly included in many serum-free media. They also list the positively charged polymers, such as poly-L-lysine, and four proteins that affect the abilities of cells to spread on and attach to surfaces. Poly-D-lysine and poly-DL-ornithine have also been used successfully for this purpose [McKeehan and Ham, 1976].

Only a few examples can be given here of factors that fulfill the specific requirements of particular cells and tissues. As with combinations of nutrients, so also with stimulatory factors, the metabolic interactions that occur within cells and in the medium enable a number of distinct and different combinations of factors to be effective. These interactions are exemplified within the growth factor category itself. For example, polypeptides have been classified as *endocrine*, which act in vivo on distant targets, and may act on these target cells in vitro; *autocrine,* in which an active endocrine hormone, e.g., an estrogen, stimulates cells to produce their own polypeptide growth factors; and *paracrine,* in which such factors act upon *adjacent* cells. All three types act upon mammary tissues [Leland et al., 1982; Ikeda et al., 1982].

Whole mouse mammary gland can undergo all of its developmental stages in vitro, in a serum-free medium supplemented with a succession of combinations of steroid and peptide hormones appropriate for each stage [Wood et al., 1975; Banerjee et al., 1982]. In a medium supplemented with 17β-estradiol, insulin, transferrin, dexamethasone, and triiodothyronine, proliferation of human breast cancer cells ceased within 7 days when either triiodothyronine or transferrin was omitted [Allegra and Lippman, 1978]. Neural cells, besides neurons, that can respond to NGF, and glial cells that can respond to GLF and GMF, respond to a variety of factors. Thus a medium that was designed for Schwann cells contains insulin, hydrocortisone, transferrin, progesterone, putrescine, and selenium [Bunge et al., 1982]. A neuroblast medium contains insulin, hydrocortisone, transferrin, the tripeptide glycyl-L-histidyl-L-lysine [Pickart and Thaler, 1973], and thyroid-stimulating hormone (TSH) [Coon and Sinback, 1982]. Astrocytes have been grown with insulin, hydrocortisone, putrescine, and prostaglandin (PG)E$_\alpha$ [Morrison and de Vellis, 1982]. A fastidious oat cell carcinoma of human lung has been grown in the RPMI 1640-based HITES medium (hydrocortisone, insulin, transferrin, 17β-estradiol, selenium) by Simms et al. [1980] and Minna et al. [1982]. The estrogen 17β-estradiol, with hydrocortisone and EGF, also stimulates the growth of human keratinocytes [Peehl and Ham, 1980a].

Among vitamins that exert specific effects, vitamin A affects differentiation of skin [Fell and Mellanby, 1953; Yuspa and Harris, 1974], and the growth, but not the differentiation, of retina [Fell and Rinaldini, 1965], and suppresses keratinization of skin and vagina [Kahn, 1954; Lasnitzki, 1961; Fell and Rinaldini, 1965]. Vitamin E, acting primarily as an antioxidant, exhibits multiple functions on cultures of Leydig cells, affecting cell survival, receptor maintenance, stimulation of testosterone synergistically with human chorionic gonadotropin (hCG), and suppression of prostaglandin secretion [Mather et al., 1982]. These are representative examples to illustrate the combinations of factors that enable cells in serum-free media to express their growth and functional potentials.

BUFFERS AND GAS PHASE

Buffers are treated separately from the other components of culture media because of their special importance for maintaining a proper environment for the metabolism, growth, and functioning of cells. Major ions (Na$^+$, K$^+$, HCO$_3^-$, and HPO$_4^{2-}$) are usually regarded as the principal components in pH control, along with H$^+$ and OH$^-$, which enter into the ion balance. Other components, including amino acids, if present in high concentrations, can contribute to the buffering power of a medium. The media of Leibovitz [1963] and of Ling et al. [1968] are examples of solutions in which organic compounds exert significant buffering power. The medium L-15 of Leibovitz [1963] eliminated bicarbonate completely. In this medium, glucose was replaced by galactose, to reduce the amount of lactic acid generated; high levels of the basic amino acids arginine, histidine, and lysine as free bases, and relatively high concentrations of inorganic phosphates (basic and acidic) were included; and 5 mM pyruvate provided an endogenous source of CO_2. Another medium, based on the principal of reducing $NaHCO_3$ and supplying pyruvate, was the CMRL 1415 of Healy and Parker [1966]. This medium also contains glucose and galactose in equimolar proportions (2.78 + 2.78 mM), instead of the 5.56 mM glucose of the earlier medium CMRL 1066, and 2 mM pyruvate. These changes, and the inclusion of high levels of arginine and glutamine, enabled Healy and Parker [1966] to omit completely the bicarbonate in medium CMRL 1415-ATM, which could be used in free gas exchange with the atmosphere.

Conventional media, containing amounts of $NaHCO_3$ comparable to those in circulating blood (e.g., 26 mM) require to be equilibrated with CO_2 at far above atmospheric concentrations, e.g., at 37°C and 760 mm Hg, with 5% CO_2 to maintain a pH of 7.66, and with 10% CO_2 for a pH of 6.33 [Waymouth, 1981b]. Such amounts of bicarbonate are, in the organism, dealt with by a sensitive homeostatic mechanism. They are not appropriate, nor do they seem to be generally required, for cell cultures, as has been demonstrated by Kelley et al. [1960a,b], by Gwatkin and Siminovitch [1960], by Moore et al. [1963, 1966], by Ling et al. [1968], and in a number of later studies.

The zwitterionic buffers described by Good et al. [1966] have become popular as replacements for bicarbonate buffers, in part because of the convenience of avoiding the necessity to maintain strict control of the gas phase. HEPES (N-2-hydroxyethyl-piperazine-N'-2-ethanesulfonic acid) is the most commonly used in culture media. Compared to the pK_a at 37°C of $NaHCO_3$ of 6.2, and that of Tris [tris(hydroxymethyl)aminomethane] of 7.9, HEPES has a pK_{a2} of 7.3. Combinations of several of the Good's buffers were recommended by Eagle [1971]. Shipman [1969] evaluated these buffers for the culture of several cell types, in a study that included measurements of the amounts of HEPES or $NaHCO_3$ required to raise the pH by one unit. High concentrations of HEPES (e.g., 100 mM) were shown by Shipman to cause diffuse cytoplasmic vacuolation in some cells, but 25 mM was in general found nontoxic and 10 mM sufficient to buffer conventional media supplemented with serum. However, Poole et al. [1982], who have used HEPES at 15 mM and TES (N-tris(hydroxymethyl)methyl-2-aminoethanesulfonic acid) and BES (N,N'-bis(2-hydroxyethyl)-2-aminomethanesulfonic acid) at 10 mM in a similar serum-supplemented medium, observed vacuolation of chondrocytes, even at these concentrations. They cite 15 papers published between 1968 and 1981 in which various adverse effects (principally membrane-related) attributed to the zwitterionic buffers are reported. Some of the Good's buffers can chelate biologically important cations (reviewed by Waymouth [1981b]). Fibroblasts appear to be more tolerant than epithelial cells to HEPES [Ham, 1981].

The possibility of such damaging effects on cells needs to be weighed against the inconvenience and expense of using high CO_2. Advantages as well as disadvantages of HEPES and TES have been recorded, e.g., the increase in number of population doublings of human lung cells [Massie et al., 1972] and chick cells [Massie et al., 1974], over the 50+ doublings regarded as the norm for human lung fibroblasts [Hayflick and Moorhead, 1961; Hayflick, 1965] in a $NaHCO_3$–CO_2-buffered Eagle's basal medium. The use of a bicarbonate-buffered, but more complete, medium (e.g., CMRL 1969) for WI-38 human lung fibroblasts was shown by Kadanka et al. [1973] to prevent heteroploid transformation at high passage. Yamane et al. [1981] achieved 76 doublings of human embryo fibroblasts in a serum-free medium containing HEPES and $NaHCO_3$, several growth factors, and trace elements.

As noted below (p. 42), a useful buffer for use in the presence of low or no bicarbonate has proven to be sodium β-glycerophosphate, first used for this purpose in a cell culture medium by Ling et al. [1968]. It buffers satisfactorily, having a pK_a at 37°C of 6.6, compared to that of $NaHCO_3$ of 6.2 [Waymouth, 1978, 1981b]. Even in the presence of liver cells, which may be expected to contain esterases, β-glycerophosphate was found not to release inorganic phosphate into the medium [Waymouth, 1981b].

Various "optimal" pH values for different cell types in vitro have been reported (e.g., by Ceccarini and Eagle [1971], Eagle [1973], and Bear and Schneider [1977]). However, that such pH optima are not necessarily absolute or characteristic of the particular cells, but may depend rather on conditions in the culture system, is exemplified by comparing the report of Eisinger et al. [1979], who found Eagle's minimum essential medium (MEM) at pH 5.6–5.8 to be optimal for the proliferation of human keratinocytes and obtained no growth at pH 7.0, with that of Peehl and Ham [1980b], who grew similar cells successfully in another medium at pH 7.4.

The requirements of cells for CO_2, when not provided for through the use of bicarbonate as the major buffer, are usually met either by a low concentration of HCO_3^-, included for metabolic rather than for buffering purposes, or by adding a keto acid, to generate CO_2 metabolically, such as oxaloacetate [Gwatkin and Siminovitch, 1960] or pyruvate [Kelley et al., 1960a,b; Moore et al., 1963, 1966; Leibovitz, 1963; Ham, 1963, 1974]. Addition of a bicarbonate to a bicarbonate-free medium (Leibovitz's L-15) for neurons stimulated nonneuronal cells, thus enabling neurons to be cultivated either with or without nonneuronal cells [Mains and Patterson, 1973].

The proper level of oxygen also needs to be considered, since 20%, about the amount in ambient air, is not necessarily optimal for systems in vitro and indeed may cause cell and chromosome damage [Sanford et al., 1978].

Mouse embryos in vitro [Whitten, 1969, 1971] and some somatic cells [Parshad and Sanford, 1971; Parshad et al., 1977; Richter et al., 1972; Taylor et al., 1974] thrive better in as little as 1%, 2%, or 5% oxygen. Even in a system with atmospheric oxygen in the gas phase, however, in large volume cultures, e.g., spinner cultures of 100–200 ml, the partial pressure of oxygen dissolved in the medium may drop from 140 mm Hg in fresh medium to less than 10 mm Hg when the population of cells becomes dense (e.g., 6×10^6 cells per 1 ml), without loss of viability or of the ability of the cells to synthesize protein [Glinos et al., 1973]. In other systems, high oxygen may prove inhibitory. The effects of a wide range of oxygen tensions on WI-38 cells was explored by Balin et al. [1976], who also reviewed the literature on such effects. Mitotic inhibition by high oxygen, in human diploid cells in an autoclavable medium, can be reversed [Balin et al., 1978].

ANTIBIOTICS

It has become a common practice to include antibiotics in culture media. Such widespread use started when most media were supplemented with serum, and when it was not realized that the presence of serum can conceal, or protect against, biological activities that affect somatic cells as well as microorganisms. Even in media containing relatively high concentrations of serum, cytotoxic effects of the antibiotics most commonly used in culture media (penicillin, streptomycin, gentamicin) were recognized [Heilman, 1945; Metzger et al., 1954]. In some of the early studies, toxic effects were attributed to impurities in the antibiotic preparations then available [Bucher, 1947]. However, modern studies have demonstrated effects attributable unequivocally to the antibiotics themselves. For example, protein synthesis and protein degradation, in short-term cultures of rat hepatocytes in a serum-free medium, are inhibited by the three common antibiotics [Schwarze and Seglen, 1981], and the yield of fibroblasts in long-term cultures (up to 70 days) is reduced in antibiotic-containing media, even in the presence of 20% serum [Goetz et al., 1979]. This is, however, a property that may be used during the initial establishment of a culture as a means of reducing the population of unwanted fibroblasts and securing the selection of other cells, e.g., epithelial cells, from a mixed population.

Specific functions, including those associated with the synthesis of particular proteins, may also be affected. Thus, the output of insulin from isolated rat islets of Langerhans is reported to be reduced by 24% in the presence of 10^{-5} M, and almost completely inhibited by 10^{-3} M, gentamicin [De Lattre et al., 1982]. Long-term use of an antibiotic can, in culture as in other applications, lead to the selection of resistant microorganisms [Coriell, 1973].

For these reasons, great care should be taken in the use of antibiotics [Perlman, 1979], and especially, because of the protective role of serum, in serum-free media. Antibiotics should not be used in strict nutritional studies [Waymouth, 1965a]. In spite of warnings about the metabolic effects of antibiotics in cell cultures (e.g., by Coriell [1973], Merchant [1973], and Armstrong [1973]) their indiscriminate use has persisted. Ten years ago Armstrong [1973] stated that "the general use of a combination of penicillin and streptomycin is as antiquated in a tissue culture laboratory, as it is in clinical infectious diseases" (a statement that should perhaps better have read "*should be* as antiquated"), and further that "the effects of [an] antibiotic regimen on a particular cell metabolism must be established by trial."

This remains good advice for those cases where antibiotics have a legitimate place, e.g., for cultivation of skin, intestine, or tumors, especially of human biopsy material, where it is difficult or impossible to obtain sterile specimens. In such cases, care must be used in selecting antibiotics compatible with the growth or function that is to be studied. The relative stabilities of the antibiotics should also be kept in mind. Stability is usually pH-dependent. At physiological pHs, penicillin may lose 50% of its activity in 1 day or less, unless the pH is rather low (e.g., pH 6.5), in which case more than 25% of the activity may persist for 5 days. Streptomycin retains 30–50% of its activity for up to 14 days at pH 6.5–8.0 [Schafer et al., 1972]. Gentamicin has the advantage of stability over a wide range of pH for at least 15 days at 37°C, and it can be sterilized by autoclaving [Schafer et al., 1972; Casemore, 1967]. Gentamicin also has mycoplasmacidal, as well as bacteriostatic, properties [Braun et al., 1970], but in taking advantage of this property, one needs to be aware that intracellular mycoplasmas are not affected [Gori et al., 1964; Boatman et al., 1976], and are therefore capable of reinfecting a culture as they are released.

ADVANTAGES OF MEDIA PREPARED IN THE LABORATORY VS COMMERCIAL MEDIA

A number of media, but usually not the most up-to-date or those designed for specialized or fastidious cells, can be obtained commercially. These media, because intended for general rather than specific use, are often incomplete (except for very short-term use), usually require supplementation with serum or serum fractions, and often also with hormones and growth factors. Advantages of commercially prepared media include some saving of time and avoidance of the necessity to keep on hand stocks of all the components. Such considerations are most pertinent for small users, or those who make occasional use of cell culture techniques. The disadvantages are that some media components are unstable and that media may deteriorate on storage, especially when exposed to fluorescent room light [Wang, 1975, 1976; Wang et al., 1974; Nixon and Wang, 1977]. A principal product of this effect has been identified as hydrogen peroxide, and about 40% of the toxicity can be removed by catalase, but other toxic products are also present [Wang and Nixon, 1978]. Commercial suppliers of media take time—quite legitimately— to test each batch of medium before placing it on the market, but this causes an inevitable delay between preparation and use by the purchaser. The commercial suppliers also routinely test the effectiveness of their media on one or more cell lines; but it should be borne in mind that these cells may be different from those for which the medium is purchased. Tests are also made of the osmolality of each batch, marked variation of which from the norm signaling an error in the composition of the lot, and are made for microbiological contaminations, including mycoplasmas.

Media prepared in the laboratory for a specific task at hand and to meet the requirements of particular cells for major and minor nutrients, ion balance, hormones, growth factors, osmolality, and gas phase are the ideal towards which the investigator should work. Some guidance in deciding how to choose or construct the right medium is given by Ham and McKeehan [1979]. It is characteristic of biological systems, however, including cells in culture, that they can adjust their metabolism within fairly wide limits to variations in the environment. One should therefore not expect only one set of conditions to be able to fulfill all the requirements. It is nevertheless worth the effort, in the long run, to study the needs of ones chosen cells (including especially their known functions in vivo) and to try to meet them as closely as possible.

PROCEDURES FOR PREPARING MEDIA IN THE LABORATORY

Media preparation requires a disciplined routine. It need not be very time-consuming, if stock solutions of groups of components are made up and stored (in most cases frozen). Combinations of these are used to make complete media. This method has the advantage that the development of optimal conditions can be approached by testing variations in certain components or groups of components, e.g., those in one of the several stock solutions.

Required for the preparation of media are chemicals and biologicals (e.g., hormones and growth factors) of high purity; a good analytical balance; a hot plate with magnetic stirrer; volumetric flasks of various volumes; a pH meter; an osmometer, preferably a vapor pressure type, which uses only microliter quantities of samples; and equipment for sterilizing, usually by membrane filtration, but in some cases by autoclaving. Polycarbonate membrane filters are preferred over cellulose ester filters, because they release less soluble materials into the filtrate and, if protein carriers or protein hormones are used, retain less protein [Hawker and Hawker, 1975].

Examples will be found in the literature of general procedures for preparing media, e.g., by McKeehan et al. [1977]. Two media are described below, with full details of the methods of preparing stock solutions, including sterilization and storage, and of making the complete media from these stocks. Medium MAB87/3 [Waymouth, 1965b; Gorham and Waymouth, 1965; Waymouth, 1976] has been used for cultivation of various mouse epithelial cell types, including lung [Waymouth, 1965b] and (with the addition of cortisol or dexamethasone) fetal liver hepatocytes [Waymouth et al., 1971; Breslow et al., 1973]. Further modifications have been made, including a high osmolality (340 mosM/kg) and low Na^+/K^+ (= 14) for pancreatic islets [Leiter et al., 1974]; and, with several hormones and growth factors and a sodium β-glycerophosphate buffer, an osmolality of 350 mosM/kg and Na^+/K^+ of 51.5 (medium MPG 160/3) for mouse prostatic epithelium [Waymouth et al., 1982].

The second medium here described is an autoclavable medium, AM77/B, [Waymouth, 1979]. Culture media can be autoclaved without significant destruction of active components and are convenient and time-saving. Early examples are those of Nagle [1968] and Yamane et al. [1968]. In these, however, two heat-labile components, bicarbonate and L-glutamine, were sterilized separately and added to the basic autoclaved solution. Besides its heat lability, bicarbonate has the disadvantages noted above (p. 36). Glutamine, when heated in solution, cyclizes to produce 5-carboxypyrrolidone, which is toxic to some cells [Goetz et al., 1973], and ammonia. Attempts to avoid the problem of the heat-labile substances include that of Nagle and Brown [1971], who eliminated the L-glutamine, increased the L-alanine, and added separately after autoclaving a low concentration (12 mM) of bicarbonate. A medium similar to that of Yamane et al. [1968], described by Yasamura et al. [1978], contains glutamine, but these authors note that for some cells it could be replaced by some of the so-called "nonessential" amino acids.

A similar principle has been used in medium AM77/B, which has been used for the study of sterol synthesis in a mouse strain L, clone 929, subline [Marshall and Heiniger, 1979]. In this medium, and others of similar design, 1) L-glutamine is replaced by higher levels than in medium MAB87/3 of the seven amino acids: L-alanine, L-asparagine, L-isoleucine, L-leucine, L-proline, L-serine, and L-valine [Waymouth, 1978, 1979]; 2) a succinic acid–succinate buffer pH 4.25, as recommended by Yamane et al. [1968], is included in the principal solution, so that the B vitamins are autoclaved at this, the pH of their maximum stability; 3) sodium β-glycerophosphate replaces NaHCO$_3$; and 4) choline bitartrate, which was introduced for culture media by Swim [1967], is used in place of the deliquescent choline chloride.

QUALITY CONTROL

The final measure of the quality of a medium is its ability to support the growth, differentiation, or function of the cells under investigation. However, stock solutions as well as the final medium need to be checked for errors in their composition, and the two most useful measures are pH and osmolality. Errors in weighing can often be revealed by these tests. Small variations from the normal pH can be adjusted by addition of HCl or NaOH, but variations of more than a few tenths of a pH unit suggest a problem. Osmolality also readily detects errors in amounts of osmotically active compounds [Waymouth, 1973]. When small adjustments have to be made in osmolality, a simple formula for this has been suggested by McLimans [1979]. Addition to each milliliter of medium of x ml of a 1 mg/ml solution of NaCl (32 mosM/kg) is made according to the following relationship:

$$x = \frac{(\text{desired mosM}) - (\text{observed mosM})}{32}.$$

The pH and osmolality of every stock solution and every batch of fresh medium help to ensure against the results of human error in their preparation.

The osmolality of a medium needs to be tailored to the needs of the cell to be cultured. Optimal osmolality varies with species and cell type, and may even vary slightly with sex [Waymouth, 1970]. For example, Naglee et al. [1969] demonstrated by testing on rabbit embryo development media in the range of 230–340 mosM/kg, varied by 20 mosM increments ($\pm$ NaCl), that the number of hatching blastocysts depended significantly upon optimal osmolality (or perhaps Na$^+$/K$^+$ balance). Vahouny et al. [1970] report a requirement, not only for high K$^+$ (see p. 28) for beating heart cells, but for a narrow range of tolerance to changes of osmolality (varied by addition of sucrose).

MEDIUM MAB87/3

This medium was first used for mouse cartilage and bone [Gorham and Waymouth, 1965] and for mouse lung [Waymouth, 1965b] and later, with addition of dexamethasone or cortisol [Waymouth et al., 1971] for hepato-cytes from mouse fetal liver in long-term culture. Modifications of MAB87/3 in which the bicarbonate buffer is replaced by sodium β-glycerophosphate (as first recommended by Ling et al. [1968]) have been described [Waymouth, 1978, 1981a,b].

Stock Solutions for MAB87/3

S/S87/3 A (× 10).

Component	mg per 1,000 ml	
NaCl	60,000	
KCl	1,500	
$CaCl_2 \cdot 2H_2O$	1,200	
$MgCl_2 \cdot 6H_2O$	2,400	(25 ml of solution A1)
$MgSO_4 \cdot 7H_2O$	1,000	
Dextrose	50,000	
Phenol red (Na salt), 1% aqueous	3.5	(0.35 ml of 1%)

Special instructions. First prepare solution A1 as follows: Dissolve 9,600 mg of $MgCl_2 \cdot 6H_2O$ from a freshly opened bottle of 100 ml distilled water. This is done because the salt $MgCl_2 \cdot 6H_2O$ is highly deliquescent and, in a frequently opened bottle, takes up enough water to make accurate weighing impossible.

Dissolve: The ingredients with heat on a magnetic stirrer.

Sterilize: May be autoclaved, but this is unnecessary unless the medium is to be made up aseptically from sterile stock solutions.

Store: At 5°C indefinitely.

S/S87/3 B (× 10).

Component	mg per 1,000 ml	
Thiamin·HCl	100	
Nicotinamide	10	
Pyridoxin·HCl	10	
Biotin	0.2	(10 ml of solution B1)
Calcium pantothenate	10	
i-Inositol	10	
Cyanocobalamin (vitamin B_{12})	2	(5 ml of solution B2)
Choline chloride	2,500	
Riboflavin	10	(Note 1)
Folic acid	5	(Note 2)

Special instructions. Biotin solution B1: Dissolve 10 mg biotin in 500 ml water. Store frozen.

Cyanocobalamin solution B2: Dissolve 40 mg in 100 ml water. Store frozen.

Dissolve: All the components except the riboflavin and folic acid in about 500 ml of water. *Note 1:* Dissolve 10 mg riboflavin in about 5 ml water, add 0.3 ml of 1 N NaOH. When dissolved, dilute with more water and neutralize with 0.25 ml of 1 N HCl. *Note 2:* Dissolve 5 mg folic acid in about 5 ml water, add 0.15 ml 1 N NaOH. When dissolved, dilute with more water and neutralize with 0.1 ml of 1 N HCl. Add these solutions to the rest of the vitamins and make up to volume.

Sterilize: By filtration.

Store: At 5°C or frozen protected from light.

S/S87/3 C (× 40).

Component	mg per 100 ml
Ascorbic acid	70
Cysteine·HCl·H$_2$O	400
Glutathione (reduced)	60

Special instructions. *Dissolve:* Without heat.

Sterilize: By filtration.

Store: At 5°C. Discard after 2 weeks.

S/S87/3 D (× 10).

Component	mg per 1,000 ml	
L-alanine	112	
L-arginine·HCl	750	
L-asparagine	240	
L-aspartic acid	600	
L-cystine	150	(Note 1)
L-glutamic acid	1,500	
Glycine	500	
L-histidine·HCl·H$_2$O	1,500	
L-isoleucine	250	
L-leucine	500	
L-lysine·HCl	2,400	
L-methionine	500	
L-phenylalanine	500	
L-proline	500	
L-serine	128	
L-threonine	750	

L-tryptophan	400	
L-tyrosine	400	(Note 2)
L-valine	650	
Hypoxanthine	250	(Note 3)
Thymidine	80	
Phenol red, 1%	3.5	(0.35 ml of 1%)

Special instructions. *Note 1:* Dissolve separately in 15 ml of 1 N NaOH. *Note 2:* Dissolve separately in 10 ml of 1 N. *Note 3:* Dissolve separately in water with heating.

Dissolve: Add the cystine, tyrosine, and hypoxanthine solutions to the other components and dissolve with heat on a magnetic stirrer.

Sterilize: Autoclave.

Store: At room temperature. pH is 7.7.

S/S87/3 E (× 100).

Component	mg per 100 ml (87/3E)	mg per 1,000 ml (E1)
$CuSO_4 \cdot 5H_2O$	5.0	200
$MnSO_4 \cdot H_2O$	1.6	64
$ZnSO_4 \cdot 7H_2O$	3.0	120
$(NH_4)_6Mo_7O_{24} \cdot 4H_2O$	2.5	100
$CoCl_2 \cdot 6H_2O$	2.2	88
$FeSO_4$	45.0	1,800
Ascorbic acid		100

Special instructions. First prepare the concentrated stock solution E1, at 40 times the concentration of stock solution 87/3 E. Dissolve the salts in the above order without heat. The ascorbic acid is included to keep the Fe^{2+} in the reduced condition and in solution.

Store: At room temperature, not sterilized, in the dark. Immediately before making up complete medium, prepare solution 87/3 E by diluting 5 ml of solution E1 to 200 ml.

S/S87/3 F (× 40).

Component	mg per 100 ml
L-glutamine	1,400

Special instructions. *Dissolve:* In cold water with vigorous stirring. *Sterilize:* By filtration or store unsterilized.

Store: Frozen in polycarbonate tubes (IEC autoclear No. 1648, 15 ml, 16 × 155 mm) with nontoxic rubber stoppers (West Rubber Co., No. 0).

Thaw: Immediately before inclusion in complete medium. Shake vigorously to ensure complete resolution of the glutamine.

S/S87/3 G (× 25).

Component	mg per 500 ml
Insulin (crystalline zinc)	100

Special instructions. A high-activity (about 26 IU/mg) porcine or bovine insulin may be used.

Dissolve: In a small volume of water, add 0.1 ml of 1 N HCl, dilute and neutralize as soon as the insulin has dissolved with 0.1 ml 1 N NaOH.

Sterilize: By filtration or store unsterilized.

Store: Frozen in polycarbonate tubes. Insulin may be lost on storage in glass, because of its strong adhesion to glass surfaces.

S/S87/3 H (× 10).

Component	mg per 500 ml
Na_2HPO_4	1,500
KH_2PO_4	1,040
$NaHCO_3$	11,200
Phenol red, 1%	1.5 (0.15 ml of 1%)

Special instructions. *Dissolve:* With vigorous stirring.

Sterilize: By filtration at low vacuum or under pressure, or store unsterilized.

Store: At 5°C.

Distilled Water

Distilled water is used to make up the complete medium to the correct volume. Freshly distilled water of high resistance ($1-2 \times 10^6$ ohms) should be used. Water that is stored for prolonged periods of time may pick up solutes from the atmosphere or the container, and occasionally microorganisms, if improperly stored. Small amounts of water for preparing complete medium aseptically can be autoclaved.

Preparation of Complete Medium MAB87/3 From Stock Solutions

Mix the component stock solutions in the following order. Final composition of the medium will be as described in Table I. Prepare amounts of medium that will be used up in 1-2 weeks. Complete media should not be used after storage at 5°C for more than 3 weeks.

Stock solution	For 100 ml	For 500 ml	For 1,000 ml
S/S87/3 A	10.0	50.0	100.0
B	10.0	50.0	100.0
C	2.5	12.5	25.0
D	10.0	50.0	100.0
E	1.0	5.0	10.0
F	2.5	12.5	25.0
G	4.0	20.0	40.0
H	10.0	50.0	100.0
Distilled water	50.0	250.0	500.0

Sterilize: By filtration through 0.2-μm polycarbonate filter.
Gas: Before storage with 10% CO_2 in air.
Quality control: Measure samples for pH (7.2) and osmolality (300 $\pm$ 5 mosM/kg).
Store: In tubes or bottles at 5°C, in the dark, with minimal air space above the fluid, for not more than 3 weeks.

MEDIUM AM77/B
Stock Solutions for Autoclavable Medium AM77/B

Medium AM77/B is composed of two primary solutions, medium *AM77/1* and the buffer solution *AM77/2*.

The stock solution that constitutes the major part of the medium is designated AM77/1, and is made from a series of substocks: AAM76/1 (amino acids and nucleosides); BAM75/1 (the B vitamins and succinic acid); and CAM77/1 (major and minor salts, dextrose, sodium succinate, and sodium pyruvate). Included in AAM76/1 are separate solutions of the four nucleosides; in BAM75/1, separate solutions of choline bitartrate, cyanocobalamin (B_{12}), and biotin; and in CAM77/1, a separate solution, TM74/1, of the salts of the trace elements.

Included in the buffer solution AM77/2, which is autoclaved separately from AM77/1, are the principal buffers (sodium β-glycerophosphate and disodium hydrogen phosphate), and also NaOH and KOH. These components adjust the medium to its final pH (7.0) and also the Na/K ratio to 32.5. Potassium dihydrogen phosphate is in AM77/1.

TABLE I. Composition of Complete Medium MAB87/3

Component	mg per 1000 ml	nmoles
NaCl	6,000.0	103.45
KCl	150.0	2.03
$CaCl_2 \cdot 2H_2O$	120.0	0.82
$MgCl_2 \cdot 6H_2O$	240.0	1.18
$MgSO_4 \cdot 7H_2O$	100.0	0.41
$NaHPO_4$	300.0	2.12
KH_2PO_4	208.0	1.53
$NaHCO_3$	2,240.0	26.66
$FeSO_4$	0.45	3×10^{-3}
$CuSO_4 \cdot 5H_2O$	0.05	2×10^{-4}
$MnSO_4H_2O$	0.016	9.5×10^{-5}
$ZnSO_4 \cdot 7H_2O$	0.03	1×10^{-4}
$(NH_4)_6Mo_7O_{24} \cdot 4H_2O$	0.024	2×10^{-5}
$CoCl_2 \cdot 6H_2O$	0.022	9.2×10^{-5}
Dextrose (anhydrous)	5,000.0	27.8
Ascorbic acid	17.5	0.1
Glutathione	15.0	0.05
Hypoxanthine	25.0	0.18
Thymidine	8.0	0.03
Choline chloride	250.00	1.80
Thiamin $\cdot$ HCl	10.0	0.03
Ca pantothenate	1.0	2.1×10^{-3}
Riboflavin	1.0	2.7×10^{-3}
Pyridoxin $\cdot$ HCl	1.0	4.9×10^{-3}
i-Inositol	1.0	5.5×10^{-3}
Nicotinamide	1.0	8.2×10^{-3}
Vitamin B_{12}	0.2	1.5×10^{-4}
Folic acid	0.5	1.1×10^{-3}
Biotin	0.02	8.2×10^{-5}
L-glutamine	350.0	2.40
L-lysine $\cdot$ HCl	240.0	1.32
L-histidine $\cdot$ HCl $\cdot$ H_2O	150.0	0.72
L-glutamic acid	150.0	1.02
L-cysteine $\cdot$ HCl $\cdot$ H_2O	100.0	0.58
L-arginine $\cdot$ HCl	75.0	0.36
L-threonine	75.0	0.63
L-valine	65.0	0.55
L-aspartic acid	60.0	0.45
Glycine	50.0	0.67
L-proline	50.0	0.43
L-leucine	50.0	0.38
L-methionine	50.0	0.33
L-phenylalanine	50.0	0.30
L-tryptophan	40.0	0.20
L-tyrosine	40.0	0.22

TABLE I. *(Continued)*

Component	mg per 1000 ml	nmoles
L-isoleucine	25.0	0.19
L-asparagine	24.0	0.18
L-cystine	15.0	0.06
L-serine	12.8	0.12
L-alanine	11.2	0.12
Insulin	8.0	1.4×10^{-3}
Phenol red	8.5	23×10^{-3}
NaOH	100.0	2.5

pH = 7.2; osmolality = 300 ± 5 mosM/kg; Na/K = 38.

Stock Solutions for AM77/1

AAM76/1.

Component	mg per 1,000 ml
L-alanine	890
L-arginine·HCl	210
L-asparagine	528
L-aspartic acid	54
L-cysteine·HCl·H$_2$O	176
L-glutamic acid	290
Glycine	100
L-histidine·HCl·H$_2$O	200
L-isoleucine	260
L-leucine	520
L-lysine·HCl	300
L-methionine	160
L-phenylalanine	160
L-proline	460
L-serine	730
L-threonine	240
L-tryptophan	80
L-valine	240
L-tyrosine	120
Hypoxanthine	60
Thymidine	60

(continued)

Stock Solutions for AM77/1 *(Continued)*

Deoxyadenosine·H_2O	20	(20 ml of solution dAdo/75)
Deoxycytidine·HCl	20	(20 ml of solution dCyd/75)
Uridine	2.5	(5 ml of solution Urd/75)
Cytidine	2.5	(5 ml of solution Cyd/75)
KH_2PO_4	640	
Phenol red	10	(1 ml of 1% of Na salt)

Special instructions. Dissolve tyrosine and hypoxanthine first in about 200 ml of water, with heat. The other amino acids and nucleosides are relatively more soluble. Prepare stock solutions of the nucleosides as follows:

	mg per 100 ml
dAdo/75:	
deoxyadenosine·H_2O	100
dCyd/75:	
deoxycytidine·HCl	100
Cyd/75: cytidine	50
Urd/75: uridine	50

Store: Frozen, nonsterile.

BAM75/1.

Component	*mg per 1,000 ml*	
Thiamin·HCl	20	
Ca pantothenate	20	
Riboflavin	10	(Note 1)
Folic acid	10	(Note 2)
Pyridoxin·HCl	10	
i-Inositol	20	
Nicotinamide	10	
Biotin	0.2	(1.0 ml of solution B1)
Cyanocobalamin (B_{12})	0.2	(0.5 ml of solution B2)

| Choline bitartrate | 50 | (10 ml of solution CBT) |
| Succinic acid | 750 | |

Special instructions. Choline bitartrate is used, to avoid the problems associated with the deliquescence of choline chloride. The original formulation [Waymouth, 1979] called for pyridoxal. However, pyridoxin is more stable to heat and is therefore recommended.

Biotin solution B1: Dissolve 10 mg biotin in 500 ml water. Store frozen.

Cyanocobalamin solution B2: Dissolve 40 mg in 100 ml water. Store frozen.

Choline bitartrate solution CBT: Dissolve 500 mg choline bitartrate (ICN Co.) in 100 ml water. Store frozen.

Dissolve: All components except riboflavin and folic acid in about 500 ml of water. *Note 1:* Dissolve riboflavin in about 5 ml water, add 0.4 ml 1 N NaOH, dilute with more water, neutralize immediately with 0.35 ml of 1 N HCl. *Note 2:* Dissolve folic acid separately in about 10 ml water, add 0.3 ml 1 N NaOH, when dissolved add more water and 0.25 ml of 1 N HCl.

Store: Frozen, nonsterile, protected from light.

CAM 77/1.

Component	mg per 1,000 ml	
NaCl	32,500	
KCl	1,000	
$CaCl_2 \cdot H_2O$	600	
$MgCl_2 \cdot 6H_2O$	1,125	(10 ml of solution 68/Mg)
Dextrose, anhydrous	12,000	
Na_2 succinate $\cdot 6H_2O$	500	
$CuSO_4 \cdot 5H_2O$	0.25	
$ZnSO_4 \cdot 7H_2O$	0.75	
$MnSO_4 \cdot H_2O$	0.80	(50 ml of solution TM74/1)
$CoCl_2 \cdot 6H_2O$	0.50	
$FeNH_4(SO_4)_2 \cdot 12H_2O$	5.00	
Na pyruvate	1,100	
Phenol red	25	(2.5 ml of 1% Na salt)

Special instructions. Solution 68/Mg: Dissolve 9,600 mg of $MgCl_2 \cdot 6H_2O$ from a freshly opened bottle in 100 ml distilled water. (This is the same as solution A1 used in medium MAB87/3.) Prepare stock solution of trace element salts (TM74/1) as follows:

TM74/1.

Component	mg per 100 ml
$CuSO_4 \cdot 5H_2O$	5.0
$ZnSO_4 \cdot 7H_2O$	15.0
$MnSO_4 \cdot H_2O$	1.6
$CoCl_2 \cdot 6H_2O$	10.0
$FeNH_4(SO_4)_2 \cdot 12H_2O$	100.0
Ascorbic acid	20 to keep the Fe^{2+} reduced

The antibiotic gentamicin may be included, at 60 mg/1,000 ml, in solution CAM 77/1.

Store: Frozen, non sterile.

Preparation of stock solution AM77/1. This is prepared by mixing the three solutions as follows: AM76/1, 500 ml; BAM75/1, 100; CAM77/1, 200. The pH is adjusted to pH 4.25 by addition of about 0.5 ml of 1 N NaOH. The 800 ml are pieced out in 80-ml lots in milk dilution bottles (e.g., Kimax #14915, with the caps loose and covered with aluminum foil. With this amount of fluid in bottles of these dimensions, it has been determined empirically that a 5% loss of water occurs by evaporation during the autoclaving cycle. Therefore, when the bottles have cooled, 4 ml of sterile distilled water can be added to each bottle, to return the volume to 80 ml. However, the usual procedure is to make up the complete medium immediately after autoclaving and cooling.

Stock Solution AM77/2

Buffer Solution AM77/2 (× 20).

Component	mg per 100 ml
$Na_2\beta$-glycero-phosphate $\cdot 5\frac{1}{2}H_2O$	16,400
Na_2HPO_4	800
NaOH	12 (3 ml of 0.1 N)
KOH	24 (4.3 ml of 0.1 N)

Special instructions. Because the amount of water of crystallization in sodium β-glycerophosphate is variable, the chemical should be purchased from a source that supplies a water analysis with each batch (e.g., Sigma), or it should be analyzed for water, or (for precise work) a solution should be made up and analyzed for total phosphorus. This is so that the correct molarity can be included in each batch of medium AM77/B (520 mM in AM77/2, 26 mM in AM77/B).

Preparation of Complete Autoclavable Medium AM77/B From Stock Solutions

Each 100 ml of medium AM77/B contains AM77/1, 80.0 ml; AM77/2, 5.0; sterile water 15.0. The main solution, AM77/1; the buffer solution, AM77/2; and some distilled water are sterilized separately by autoclaving for 20 minutes at 122°C.

To the sterile AM77/1 are added 5 ml of AM77/2 and 19 ml of distilled water (i.e., 4 ml to restore the volume of AM77/1 to 80 ml + 15 ml), to a total of 100 ml medium.

Quality control: The pH (7.0 ± 0.07) and osmolality (322 ± 7.4 mosM/kg) should be checked. The Na/K ratio should be 32.5.

Store: The medium AM77/B should be stored at 5°C, protected from light, and used within 3 weeks.

Final composition of medium AM77/B will be as described in Table II.

TABLE II. Composition of Complete Medium AM77/B

Component	mg per liter	mmoles
NaCl	6,500.0	112
KCl	200.0	2.7
$CaCl_2 \cdot 2H_2O$	120.0	0.8
$MgCl_2 \cdot 6H_2O$	225.0	1.1
Na_2HPO_4	400.0	2.8
KH_2PO_4	320.0	2.4
$CuSO_4 \cdot 5H_2O$	0.05	2.0×10^{-4}
$ZnSO_4 \cdot 7H_2O$	0.15	5.2×10^{-4}
$MnSO_4 \cdot H_2O$	0.16	9.5×10^{-4}
$CoCl_2 \cdot 6H_2O$	0.10	4.2×10^{-4}
$FeNH_4(SO_4)_2 \cdot 12H_2O$	1.00	2.0×10^{-3}
Na pyruvate	220.0	2.0
Na_2 succinate $\cdot 6H_2O$	100.0	0.37
Succinic acid	75.0	0.63
Na_2 β-glycero-phosphate $\cdot 5\frac{1}{2}H_2O$	8.200.0	26.0
Dextrose	2,400.0	13.3
Choline bitartrate	5.0	0.02
Thiamin $\cdot$ HCl	2.0	5.9×10^{-3}
Ca pantothenate	2.0	4.2×10^{-3}
Riboflavin	1.0	2.6×10^{-3}
Pyridoxin $\cdot$ HCl	1.0	4.9×10^{-3}

(continued)

TABLE II. Composition of Complete Medium AM77/B (Continued)

Component	mg per liter	mmoles
Folic acid	1.0	2.3×10^{-3}
i-Inositol	2.0	1.0×10^{-2}
Nicotinamide	1.0	8.2×10^{-3}
Cyanocobalamin	0.02	1.5×10^{-5}
Biotin	0.02	8.2×10^{-5}
Hypoxanthine	30.0	0.22
Thymidine	30.0	0.12
Deoxyadenosine $\cdot$ H$_2$O	10.0	3.7×10^{-2}
Deoxycytidine $\cdot$ HCl	10.0	3.8×10^{-2}
Uridine	1.25	5.0×10^{-3}
Cytidine	1.25	5.0×10^{-3}
L-alanine	445.0	5.00
L-arginine $\cdot$ HCl	105.0	0.50
L-asparagine	264.0	2.00
L-aspartic acid	27.0	0.20
L-cysteine $\cdot$ HCl $\cdot$ H$_2$O	88.0	0.48
L-glutamic acid	145.0	0.99
Glycine	50.0	0.67
L-histidine $\cdot$ HCl $\cdot$ H$_2$O	100.0	0.48
L-isoleucine	130.0	1.00
L-leucine	260.0	2.00
L-lysine $\cdot$ HCl	150.0	0.82
L-methionine	80.0	0.54
L-phenylalanine	80.0	0.48
L-proline	230.0	2.00
L-serine	365.0	3.48
L-threonine	120.0	1.00
L-tryptophan	40.0	0.20
L-tyrosine	60.0	0.33
L-valine	120.0	1.02
NaOH	6.0	0.15
KOH	12.0	0.214
Phenol red	10.0	0.028

Osmolality = 322.5 ± 7.4 mosM/kg (n = 92). pH = 7.0 ± 0.07 (n = 28). Na = 172.49; K = 5.31. Na/K = 32.46.

REFERENCES

Alarcon RA (1964): Isolation of acrolein from incubated mixtures of spermine with calf serum and its effects on mammalian cells. Arch Biochem Biophys 106:240–242.

Allegra JC, Lippman ME (1978): Growth of a human breast cancer line in serum-free hormone-supplemented medium. Cancer Res 38:3823-3829.

Armstrong D (1973): Contamination of tissue culture by bacteria and fungi. In Fogh J (ed): "Contamination in Tissue Culture." New York: Academic, Chap 3, pp 51-64.

Aronowicz JL (1981): How pure is "pure"? In Waymouth C, Ham RG, Chapple PJ (eds): "The Growth Requirements of Vertebrate Cells In Vitro." New York: Cambridge University Press, Chap 6, pp 82-93.

Baker JB, Brash GS, Carney DH, Cunningham DD (1978): Dexamethasone modulates binding and action of epidermal growth factor in serum-free cell culture. Proc Natl Acad Sci USA 75:1882-1886.

Baker LE, Ebeling AH (1939): Artificial maintenance media for cell and organ cultivation. I. The cultivation of fibroblasts in artificial and serumless media. J Exp Med 69:365-378.

Balin AK, Goodman DBP, Rasmussen H, Cristofalo VJ (1976): The effect of oxygen tension on the growth and metabolism of WI-38 cells. J Cell Physiol 89:235-250.

Balin AK, Goodman DBP, Rasmussen H, Cristofalo VJ (1978): Oxygen-sensitive stages of the cell cycle of human diploid cells. J Cell Biol 78:390-400.

Balk SD, Whitfield JF, Youdale T, Braun AC (1973): Roles of calcium, serum, plasma, and folic acid in the control of proliferation of normal and Rous sarcoma virus-infected chicken fibroblasts. Proc Natl Acad Sci USA 70:675-679.

Banerjee MR, Mehta NM, Ganguly R, Majumder PK, Ganguly N, Joshi J (1982): Selective gene expression in an isolated whole mammary organ in vitro. In Sato GH, Pardee AB, Sirbasku DA (eds): "Growth of Cells in Hormonally Defined Media." Cold Spring Harbor, New York: Cold Spring Harbor Laboratory, Book B, pp 789-805.

Bannai S, Tsukeda H, Okumura H (1977): Effect of antioxidants on cultured human diploid fibroblasts exposed to cystine-free medium. Biochem Biophys Res Commun 74:1582-1588.

Barnes D, Sato G (1980a): Methods for growth of cultured cells in serum-free medium. Anal Biochem 102:255-270.

Barnes D, Sato G (1980b): Serum-free cell culture: A unifying approach. Cell 22:649-655.

Baugh CL, Fitzgerald J, Tytell AA (1967): Growth of the L-cell in galactose medium. J Cell Physiol 69:257-261.

Bear MP, Schneider FH (1977): The effect of medium pH on rate of growth, neurite formation and acetylcholinesterase activity in mouse neuroblastoma cells. J Cell Physiol 91:63-68.

Benke PJ, Dittmar D (1976): Human lymphocyte response to D-valine. Biochim Biophys Acta 451:635-637.

Bettger WJ, Ham RG (1982a): The nutritional requirements of cultured mammalian cells. In Draper HH (ed): "Advances in Nutritional Research." New York: Plenum, Vol 4, Chap 9, pp 249-286.

Bettger WJ, Ham RG (1982b): The critical role of lipids in supporting clonal growth of human diploid fibroblasts in a defined medium. In Sato GH, Pardee AB, Sirbasku DA (eds): "Growth of Cells in Hormonally Defined Media." Cold Spring Harbor, New York: Cold Spring Harbor Laboratory, Book A, pp 61-64.

Bettger WJ, Boyce ST, Walthall BJ, Ham RG (1981): Rapid clonal growth and serial passage of human diploid fibroblasts in a lipid-enriched synthetic medium supplemented with epidermal growth factor, insulin, and dexamethasone. Proc Natl Acad Sci USA 78:5588-5592.

Blaschko H, Bonney R (1962): Spermine oxidase and benzylamine oxidase. Distribution, development and substrate specificity. Proc R Soc Lond Ser B 156:268-279.

Blaschko H, Hawes R (1959): Observations on spermine oxidase of mammalian plasmas. J Physiol (Lond) 145:124-131.

Block RJ, Bolling D (1945): "The Amino Acid Composition of Proteins and Foods." Springfield, Illinois: C C Thomas.

Blundell TL, Humbel RE (1980): Hormone families: Pancreatic hormones and homologous growth factors. Nature 287:781-787.

Boatman E, Cartwright F, Kenny F (1976): Morphometry and electron microscopy of HeLa cells infected with bovine *Mycoplasma*. Cell Tissue Res 170:1-16.

Boone CW, Mantel N, Caruso TD, Kazam E, Stevenson RE (1972): Quality control studies of fetal bovine serum used in tissue culture. In Vitro 7:174-189.

Boynton AL, Whitfield JF (1976): Different calcium requirements for proliferation of conditionally and unconditionally tumorigenic mouse cells. Proc Natl Acad Sci USA 73:1651-1654.

Boynton AL, Whitfield JF, Isaacs RJ, Morton HJ (1974): Control of 3T3 cell proliferation by calcium. In Vitro 10:12-17.

Bradshaw RA, Young M (1976): Commentary: Nerve growth factor — Recent developments and perspectives. Biochem Pharmacol 25:1445-1449.

Braun P, Klein JO, Kass EH (1970): Susceptibility of genital mycoplasmas to antimicrobial agents. Appl Micrbiol 19:62-70.

Breslow JL, Sloan HR, Ferrans VJ, Anderson JL, Levy RI (1973): Characterization of the mouse liver cell line FL83B. Exp Cell Res 78:441-453.

Bucher O (1947): Die Wirkung von Penicillin auf Gewebekulturen. IV. Cytopharmakologische Untersuchung mit reinen Penicillin G. Schweiz Med Wochenschr 77:849-852.

Bunge RP, Bunge MB, Carey DJ, Cornbrooks C, Higgins D, Johnson MI, Kleinschmidt DC, Wood PM, Iacovitti L, Moya F (1982): Functional expression in primary nerve tissue cultures maintained in defined medium. In Sato GH, Pardee AB, Sirbasku DA (eds): "Growth of Cells in Hormonally Defined Media." Cold Spring Harbor, New York: Cold Spring Harbor Laboratory, Book B, pp 1017-1031.

Burrows MT (1910): The cultivation of tissues of the chick embryo outside the body. J Am Med Assoc 55:2057-2058.

Burrows MT (1911): The growth of tissues of the chick embryo outside the animal body, with special reference to the nervous system. J Exp Zool 10:63-83.

Buzard RL, Horn D (1982): Growth response to dexamethasone in clonal variants of hamster melanoma cells: Role of medium components. In Sato GH, Pardee AB, Sirbasku DA (eds): "Growth of Cells in Hormonally Defined Media." Cold Spring Harbor, New York: Cold Spring Harbor Laboratory, Book A, pp 423-427.

Carpenter G (1980): Epidermal growth factor: Biology and mechanism of action. Birth Defects 16:61-72.

Carpenter G, Cohen S (1979): Epidermal growth factor. Annu Rev Biochem 48:193-216.

Carpenter G, Stoscheck GM, Soderquist AM (1982): Epidermal growth factor. Ann NY Acad Sci 397:11-17.

Casemore DP (1967): Gentamicin as a bacteriocidal agent in virological tissue culture. J Clin Pathol 20:298-299.

Castor CW, Cobel-Geard SR (1980): Connective tissue activation: Evidence for a second human platelet growth factor. Clin Res 28:139A.

Castor CW, Fremuth TD (1982): Factors modifying DNA synthesis by lung fibroblasts in vitro. Proc Soc Exp Biol Med 171:109-113.

Castor CW, Cobel-Geard SR, Hossler PA, Kelch RP (1981): Connective tissue activating peptide III. XXII. A platelet growth factor in human growth hormone deficient patients. J Clin Endocrinol Metab 52:128-132.

Ceccarini C, Eagle H (1971): Induction and reversal of contact inhibition of growth by pH modification. Nature New Biol 233:271-273.

Chen HW, Kandutsch AA (1981): Cholesterol requirements for cell growth: Endogenous synthesis versus exogenous sources. In Waymouth C, Ham RG, Chapple RJ (eds): "The Growth Requirements of Vertebrate Cells In Vitro." New York: Cambridge University Press, Chap 21, pp 327-342.

Christensen HN (1982): Interorgan amino acid nutrition. Physiol Rev 62:1193-1233.

Cohen S, Taylor JM (1974): Epidermal growth factor: Chemical and biological characterization. Recent Prog Hormone Res 30:533-550.

Coon HG, Sinback CN (1982): Cultures of rat neuroblasts that divide and differentiate in vitro. In Sato GH, Pardee AB, Sirbasku DA (eds): "Growth of Cells in Hormonally Defined Media." Cold Spring Harbor, New York: Cold Spring Harbor Laboratory, Book B, pp 1007-1016.

Corfield VA, Hay RJ (1978): Effects of cystine or glutamine restriction on human diploid fibroblasts in culture. In Vitro 14:787-794.

Coriell LL (1973): Methods of prevention of bacterial, fungal and other contaminations. In Fogh J (ed): "Contamination in Tissue Culture." New York: Academic, Chap 2, pp 29-49.

Crocker FS, Vernier RL (1970): Fetal kidney in organ culture: Abnormalities of development induced by decreased amounts of potassium. Science 169:485-487.

Cuatrecasas P, Hollenberg MD (1976): Membrane receptors and hormone action. Adv Protein Chem 30:251-451.

Das M (1982): Epidermal growth factor: Mechanisms of action. Int Rev Cytol 78:233-256.

De Haan RL (1967): Regulation of spontaneous activity and growth of embryonic chick heart cells in tissue culture. Develop Biol 16:216-249.

De Haan RL (1970): The potassium-sensitivity of isolated embryonic heart cells increases with development. Dev Biol 23:226-240.

De Lattre E, Santos ML, Boschero AC (1982): Effect of gentamycin on insulin release and ^{45}Ca net uptake by isolated islets. Experientia 38:1320-1321.

De Luca C, Carruthers C, Tritsch GL (1966a): Extracts of plasma and serum toxic for mammalian cells cultured in vitro. Exp Cell Res 43:403-412.

De Luca C, Habeeb AFSA, Tritsch GL (1966b): The stimulation of growth of mammalian cells in vitro by a peptide fraction from enzymatic digests of serum. Exp Cell Res 43:98-106.

Eagle H (1971): Buffer combinations for mammalian cell culture. Science 174:500-503.

Eagle H (1973): The effect of environmental pH on the growth of normal and malignant cells. J Cell Physiol 82:1-8.

Eagle H, Barban S, Levy M, Schulze HO (1958): The utilization of carbohydrates by human cell cultures. J Biol Chem 233:551-558.

Eisinger M, Lee JS, Hefton JM, Darzynkiewicz Z, Chiao JW, de Harven E (1979): Human epidermal cell cultures: Growth and differentiation in the absence of dermal components or medium supplements. Proc Natl Acad Sci USA 76:5340-5344.

Erickson GF, Hofeditz C, Casper R (1982): Obligatory role of defined medium in the hormone-dependent differentiation of rat granulosa cells. In Sato GH, Pardee AB, Sirbasku DA (eds): "Growth of Cells in Hormonally Defined Media." Cold Spring Harbor, New York: Cold Spring Harbor Laboratory, Book B, pp 1129-1140.

Esber HJ, Payne IJ, Bogden AE (1973): Variability of hormone concentrations and ratios in commercial sera used for tissue culture. J Natl Cancer Inst 50:559-562.

Evans VJ, Sanford KK (1978): Development of defined media for studies on malignant transformation in culture. In Katsuta H (ed): "Nutritional Requirements of Cultured Cells." Baltimore: University Park Press, pp 149-192.

Evans VJ, Bryant JC, Fioramonti MC, McQuilkin WT, Sanford KK, Earle WR (1956): Studies of nutrient media for tissue cells in vitro. I. A protein-free chemically defined medium for cultivation of strain L cells. Cancer Res 16:77–86.

Evans VJ, Bryant JC, Kerr HA, Schilling EL (1964): Chemically defined media for cultivation of long-term cell strains from four mammalian species. Exp Cell Res 36:439–474.

Faris B, Snider R, Levine A, Moscaritolo R, Salcedo L, Franzblau C (1978): Effect of ascorbate on collagen synthesis by lung embryonic fibroblasts. In Vitro 14:1022–1027.

Fell HB, Mellanby E (1953): Metaplasia produced in cultures of chick ectoderm by high vitamin A. J Physiol (Lond) 119:470–488.

Fell HB, Rinaldini L (1965): The effects of vitamins A and C on cells and tissues in culture. In Willmer EN (ed): "Cells and Tissues in Culture." New York: Academic, Chap 17, pp 659–699.

Feng J, Melcher AH, Brunette DM, Moe HK (1977): Determination of L-ascorbic acid levels in culture medium: Concentrations in commercial media and maintenance of levels under conditions of organ culture. In Vitro 13:91–99.

Fillerup DL, Migliore JV, Mead JF (1958): The uptake of lipoproteins by ascites tumor cells. J Biol Chem 233:98–101.

Fisher KRS, Liversage RA (1973): A high potassium organ culture medium for chick embryo gonad primordia. In Vitro 8:419.

Frost DV, McIntire FC (1944): Hydrolysis of pantothenate: A first order reaction. Relation to thiamine stability. J Am Chem Soc 66:425–427.

Gilbert SF, Migeon BR (1975): D-Valine as a selective agent for normal human and rodent epithelial cells in culture. Cell 5:11–17.

Glinos AD, Vail JM, Taylor B (1973): Density-dependent regulation of growth in L cell suspension cultures. II. Synthesis of total protein and collagen in presence of rapidly declining oxygen tensions. Exp Cell Res 78:319–328.

Goetz IE, Weinstein C, Roberts E (1973): Inhibition of growth of hamster tumor cells in vitro by pyrrolidone carboxylic acid in a glutamine-dependent system. In Vitro 8:279–282.

Goetz IE, Moklebust R, Warren CJ (1979): Effects of some antibiotics on the growth of human diploid skin fibroblasts in cell culture. In Vitro 15:114–119.

Good NE, Winget GD, Winter W, Connolly TN, Izawa S, Singh RMM (1966): Hydrogen ion buffers for biological research. Biochemistry 5:467–477.

Gorham LW, Waymouth C (1965): Differentiation in vitro of embryonic cartilage and bone in a chemically defined medium. Proc Soc Exp Biol Med 119:287–290.

Gori GB, Lee DY (1964): A method for eradication of mycoplasma infections in cell cultures. Proc Soc Exp Biol Med 117:918–921.

Gospodarowicz D (1974): Localisation of a fibroblast growth factor and its effect alone and with hydrocortisone on 3T3 cell growth. Nature 249:123–127.

Gospodarowicz D (1975): Purification of a fibroblast growth factor from bovine pituitary. J Biol Chem 250:2515–2520.

Gospodarowicz D, Moran JS (1976): Growth factors in mammalian cell culture. Annu Rev Biochem 45:531–558.

Gospodarowicz D, Vlodavsky I, Greenburg G, Birdwell CR (1982): The effects of the epidermal and fibroblast growth factors upon cell proliferation and differentiation using vascular and corneal endothelial cells as a model. In Waymouth C, Ham RG, Chapple PJ (eds): "The Growth Requirements of Vertebrate Cells In Vitro." New York: Cambridge University Press, Chap 36, pp 492–542.

Gregg CT (1972): Some aspects of the energy metabolism of mammalian cells. In Rothblat GH, Cristofalo VJ (eds): "Growth, Nutrition, and Metabolism of Cells in Culture." New York: Academic, Vol 1, Chap 4, pp 83-136.

Griffiths JB, Pirt SJ (1967): The uptake of amino acids by mouse cells (strain LS) during growth by batch culture and chemostat culture: The influence of cell growth rate. Proc R Soc (Lond) Ser B 168:421-438.

Groelke JW, Baseman JB, Amos H (1979): Regulation of the $G_1 \rightarrow S$ phase transition in chick embryo fibroblasts with α-keto acids and L-alanine. J Cell Physiol 101:391-398.

Grosswicz N, Arnovitch J, Rachmilewicz M, Izak O, Sadovsky A, Bercovici B (1960): Folic and folinic acid in maternal and foetal blood. Br J Haematol 6:296-302.

Guskey AE, Jenkin HM (1976): The serial cultivation of suspended BHK-21/13 cells in serum-free Waymouth medium. Proc Soc Exp Biol Med 151:221-224.

Gwatkin RB, Siminovitch L (1960): Multiplications of single mammalian cells in a nonbicarbonate medium. Proc Soc Exp Biol Med 103:718-721.

Ham RG (1963): An improved nutrient solution for diploid Chinese hamster and cell lines. Exp Cell Res 29:515-526.

Ham RG (1964): Putrescine and related amines as growth factors for a mammalian cell line. Biochem Biophys Res Commun 14:34-38.

Ham RG (1965): Clonal growth of mammalian cells in a chemically defined, synthetic medium. Proc Natl Acad Sci USA 53:288-293.

Ham RG (1974): Nutritional requirements of primary cultures: A neglected problem of modern biology. In Vitro 10:119-129.

Ham RG (1981): Survival and growth requirements of nontransformed cells. In Baserga R (ed): "Handbook of Experimental Pharmacology." New York: Springer-Verlag, Vol 57, Chap 2, pp 13-88.

Ham RG, McKeehan WL (1979): Media and growth requirements. Methods Enzymol 58:44-93.

Hamilton WG, Ham RG (1977): Clonal growth of Chinese hamster cell lines in protein-free media. In Vitro 13:537-547.

Harris M, Kutsky R (1953): Utilization of added sugar by chick heart fibroblasts in vitro. J Cell Comp Physiol 42:449-470.

Harrison RG (1908): Embryonic transplantation and development of the nervous system. Anat Rec 2:385-410.

Hawker RJ, Hawker LM (1975): Protein losses during sterilizing by filtration. Lab Practice 24:805-818.

Hayflick L (1965): The limited in vitro lifetime of human diploid cell strains. Exp Cell Res 37:614-636.

Hayflick L, Moorhead PS (1961): The serial cultivation of human diploid cell strains. Exp Cell Res 25:585-621.

Healy GM, Parker RC (1966): An improved chemically defined basal medium (CMRL 1415) for newly explanted mouse embryo cells. J Cell Biol 30:531-538.

Healy GM, Morgan JF, Parker RC (1952): Trace metal content of some natural and synthetic media. J Biol Chem 198:305-312.

Heilman CH (1945): Cytotoxicity of streptomycin and streptothricin. Proc Soc Exp Biol Med 60:365-367.

Hertz L, Dittman L, Mandel P (1973): K^+ induced stimulation of oxygen uptake in cultured cerebral glial cells. Brain Res 60:517-520.

Higuchi K (1970): An improved chemically defined culture medium for strain L mouse cells based on growth responses to graded levels of nutrients including iron and zinc ions. J Cell Physiol 75:65-72.

Hill PMM, Young M (1973): Net placental transfer of free amino acids against varying concentrations. J Physiol (Lond) 235:409-422.

Hinman JW (1972): Prostaglandins. Annu Rev Biochem 41:161-178.

Hogenkamp HPC (1966): The photolysis of methylcobalamin. Biochemistry 5:417-422.

Hollenberg MD, Cuatrecasas P (1973): Epidermal growth factor: Receptors in human fibroblasts and modulation of action by cholera toxin. Proc Natl Acad Sci USA 70:2964-2968.

Honn KV, Singley JA, Chavin W (1975): Febal bovine serum: A multivariate standard. Proc Soc Exp Biol Med 149:344-347.

Hopkins L, McFadyen IR, Young M (1971): The free amino acids concentrations in maternal and foetal plasmas in the pregnant ewe. J Physiol (Lond) 215:9P-10P.

Hornsby PJ, Simonian MH, Gill GN (1979): Aging of adrenocortical cells in culture. Int Rev Cytol (Suppl) 10:131-162.

Hu PC, Collier AM, Baseman JB (1975): Alternations in the metabolism of hamster tracheas in organ culture after infection by virulent *Mycoplasma pneumoniae*. Infect Immunol 11:704-710.

Ikeda T, Liu Q-F, Danielpour D, Officer JB, Iio M, Leland FE, Sirbasku DA (1982): Identification of estrogen-inducible growth factors (estromedins) for rat and human mammary tumor cells in culture. In Vitro 18:961-979.

Jackson MJ, Shin S (1982): Inositol as a growth factor for mammalian cells in culture. In Sato GH, Pardee AB, Sirbasku DA (eds): "Growth of Cells in Hormonally Defined Media." Cold Spring Harbor, New York: Cold Spring Harbor Laboratory, Book A, pp 75-86.

Jenkin HM, Anderson LE (1970): The effect of oleic acid on the growth of monkey kidney cells (LLC-MK$_2$). Exp Cell Res 59:6-10.

Jensen PKA, Therkelsen AJ (1982): Selective inhibition of fibroblasts by spermine in primary cultures of normal human skin epithelial cells. In Vitro 18:867-871.

Jiménez de Asúa L, Surian ES, Flawia MN, Torres HN (1973): Effect of insulin on the growth pattern and adenylate cyclase activity of BHK fibroblasts. Proc Natl Acad Sci USA 70:1388-1392.

Jones-Villeneuve EMV, McBurney MW, Rogers KA, Kalnins VI (1982): Retinoic acid induces embryonal carcinoma cells to differentiate into neurons and glial cells. J Cell Biol 94:253-262.

Kadanka ZK, Sparkes JD, Macmorine HG (1973): A study of the cytogenetics of the human cell strain WI-38. In Vitro 8:353-361.

Kahn RH (1954): Effect of oestrogen and of vitamin A on vaginal cornification in tissue culture. Nature 174:317.

Kamely D, Rudland PS (1976): Induction of DNA synthesis and cell division in human diploid skin fibroblasts by fibroblast growth factor. Exp Cell Res 97:120-126.

Kano-Sueoka T, Errick JE (1981): Effects of phosphoethanolamine and ethanolamine on growth of mammary carcinoma cells in culture. Exp Cell Res 136:137-145.

Kao WW-Y, Prockop DJ (1977): Proline analogue removes fibroblasts from cultured mixed cell populations. Nature 266:63-64.

Kelley GG, Adamson DJ, Oliver KL (1960a): Growth of human tissue cells in the absence of added gas phase. Am J Hyg 71:9-14.

Kelley GG, Adamson DJ, Vail MH (1960b): Further studies on the growth of human tissue cells in the absence of an added gas phase. Am J Hyg 72:275-278.

Kelley WN (1972): Purine and pyrimidine metabolism of cells in culture. In Rothblat GH, Cristofalo VJ (eds). "Growth, Nutrition, and Metabolism of Cells in Culture." New York: Academic, Vol 1, Chap 7, pp 211-256.

Kihara H, de la Flor SD (1968): Arginase in fetal calf serum. Proc Soc Exp Biol Med 129:303-304.

King ME, Spector AA (1981): Lipid metabolism in cultured cells. In Waymouth C, Ham RB, Chapple PJ (eds): "Growth Requirements of Vertebrate Cells In Vitro." New York: Cambridge University Press, Chap 19, pp 293-312.

Kitos PA, Sinclair R, Waymouth C (1962): Glutamine metabolism by animal cells growing in a synthetic medium. Exp Cell Res 27:307-316.

Kohler N, Lipton A (1974): Platelets as a source of fibroblast growth-promoting activity. Exp Cell Res 87:197-301.

Lasher RS, Zagon IS (1972): The effect of potassium on neuronal differentiation in cultures of dissociated newborn rat cerebellum. Brain Res 41:482-488.

Lasnitzki I (1961): Effect of excess vitamin A on the normal and oestrone-treated mouse vagina grown in chemically defined media. Exp Cell Res 24:37-45.

Latta JS, Buchholz DJ (1939): Effects of insulin on growth of fibroblasts in vitro. Arch Exp Zellforsch Gewebezücht 23: 146-156.

Leffert HL, Koch KS (1977): Control of animal cell proliferation. In Rothblat GH, Cristofalo VJ (eds): "Growth, Nutrition and Metabolism of Cells In Vitro." New York: Academic, Vol 3, pp 225-294.

Leffert HL, Paul D (1973): Serum dependent growth of primary cultured differentiated fetal rat hepatocytes in arginine-deficient medium. J Cell Physiol 81:113-124.

Leibovitz A (1963): The growth and maintenance of tissue-cell cultures in free gas exchange with the atmosphere. Am J Hyg 78:173-180.

Leiter EH, Coleman DL, Waymouth C (1974): Cell culture of the endocrine pancreas of the mouse in chemically defined media. In Vitro 9:421-433.

Leland FE, Danielpour D, Sirbasku DA (1982): Studies of the endocrine, paracrine, and autocrine control of mammary tumor cell growth. In Sato GH, Pardee AB, Sirbasku DA (eds): "Growth of Cells in Hormonally Defined Media." Cold Spring Harbor, New York: Cold Spring Harbor Laboratory, Book B, pp 741-750.

Levander OA, Cheng L (eds) (1980): Micronutrient interactions: Vitamins, minerals and hazardous elements. Ann NY Acad Sci 355:1-372.

Levi-Montalcini R (1952): Effect of mouse tumor transplantation on the nervous system. Ann NY Acad Sci 55:330-343.

Levi-Montalcini R (1966): The nerve growth factor: Its mode of action on sensory and sympathetic nerve cells. Harvey Lect 60:217-259.

Levi-Montalcini R, Angeletti PU (1968): Nerve growth factor. Physiol Rev 48:534-569.

Lewis MR, Lewis WH (1911a): The cultivation of tissues from chick embryos in solutions of NaCl, $CaCl_2$, KCl and $NaHCO_3$. Anat Rec 5:277-293.

Lewis MR, Lewis WH (1911b): On the growth of embryonic chick tissues in artificial media, nutrient agar and bouillon. Bull Johns Hopkins Hosp 22:126-127.

Lewis WH, Lewis MR (1912): The cultivation of chick tissues in media of known chemical composition. Anat Rec 6:207-211.

Lim R, Turriff DE, Troy SS, Kato T (1977): Differentiation of glioblasts under the influence of glia maturation factor. In Fedoroff S, Hertz L (eds): "Cell, Tissue and Organ Cultures in Neurobiology." New York: Academic, pp 223-235.

Ling CT, Gey GO, Richters V (1968): Chemically characterized concentrated corodies for continuous cell culture (the 7C's culture media). Exp Cell Res 52:469-489.

Mains RE, Patterson PH (1973): Primary cultures of dissociated sympathetic neurons. I. Establishment of long-term growth in culture and studies of differentiated properties. J Cell Biol 59:329-345.

Makino S, Jenkin HM (1975): Effect of fatty acids on growth of Japanese encephalitis virus in BHK-21 cells and phospholipid metabolism of the infected cells. J Virol 15:515-525.

Marceau N, Goyette R, Valet JP, Deschênes J (1980): The effect of dexamethasone on formation of a fibronectin extracellular matrix by rat hepatocytes in vitro. Exp Cell Res 125:497-502.

Marshall JD, Heiniger HJ (1979): High affinity concanavalin A binding to sterol-depleted L cells. J Cell Physiol 100:539-548.

Massie HR, Samis HV, Baird MB (1972): The effects of the buffer HEPES on the division potential of WI-38 cells. In Vitro 7:191-194.

Massie HR, Baird MB, Samis HV (1974): Prolonged cultivation of primary chick cultures using organic buffers. In Vitro 9:441-444.

Mather J, Wu R, Sato G (1981): The role of hormones in the growth and regulation of cells in a serum-free medium. In Waymouth C, Ham RG, Chapple PJ (eds): "The Growth Requirements of Vertebrate Cells In Vitro." New York: Cambridge University Press, Chap 16, pp 244-257.

Mather JP, Saez JM, Haour F, Dray F (1982): Hormone-hormone and hormone-vitamin interactions in the control of growth and function of Leydig cells in vitro. In Sato GH, Pardee AB, Sirbasku DA (eds): "Growth of Cells in Hormonally Defined Media." Cold Spring Harbor, New York: Cold Spring Harbor Laboratory, Book B, pp 1117-1128.

McCarty KS Jr, McCarty KS Sr (1977): Steroid hormone receptors in the regulation of differentiation. Am J Pathol 86:705-744.

McKeehan WL (1982a): Glycolysis, glutaminolysis and cell proliferation. Cell Biol Intern Rept 6:635-650.

McKeehan WL (1982b): Growth-factor-nutrient interrelationships in control of normal and transformed cell proliferation. In Sato, GH, Pardee AB, Sirbasku DA (eds): "Growth of Cells in Hormonally Defined Media." Cold Spring Harbor, New York: Cold Spring Harbor Laboratory, Book A, pp 65-74.

McKeehan WL, Ham RG (1976): Stimulation of clonal growth of normal fibroblasts with substrata coated with basic polymers. J Cell Biol 71:727-734.

McKeehan WL, McKeehan KA (1979): Oxocarboxylic acids, pyridine nucleotide-linked oxidoreductase and serum factors in regulation of cell proliferation. J Cell Physiol 101:9-16.

McKeehan WL, Hamilton WG, Ham RG (1976): Selenium is an essential trace nutrient for growth of WI-38 diploid human fibroblasts. Proc Natl Acad Sci USA 73:2023-2027.

McKeehan WL, McKeehan KA, Hammond SL, Ham RG (1977): Improved medium for the clonal growth of human diploid fibroblasts at low concentrations of serum. In Vitro 13:399-416.

McKeehan WL, McKeehan KA, Calkins D (1982): Epidermal growth factor modifies Ca^{2+}, Mg^{2+}, and 2-oxocarboxylic acid, but not K^+ and phosphate ion, requirements for multiplication of human fibroblasts. Exp Cell Res 140:25-30.

McLimans WF (1979): Mass culture of mammalian cells. Methods Enzymol 58:194-211.

Merchant DJ (1973): Summary. In Fogh J (ed): "Contamination in Tissue Culture." New York: Academic, Chap 12, pp 257-269.

Metzger JF, Fusillo MH, Cornman I, Kuhns DM (1954): Antibiotics in tissue culture. Exp Cell Res 6:337-344.

Minna JD, Carney DN, Oie H, Bunn PA Jr, Gazdar AF (1982): Growth of human small-cell lung cancer in defined medium. In Sato GH, Pardee AB, Sirbasku DA (eds): "Growth of Cells in Hormonally Defined Media." Cold Spring Harbor, New York: Cold Spring Harbor Laboratory, Book B, pp 627-639.

Monard D, Stöckel K, Goodman R, Thoenen H (1975): Distinction between nerve growth factor and glial factor. Nature 258:444-445.

Moore GE, Mount DD, Tara G, Schwartz N (1963): Growth of human tumor cells in suspension culture. Cancer Res 23:1735-1741.

Moore GE, Sandberg AA, Ulrich K (1966): Suspension cell culture and in vivo and in vitro chromosome constitution of mouse leukemia. J Natl Cancer Inst 36:405-421.

Morgan JF, Morton HJ (1960): Carbohydrate utilization by chick embryonic heart cultures. Can J Biochem Physiol 38:69-76.

Morgan JF, Morton HJ, Parker RC (1950): Nutrition of animal cells in tissue culture. I. Initial studies on a synthetic medium. Proc Soc Exp Biol Med 73:1-8.

Morrison RG, de Vellis J (1982): Growth and differentiation of purified astrocytes in a chemically defined medium. In Sato GH, Pardee AB, Sirbasku DA (eds): "Growth of Cells in Hormonally Defined Media." Cold Spring Harbor, New York: Cold Spring Harbor Laboratory, Book B, pp 973-985.

Morrison SJ, Jenkin HM (1972): Growth of Chlamydia psittaci strain meningopneumonitis in mouse L cells cultivated in a defined medium in spinner cultures. In Vitro 8:94-100.

Morton HJ (1981): Known cellular growth requirements and the composition of currently available defined media. In Waymouth C, Ham RG, Chapple PJ (eds): "The Growth Requirements of Vertebrate Cells In Vitro." New York: Cambridge University Press, Chap 2, pp 16-32.

Murakami H, Masui H, Sato GH (1982): Suspension culture of hybridoma cells in serum-free medium: Soybean phospholipids as the essential components. In Sato GH, Pardee AB, Sirbasku DA (eds): "Growth of Cells in Hormonally Defined Media." Cold Spring Harbor, New York: Cold Spring Harbor Laboratory, Book B, pp 711-715.

Nagle SC (1968): Heat-stable chemically defined medium for growth of animal cells in suspension. Appl Microbiol 16:53-55.

Nagle SC, Brown BL (1971): An improved heat-stable glutamine-free chemically defined medium for growth of mammalian cells. J Cell Physiol 77:259-264.

Naglee DL, Maurer RR, Foote RH (1969): Effect of osmolarity on in vitro development of rabbit embryos in a chemically defined medium. Exp Cell Res 58:331-333.

Nielsen FH (1981): Consideration of trace element requirements for preparation of chemically defined media. In Waymouth C, Ham RG, Chapple RJ (eds): "The Growth Requirements of Vertebrate Cells In Vitro." New York: Cambridge University Press, Chap 5, pp 68-81.

Nielsen JH, Brunstedt K (1982): Growth hormone as a growth factor for normal pancreatic islet cells in primary culture. In Sato GH, Pardee AB, Sirbasku DA (eds): "Growth of Cells in Hormonally Defined Media." Cold Spring Harbor, New York: Cold Spring Harbor Laboratory, Book A, pp 501-506.

Nixon BT, Wang RJ (1977): Formation of photoproducts lethal for human cells in culture by daylight fluorescent light and bilirubin light. Photochem Photobiol 26:589-593.

Noyes WF (1973): Culture of human fetal liver. Proc Soc Exp Biol Med 144:245-248.

Olmsted CA (1967): A physico-chemical study of fetal calf sera used as tissue culture nutrient correlated with biological tests for toxicity. Exp Cell Res 48:283-299.

Pappano AJ, Sperelakis N (1969): Spontaneous contractions of cultured heart cells in high K^+ media. Exp Cell Res 54:58-68.

Parshad R, Sanford KK (1971): Oxygen supply and stability of chromosomes in mouse embryo cells in vitro. J Natl Cancer Inst 47:1033-1035.

Parshad R, Sanford KK, Jones GM, Price FW, Taylor WG (1977): Oxygen and light effects on chromosomal aberrations in mouse cells in vitro. Exp Cell Res 104:199-205.

Pasieka AE, Morton HJ, Morgan JF (1958): The metabolism of animal tissues cultivated in vitro. III. Amino acid metabolism of strain L cells in completely synthetic media. Can J Biochem Physiol 36:771-782.

Patterson MK (1972): Uptake and utilization of amino acids by cell cultures. In Rothblat GH, Cristofalo VJ (eds): "Growth, Nutrition and Metabolism of Cells in Culture." New York: Academic, Vol 1, pp 171-209.

Pawelek J, Murray M, Fleischmann R (1982): Genetic studies of insulin action in Cloudman melanoma cells. In Sato GH, Pardee AB, Sirbasku DA (eds): "Growth of Cells in Hormonally Defined Media." Cold Spring Harbor, New York: Cold Spring Harbor Laboratory, Book B, pp 911-919.

Peehl DM, Ham RG (1980a): Growth and differentiation of human keratinocytes without a feeder layer or conditioned medium. In Vitro 16:516-525.

Peehl DM, Ham RG (1980b): Clonal growth of human keratinocytes with small amounts of dialysed serum. In Vitro 16:526-540.

Perlman D (1979): Use of antibiotics in cell culture. Methods Enzymol 58:110-116.

Perris AD, Whitfield JF (1967): Calcium and the control of mitosis in the mammal. Nature 216:1350-1351.

Perris AD, Whitfield JF, Tölg PK (1968): Role of calcium in the control of growth and cell division. Nature 219:527-529.

Peterkofsky B (1972): The effect of ascorbic acid on collagen polypeptide synthesis and proline hydroxylation during the growth of cultured fibroblasts. Arch Biochem Biophys 152:318-328.

Pfeiffer SE, Betschart B, Cook J, Mancini P, Morris R (1977): Glial cell lines. In Fedoroff S, Hertz L (eds): "Cell, Tissue and Organ Cultures in Neurobiology." New York: Academic, pp 287-346.

Pickart L, Thaler MM (1973): Tripeptide in human serum which prolongs survival of normal liver cells and stimulates growth in neoplastic liver. Nature New Biol 243:85-87.

Pledger WJ, Howe PH, Leof EB (1982): The regulation of cell proliferation by serum growth factors. Ann NY Acad Sci 397:1-10.

Poiley JA, Shuman RF, Pienta RJ (1978): Characterization of normal human embryo cells grown to over 100 population doublings. In Vitro 14:405-412.

Poole CA, Reilly HC, Flint MH (1982): The adverse effects of HEPES, TES, and BES zwitterion buffers on the ultrastructure of cultured chick embryo epiphyseal chondrocytes. In Vitro 18:755-765.

Poste G, Papahadjopoulos D, Vail WJ (1976): Lipid vesicles as carriers for introducing biologically active materials into cells. Methods Cell Biol 14:33-71.

Powers S, Alcorta D, Nicholson N, Pollack R (1982): Effects of calcium and hormones on the cytoskeleton and cell proliferation. In Sato GH, Pardee AB, Sirbasku DA (eds): "Growth of Cells in Hormonally Defined Media." Cold Spring Harbor, New York: Cold Spring Harbor Laboratory, Book A, pp 243-258.

Pumper RW (1973): Purification and standardization of water for tissue culture. In Kruse PF Jr, Patterson MK Jr (eds): "Tissue Culture: Methods and Applications." New York: Academic, Section XIV, Chap 2, pp 674-677.

Rheinwald JG, Green H (1977): Epidermal growth factor and the multiplication of cultured human epidermal keratinocytes. Nature 265:421-424.

Richter A, Sanford KK, Evans VJ (1972): Influence of oxygen and culture media on plating efficiency of some mammalian cell lines. J Natl Cancer Inst 49:1705-1712.

Roberts AB, Anzano MA, Frolik CA, Sporn MB (1982): Transforming growth factors: Characterization of two classes of factors from neoplastic and nonneoplastic tissues. In Sato GH, Pardee AB, Sirbasku DA (eds): "Growth of Cells in Hormonally Defined Media." Cold Spring Harbor, New York: Cold Spring Harbor Laboratory, Book A, pp 319-332.

Ross R, Glomset J, Kariya B, Harker L (1974): A platelet-dependent serum factor that stimulates the proliferation of arterial smooth muscle cells in vitro. Proc Natl Acad Sci USA 71:1207-1210.

Ross R, Raines E, Bowen-Pope D, Glenn K (1982): General concepts of growth factors: A progress report on platelet-derived growth factor. In Sato GH, Pardee AB, Sirbasku (eds): "Growth of Cells in Hormonally Defined Media." Cold Spring Harbor, New York: Cold Spring Harbor Laboratory, Book A, pp 27-35.

Rothstein H (1982): Regulation of the cell cycle by somatomedins. Intern Rev Cytol 78:127-232.

Rudland PS, Jiménez de Asúa L (1979): Action of growth factors in the cell cycle. Biochim Biophys Acta 560:91-133.

Salas-Prato M (1982): Growth of fetal mouse liver cells in hormone-supplemented serum-free medium. In Sato GH, Pardee AB, Sirbasku DA (eds): "Growth of Cells in Hormonally Defined Media." Cold Spring Harbor, New York: Cold Spring Harbor Laboratory, Book A, pp 615-624.

Sanford KK, Parshad R, Gantt R (1978): Light and oxygen effects on chromosomes, DNA and neoplastic transformation of cells in culture. In Katsuta H (ed): "Nutritional Requirements of Cultured Cells." Baltimore: University Park Press, pp 117-148.

Schafer TW, Pascale A, Shimonski G, Came PE (1972): Evaluation of gentamicin for use in virology and tissue culture. Appl Microbiol 23:565-570.

Schneider BL, Braunschweiger M, Mitsui Y (1978): The effect of serum batch on the in vitro lifespans of cell cultures derived from old and young donors. Exp Cell Res 115:47-52.

Schwarze PE, Seglen PO (1981): Effects of antibiotics in protein synthesis and degradation in primary cultures of rat hepatocytes. In Vitro 17:71-78.

Scott BS (1971): Effect of potassium on neuron survival in cultures of dissociated human nervous tissue. Exp Neurol 30:297-308.

Scott BS, Fisher KC (1970): Potassium concentration and number of neurons in cultures of dissociated ganglia. Exp Neurol 27:16-22.

Scott BS, Fisher KC (1971): Effect of choline, high potassium and low sodium on the number of neurons in cultures of dissociated chick ganglia. Exp Neurol 31:183-188.

Server AC, Shooter EM (1977): Nerve growth factor. Adv Protein Chem 31:339-409.

Shapiro D, Schrier BK (1973): Cell cultures of fetal rat brain: Growth and marker enzyme development. Exp Cell Res 77:239-247.

Shipman C Jr (1969): Evaluation of 4-(2-hydroxyethyl)-1-piperazineëthanesulfonic acid (HEPES) as a tissue culture buffer. Proc Soc Exp Biol Med 130:305-310.

Shooter RA, Gey GO (1952): Studies of the mineral requirements of mammalian cells. Br J Exp Pathol 33:98-103.

Simms E, Gazdar AF, Abrams PG, Minna JD (1980): Growth of human small cell (oat cell) carcinoma of the lung in serum-free growth factor-supplemented medium. Cancer Res 40:4356-4363.

Smith JR (1981): The fat soluble vitamins. In Waymouth C, Ham RG, Chapple PJ (eds): "The Growth Requirements of Vertebrate Cells In Vitro." New York: Cambridge University Press, Chap 22, pp 343-352.

Sporn MB, Clamon GH, Dunlop NM, Newton DL, Smith JM, Saffiotti U (1975): Activity of vitamin A analogues in cell cultures of mouse epidermis and organ cultures of hamster trachea. Nature 252:47-49.

Steele W, Jenkin HM (1977a): The growth characteristics of Novikoff hepatoma cells in the presence of different fatty acid:albumin ratios. Proc Soc Exp Biol Med 155:405-409.

Steele W, Jenkin HM (1977b): Effect of petroselinic and stearic acids on the alkyl diacyl glycerides of Novikoff hepatoma cells. Proc Soc Exp Biol Med 155:410-415.

Steiner DF, Chan SJ, Terris S, Hoffmann C (1978): Insulin as a cellular growth regulator. In Porter R, Whelan J (eds): "Hepatotrophic Factors." Amsterdam: Elsevier (CIBA Symposium #55), pp 217-228.

Stoien JD, Wang JD (1974): Effect of near-ultraviolet and visible light on mammalian cells in culture. II. Formation of toxic photoproducts in tissue culture medium by black light. Proc Natl Acad Sci USA 71:3961-3965.

Stoner GD, Harris CC, Myers GA, Trump BE, Connor RD (1980): Putrescine stimulates growth of human bronchial epithelial cells in primary culture. In Vitro 16:399-406.

Swierenga SHH, Whitfield JF, Gillan DJ (1976): Alteration by malignant transformation of the calcium requirements for cell proliferation in vitro. J Natl Cancer Inst 57:125-129.

Swim HE (1967): Nutrition of cells in culture. A review. Wistar Inst Symp 6:1-14.

Takaoka T, Katsuta H (1971): Long-term cultivation of mammalian cell strains in protein- and lipid-free chemically defined synthetic media. Exp Cell Res 67:295-304.

Taub M (1982): Hormones control the growth and function of cultured kidney cells. In Sato GH, Pardee AB, Sirbasku DA (eds): "Growth of Cells in Hormonally Defined Media." Cold Spring Harbor, New York: Cold Spring Harbor Laboratory, Book A, pp 581-592.

Taub M, Livingston D (1981): The development of serum-free hormone-supplemented media for primary kidney cultures and their use in examining renal function. Ann NY Acad Sci 372:406-421.

Taylor WG, Richter A, Evans VJ, Sanford KK (1974): Influence of oxygen and pH on plating efficiency and colony development of WI-38 cells and VERO cells. Exp Cell Res 86:152-156.

Thomas JA, Johnson MJ (1967): Trace-metal requirements of NCTC clone 929 strain L cells. J Natl Cancer Inst 39:337-345.

Tsao MC, Walthall BJ, Ham RG (1982): Clonal growth of normal human epidermal keratinocytes in a defined medium. J Cell Physiol 110:219-229.

Tupper J, Zorgniotti F (1977): Calcium content and distribution as a function of growth and transformation in the mouse 3T3 cell. J Cell Biol 75:12-22.

Twardzik DR, Sherwin SA, Ranchalis J, Todaro GJ (1982): Transforming growth factors in the urine of normal, pregnant and tumor-bearing humans. J Natl Cancer Inst 69:793-798.

Vahouny GV, Wei R, Starkweather R, Davis C (1970): Preparation of beating heart cells from adult rats. Science 167:1616-1618.

Van Wyk JJ, Furlanetto RW (1982): The somatomedins and other peptide growth factors. In Waymouth C, Ham RG, Chapple PJ (eds): "The Growth Requirements of Vertebrate Cells In Vitro." New York: Cambridge University Press, Chapt 29, pp 411-424.

Van Wyk JJ, Underwood LE, Hintz RL, Glemmons DR, Voina S, Weaver RP (1974): The somatomedins: A family of insulin-like hormones under growth hormone control. Rec Prog Hormone Res 30:259-318.

Vasiliev JM, Gelfand IM (1981): "Neoplastic and Normal Cells in Culture." Part III. Regulation of growth in normal and transformed cultures. Cambridge: Cambridge University Press, pp 177-241.

Vogelaar JPM, Erlichman E (1933): A feeding solution for cultures of human fibroblasts. Am J Cancer 18:28-48.

Walthall BJ, Ham RG (1981): Multiplication of human diploid fibroblasts in a synthetic medium supplemented with EGF, insulin and dexamethasone. Exp Cell Res 134:301-309.

Wang R (1975): Lethal effect of "daylight" fluorescent light on human cells in tissue culture media. Photochem Photobiol 21:373-375.

Wang RJ (1976): Effect of room fluorescent light on the deterioration of tissue culture medium. In Vitro 12:19-22.

Wang RJ, Nixon BT (1978): Identification of hydrogen peroxide as a photoproduct toxic to human cells in tissue-culture medium irradiated with "daylight" fluorescent light. In Vitro 14:715-722.

Wang RJ, Stoien JD, Landa D (1974): Lethal effects of near-ultraviolet irradiation on mammalian cells in culture. Nature 247:43-45.

Waymouth C (1954): The nutrition of animal cells. Intern Rev Cytol 3:1-68.

Waymouth C (1959): Rapid proliferation of sublines of NCTC clone 929 (strain L) mouse cells in a simple chemically defined medium (MB 752/1). J Natl Cancer Inst 22:1003-1017.

Waymouth C (1965a): Construction and use of synthetic media. In Willmer EN (ed): "Cells and Tissues in Culture." London, New York: Academic, Vol 1, Chap 3, pp 99-142.

Waymouth C (1965b): The cultivation of cells in chemically defined media and the malignant transformation of cells in vitro. In Ramakrishnan CV (ed): "Tissue Culture." The Hague: Dr W Junk, pp 168-179.

Waymouth C (1970): Osmolality of mammalian blood and of media for culture of mammalian cells. In Vitro 6:109-127.

Waymouth C (1972): Construction of tissue culture media. In Rothblat GH, Cristofalo VJ (eds): "Growth, Nutrition and Metabolism of Cells in Culture." New York, London: Academic, Vol 1, Chap 2, pp 11-47.

Waymouth C (1973): Determination and survey of osmolality in culture media. In Kruse PF Jr, Patterson MK Jr (eds): "Tissue Culture: Methods and Applications." New York: Academic, Sec XIV, Chap 6, pp 703-709.

Waymouth C (1976): Preparation of medium MAB87/3 for primary cultures of epithelial cells. Tissue Culture Assoc Manual 3:521-525.

Waymouth C (1977): Nutritional requirements of cells in culture, with special reference to neural cells. In Fedoroff S, Hertz L (eds): "Cell, Tissue and Organ Cultures in Neurobiology." New York: Academic, pp 631-648.

Waymouth C (1978): Studies on chemically defined media and the nutritional requirements of cultures of epithelial cells. In Katsuta H (ed): "Nutritional Requirements of Cultured Cells." Baltimore: University Park Press, pp 39-61.

Waymouth C (1979): Autoclavable medium AM 77B. J Cell Physiol 100:548-550.

Waymouth C (1981a): Requirements for serum-free growth of cells: Comparison of currently available defined media. In Waymouth C, Ham RG, Chapple PJ (eds): "The Growth of Vertebrate Cells In Vitro." New York: Cambridge University Press, Chap 3, pp 33-47.

Waymouth C (1981b): Major ions, buffer systems, pH, osmolality, and water quality. In Waymouth C, Ham RG, Chapple PJ (eds): "The Growth of Vertebrate Cells In Vitro." New York: Cambridge University Press, Chap 8, pp 105-117.

Waymouth C, Chen HW, Wood BG (1971): Characteristics of mouse liver parenchymal cells in chemically defined media. In Vitro 6:371.

Waymouth C, Ward PF, Blake SL (1982): Mouse prostatic epithelial cells in defined culture media. In Sato GH, Pardee AB, Sirbasku DA (eds): "Growth of Cells in Hormonally Defined Media." Cold Spring Harbor, New York: Cold Spring Harbor Laboratory, Book B, pp 1097-1108.

Webber MM, Chaproniere-Rickenberg D (1980): Spermine oxidation products are selectively toxic to fibroblasts in cultures of normal human prostatic epithelium. Cell Biol Intern Rep 4:185-193.

Westermark B, Fryklund L, Wasteson A (1981): The somatomedins as growth factors for cultured cells. In Waymouth C, Ham RG, Chapple PJ (eds): "The Growth of Vertebrate Cells In Vitro." New York: Cambridge University Press, Chap 30, pp 425-435.

Whitfield JF, Perris AD, Youdale T (1968): The role of calcium in the mitotic stimulation of rat thymocytes by detergents, agmatine and poly-L-lysine. Exp Cell Res 53:155-165.

Whitfield JF, Perris AD, Rixon RH (1969): Stimulation of mitotic activity and the initiation of deoxyribonucleic acid synthesis in populations of rat thymic lymphocytes by magnesium. J Cell Physiol 74:1-8.

Whitfield JF, McManus JP, Youdale T, Franks DJ (1971): The roles of calcium and cyclic AMP in the stimulatory action of parathyroid hormone on thymic lymphocyte proliferation. J Cell Physiol 78:355-368.

Whitfield JF, Boynton AL, MacManus JP, Rixon RH, Swierenga SHH, Walker PR (1981): Interactions of endogenous and exogenous factors in proliferative control. In Waymouth C, Ham RG, Chapple PJ (eds): "The Growth of Vertebrate Cells In Vitro." New York: Cambridge University Press, Chap 12, pp 160-196.

Whitten WK (1969): The effect of oxygen on cleavage of mouse eggs in vitro. Biol Reprod, Abstract #58, p 29.

Whitten WK (1971): Nutrient requirements for the culture of preimplantation embryos in vitro. Adv Biosci 6:129-141.

Willmer EN (1970): "Cytology and Evolution." 2nd ed. New York: Academic.

Wood BG, Washburn LL, Mukherjee AS, Banerjee MR (1975): Hormonal regulation of lobulo-alveolar growth, functional differentiation and regression of whole mouse mammary gland in organ culture. J Endocrinol 65:1-6.

Wu R, Smith D (1982): Continuous multiplication of rabbit tracheal epithelial cells in a defined, hormone-supplemented medium. In Vitro 18:800-812.

Yamane I (1978a): Role of bovine albumin in a serum-free culture medium and its application. Natl Cancer Inst Monogr 48:131-133.

Yamane I (1978b): Development and application of serum-free culture medium for primary culture. In Katsuta H (ed): "Nutritional Requirements of Cultured Cells." Baltimore: University Park Press, pp 1-21.

Yamane I, Matsuya Y, Jimbo K (1968): An autoclavable powdered cultured medium for mammalian cells. Proc Soc Exp Biol Med 127:335-336.

Yamane I, Murakami O, Kato M (1975): Role of bovine albumin in a serum-free suspension of cell culture. Proc Soc Exp Biol Med 149:439-442.

Yamane I, Kan M, Hoshi H, Minamoto Y (1981): Primary cultures of human diploid cells and its long-term transfer in a serum-free medium. Exp Cell Res 134:470-474.

Yasamura Y, Niwa A, Yamamoto K (1978): Phenotypic requirement for glutamine of kidney cells and for glutamine and arginine of liver cells in culture. In Katsuta H (ed): "Nutritional Requirements of Cultured Cells." Baltimore: University Park Press, pp 223-255.

Young M, McFayden IR (1973): Placental transfer and fetal uptake of amino acids in the pregnant ewe. J Perinat Med 1:174-182.

Yuspa SH, Harris CC (1974): Altered differentiation of mouse epidermal cells treated with retinyl acetate in vitro. Exp Cell Res 86:95-105.

Methods for Preparation of Media, Supplements, and Substrata
for Serum-Free Animal Cell Culture, pages 69–86
© 1984 Alan R. Liss, Inc., 150 Fifth Avenue, New York, NY 10011

3
Preparations and Uses of Lipoproteins to Culture Normal Diploid and Tumor Cells Under Serum-Free Conditions

Denis Gospodarowicz

The principal function of plasma lipoproteins is either to deliver or to remove lipids from the cells. They could also have other biological functions yet to be discovered. They consist of noncovalent aggregates of phospholipids, cholesterol, and apoproteins whose composition/ratio and diversity differ, depending on the lipoproteins considered. Four classes are currently recognized: the chylomicrons, very low density lipoproteins (VLDL), low-density lipoprotein (LDL), and high-density lipoprotein (HDL) [Lewis, 1976]. Since only the two last classes of lipoproteins have been extensively studied in terms of their action on cell growth, we will describe the purification of those two classes of lipoproteins by ultracentrifugal flotation techniques initially developed by Havel et al. [1955].

LDL is defined operationally as the lipoprotein class that can be isolated in the density range 1.019–1.063. It has a molecular weight of $(2.2-2.3) \times 10^6$ and consists of particles with a diameter of 19–25 nm. With a concentration of about 3–4 mg/ml, it is the most abundant of the lipoproteins in human plasma [Lewis, 1976]. It also has a higher "payload" of lipid than HDL: lipid composes 76–78% of LDL. The main lipid cargo is cholesterol (8%) and cholesteryl ester (37%), with lesser amounts of phospholipid (20%) and a small triglyceride content (10%); 60–70% of the cholesterol content of whole plasma is in LDL. Lindgren et al. [1959] and Ewing et al. [1965] noted the following percentage composition of the lipid components: cholesteryl ester

Cancer Research Institute and the Departments of Medicine and Ophthalmology, University of California Medical Center, San Francisco, California 94143

46, cholesterol 14, triglyceride 14, and phospholipid 25. The carbohydrate content of LDL is at least 3%, including sialic acid in a terminal position in the polysaccharide chain, with glucosamine and several monosaccharides. Proteins account for 23% of the weight of the particles. Apoprotein B is the main apoprotein to be found in LDL particles. Small amounts of ApoC ($<$ 10%) can also be present. The principal function of LDL, as outlined in the elegant studies of Brown and Goldstein [1978], is to deliver cholesterol to the cells.

In most mammalian species, HDL is the most abundant of the serum lipoproteins, but in human serum LDL is present at higher concentration—at all ages and in both sexes. Despite its relatively low concentration in human plasma (1 mg/ml), HDL is of fundamental importance in the mobilization of cholesterol from the tissues. Its significance is illustrated by the rare but theoretically important genetic disorder Tangier disease, in which HDL is low in concentration and abnormal in composition, and cholesterol storage occurs particularly in reticuloendothelial tissues.

Human HDL floats when submitted to ultracentrifugation in the density range of 1.07-1.21. In terms of density HDL is heterogeneous; HDL_2 has a density range 1.07-1.125, is larger in diameter, and has a higher proportion of lipid than does HDL_3 (d = 1.13-1.21). Their molecular weights are 360,000 and 175,000, respectively, and their diameter ranges 7-10 nm and 4-7 nm.

HDL particles contain 55% proteins, composed of the following apoproteins: ApoAI and II (80%), ApoB (5%), ApoC (15%), ApoD ($<$ 5%). The lipid moiety of HDL is composed mostly of glycerides (4%) and phospholipids (24%), phosphatidyl choline accounting for 80% of the phospholipid. Cholesterol and cholesteryl esters account for 2% and 15%, respectively, of the amount of lipids.

In this short review, we will concentrate on the techniques involved in purifying and characterizing LDL and HDL from human plasma, and on the preparation of unilamellar phosphatidyl choline (PC) liposomes, since this is the main phospholipid present in HDL particles. The use of lipoprotein and PC liposomes for growing cells in defined medium will be briefly described.

ROTORS AND CENTRIFUGES

The most common technique for purifying lipoproteins is ultracentrifugal flotation techniques. The following rotors and ultracentrifuges can be used: Beckman rotor 50.2-Ti (capacity 300 ml), rotor 42.1 (capacity 200 ml), or rotor 45 Ti (capacity 600 ml) run in Beckman ultracentrifuge models L-5-50E, L, or L8-70, respectively. All run at a temperature of 4°C.

PREPARATION BY FLOTATION OF HUMAN LDL

LDL particles that have a density between 1.019 and 1.063 can be purified from plasma by preparative ultracentrifugation. Two stock solutions of potassium bromide (KBr) are first prepared and kept at room temperature. Solution A is composed of 4.4 g of NaCl dissolved in 500 ml of double-distilled water (d = 1.005). The second solution (B) is composed of 76.5 g NaCl and 177 g KBr dissolved in 500 ml of double-distilled water (d = 1.346). Since the methodology of preparing purified lipoproteins rests mostly on the density of the solution used, it is important to check the density at various points by weighing aliquots of plasma or various solutions (i.e., 1 ml of a solution of a density of 1.346 g/cm^3 should weigh 1.346 gm).

A 120 ml quantity of solution A (d = 1.005) and 36.8 ml of solution B (d = 1.346) are mixed, giving 156.8 ml of a solution with a density of 1.085. Then 85 ml of this solution is added to 425 ml of plasma, raising its density to 1.019. The plasma solution is then spun for 48 h in a 50.2 Ti rotor (total capacity 300 ml) at 35,000 rpm in a Beckman L-5-50E ultracentrifuge. What is left of the solution is spun for 48 h in a 42.1 rotor (total capacity 200 ml) at 40,000 rpm in a Beckman Model L centrifuge. The top plasma layer of the tubes, containing mostly chylomicrons and VLDL, is then discarded. The buffer present in the upper third of the tube is also discarded. The bottom two-thirds is collected (total volume around 140–200 ml). A KBr solution is composed of 20 ml of a solution A (d = 1.005) and 26.7 ml of solution B (d = 1.346) is then added to the collected plasma at a ratio of (ml collected)/3, raising its density to 1.063. The plasma is then spun again in a 50.2 Ti rotor at 35,000 rpm in a Beckman L-5-50E ultracentrifuge for 48 h. The top layer, which is yellow, is collected from each tube, pooled, and spun for 24 h at 35,000 rpm in a 50.2 rotor, as described above. This step is a washing step in which most, if not all, contaminating plasma proteins will be eliminated. After centrifugation, the top yellow layers are again collected, put in a single 25-ml tube, and spun again at 35,000 rpm in a Beckman 50.2 Ti rotor for 24 h. This step serves as both a washing and concentrating step. The top yellow layer, whose protein content is 15 mg/ml (or 75 mg/ml lipoprotein), is then collected (total volume 15 ml) and stored at 4°C for up to a month. Before being used for cell culture, the aliquots of the LDL stock solution are dialyzed overnight against a solution composed of 0.9% NaCl, 25 mM EDTA, and 5 mM Tris-HCl (pH 7.2). The protein concentration is determined by the modification of Maxwell et al. [1978] of the method of Lowry et al. [1951].

PREPARATION OF HUMAN AND BOVINE HDL

Human HDL particles have a density (between 1.07 and 1.21) that differs from that of other lipoproteins. They can therefore be purified from plasma and other lipoproteins by differential ultracentrifugal flotation. Solid KBr (103.4 g) is added to 1,100 ml of plasma to raise its final density to 1.07. If a different volume of plasma is used, the quantity of KBr added to it in order to reach the same final density is given by the following formula:

$$\frac{\text{ml plasma} \times (1.07-1.005)}{1-0.312} = \text{g KBr}$$

The plasma to which solid KBr has been added should be stirred in the cold for 30 min in order to dissolve the salt fully. The plasma is then spun for 48 h in a type 45 Ti rotor (total volume 600 ml) at 40,000 rpm in a Beckman ultracentrifuge L.8-10. Other rotors and centrifuges that can be used are rotor 50.2 Ti (total volume 300 ml) run at 35,000 rpm in a Beckman ultracentrifuge L-5-50 or the rotor 42.1 (total volume 200 ml) run at 40,000 rpm in a Beckman ultracentrifuge Model L. The top layers (upper half of the tubes), containing chylomicrons, VLDL, and LDL, are then discarded, and the bottom layers are collected, the total volume is determined, and crystalline KBr is added according to the following formula,

$$\frac{\text{ml plasma} \times (1.21-1.07)}{1-0.312} = \text{g KBr}$$

to raise the plasma density to 1.21. Once the KBr is totally dissolved (4°C with stirring), the plasma is spun for 48 h in a Beckman ultracentrifuge L-5-50E (rotor 50.2, tube volume 25 ml, total volume 300 ml) and in a Beckman ultracentrifuge model L (rotor 42.1, tube volume 25 ml, total volume 200 ml). After centrifugation, the top layer, which is about 5 mm wide and intensely yellow, is collected with a syringe from each tube. In order to wash and concentrate the HDL preparation, the HDL solutions are centrifuged once more for 24 h, as described above. The top layers containing the HDL are then pooled (15 ml total volume, average protein concentration of 30-40 mg/ml, or 60-80 mg/ml lipoprotein). The HDL solution can be stored at 4°C for up to, but for *no longer* than, a month. Before using them for cell culture, aliquots of the HDL solution are dialyzed overnight against a solution composed of 0.9% NaCl, 25 mM EDTA, and 5 mM Tris-HCl, pH 7.2. Protein concentration is determined by the modification of Maxwell et al. [1978] of the method of Lowry et al. [1951].

   HDL can also be prepared easily from bovine plasma, in which it constitutes about 80% (by weight) of the total lipoproteins of d = 1.21 g/ml. In contrast to human HDL, bovine HDL has a density interval of 1.083-1.125 g/ml

[Jonas, 1972]. The density of plasma is adjusted to 1.083, in order to eliminate chylomicrons, VLDL, and LDL, and then is raised to 1.125 after spinning. Bovine HDL is then washed and concentrated, as described for human HDL, in a solution of density 1.125 g/cm^3.

PREPARATION OF SUBCLASSES OF HUMAN HDL

Human HDL is composed of subclasses of particles with apoproteins and lipids present at different ratios. These can be separated on the basis of their different densities. While human HDL_2 had a density of 1.075-1.125 g/cm^3, that of HDL_3 is 1.130-1.21. As in the case of HDL, the density of the plasma is raised to 1.07 and the plasma is spun (48 h) as previously described. The top layers (upper half of the tubes) containing chylomicrons, VLDL, and LDL are then discarded and the bottom layers are collected, the total volume is determined, and crystalline KBr is added according to the following formula:

$$\frac{\text{ml plasma} \times (1.125-1.07)}{1-0.312} = \text{g KBr}$$

to raise the plasma density to 1.125. Once the KBr is totally dissolved, the plasma is spun for 48 h, as previously described. After centrifugation, the top layers containing HDL_2 are collected and recentrifuged for 24 h in order to wash and concentrate the preparation.

To purify HDL_3, the plasma density is raised to 1.13 following the formula:

$$\frac{\text{ml plasma} \times (1.13-1.005)}{1-0.312} = \text{g KBr}$$

After the KBr is totally dissolved, the plasma is spun for 48 h as described. After centrifugation, the top yellow layers are discarded and the bottom layers are collected, the total volume is determined, and the crystalline KBr is added according to the following formula:

$$\frac{\text{ml plasma} \times (1.21-1.13)}{1-0.312} = \text{g KBr}$$

to raise the plasma density to 1.21. It is then spun for 48 h, and the top layers are collected and respun for 24 h. The top yellow layers containing HDL_3 are then collected and kept at 4°C. Before use, HDL_2 and HDL_3 should be dialyzed as described for HDL.

CHARACTERIZATION OF LDL AND HDL PARTICLES

In order to characterize LDL and HDL particles and assess the relative degree of contamination of lipoproteins by each other or by plasma protein, the lipoprotein preparation must be analyzed by gel electrophoresis and by immunodiffusion.

Gel Electrophoresis

An easy test consists in analyzing the homogeneity of each lipoprotein fraction on agarose electrophoretic gels (Fig. 1). LDL and HDL should migrate as distinct bands. More complex, but also more precise analytically, is the examination of each lipoprotein fraction by slab gel electrophoresis (Figs. 2 and 3) [Tauber et al., 1981a,b,c]. To eliminate the possibility of contamination by plasma proteins, the purity of the HDL preparations should be analyzed by slab gel electrophoresis (10-18% and 5-18%, respectively, exponential poly-

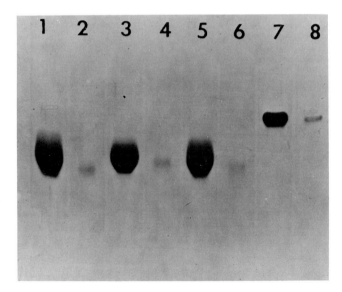

Fig. 1. Agarose gel electrophoresis of HDL_3, HDL_2, HDL, and LDL. Human lipoproteins were loaded onto an agarose gel (Corning ACI) as follows: lanes 1 and 2, HDL_3 at 100 μg and 1 μg protein, respectively; lanes 3 and 4, HDL_2 at 100 μg and 1 μg protein, respectively; lanes 5 and 6, HDL at 100 μg and 1 μg, protein respectively, and lanes 7 and 8, LDL at 50 μg and 1 μg protein, respectively. The gel was processed in a Corning ACI cassette system with a barbital buffer (0.05 M, pH 8.6) for 35 min. The lipoproteins were fixed and stained by 15 min of exposure to amido black 10B (Corning ACI). The gel was then cleared by washing in a solution of 5% acetic acid.

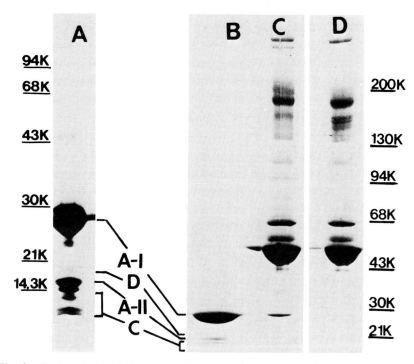

Fig. 2. Sodium dodecyl sulfate polyacrylamide gel electrophoresis of human HDL, HS, and human LPDS. Human HDL preparations were delipidated with tetramethyl urea and reduced with 0.1 N dithiothreitol. Samples of human HDL (B), serum (C), and human LPDS (D) were then boiled for 5 min, and 50-μg protein aliquots were applied to a 5-18% exponential gradient polyacrylamide gel containing 0.1% SDS. High-density lipoprotein was also analyzed separately (A, 30 μg protein) after delipidation with tetramethyl urea with a 10-18% gradient gel to enhance the separation of the different apolipoproteins. Identification of apolipoproteins ApoAI (A-I), ApoAII (A-II), ApoD (D), and ApoC (C) is based on reported molecular weights and comparison to molecular weight standards.

acrylamide gel gradient containing 0.1% sodium dodecyl sulfate) with or without prior delipidation with tetramethyl urea (Fig. 2). When the electrophoretic patterns of HDL preparations are compared to those of plasma or lipoproteins deficient serum (LPDS), no obvious contamination by plasma proteins should be observable (Fig. 2).

When HDL_2 and HDL_3 preparations are prepared, their purity should be compared to that of HDL by similar techniques. HDL, HDL_2, and HDL_3 are delipidated with tetramethyl urea and applied to an exponential gradient (10-18%) polyacrylamide/SDS slab gel with a 3% stacking gel (Fig. 2). Samples are run reduced or unreduced (Fig. 3). To reduce the apoprotein AII-E

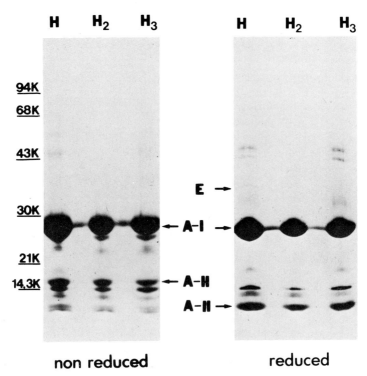

Fig. 3. Sodium dodecyl sulfate/polyacrylamide gel electrophoresis of HDL, HDL$_2$, and HDL$_3$. Human HDL (H), HDL$_2$ (H$_2$), and HDL$_3$ (H$_3$) were delipidated with tetramethyl urea and reduced or not reduced with 0.1 M dithiothreitol. Samples were then boiled for 5 min, and an aliquot of each lipoprotein (30 µg protein) was applied to a 10–18% exponential gradient sodium dodecyl sulfate/polyacrylamide gel. Identification of apolipoprotein apoprotein AI (A-I), apoprotein AII (A-II), and apoprotein E (E) is based on reported molecular weights and comparison to molecular weight standards.

complex present in human HDL$_2$, dithiothreitol is added at a concentration of 0.1 M before the samples are boiled. As shown in Figure 3, apoprotein E (M$_r$ 37,000) is found only in HDL and HDL$_2$, and not in HDL$_3$.

Double Immunodiffusion

Labeled and native high-density and low-density lipoproteins are tested by double immunodiffusion against either rabbit anti-human β-lipoprotein (anti-LDL; Cappel Laboratories, Cochranville, PA) or rabbit anti-human α-lipoprotein (anti-HDL; Behring Diagnostics, American Hoechst Co., Somerville, NJ) [Tauber et al., 1981c]. To enhance the level of detectability of contaminants, one can iodinate LDL and HDL by a modification of the iodine monochloride method, as described previously [MacFarlane, 1958; Vlodavsky et al., 1978; Fielding et al., 1979]. After labeling, unbound ^{125}I is removed by exclusion gel chromatography on a Sephadex G-50 (Pharmacia) column followed by dialysis of the labeled lipoproteins against 2,000 volumes of saline containing 0.01T EDTA and 10 mM Tris-HCl, pH 7.4. Specific activity of the labeled LDL is 265–385 cpm per 1 ng protein and for labeled HDL is 252–580 cpm per 1 ng protein. Less than 5% of the radioactivity in labeled LDL and labeled HDL should be soluble in 10% trichloroacetic acid, and lipid-bound radioactivity, determined by extraction of the iodinated lipoproteins with chloroform/methanol (1:2), is 5–7%.

Plates containing 0.6% agarose in phosphate-buffered saline (PBS) are incubated with the appropriate samples for 18 h at room temperature and then washed extensively with PBS (six changes over a 96-h period), fixed with 7% acetic acid (wt./vol.), stained with 0.1% amido black, and photographed. After drying, stained gels are subjected to autoradiography (3-h exposure). When labeled or native LDL and HDL preparations are analyzed by double immunodiffusion, 1 μg of unlabeled LDL protein and 0.1 μg of ^{125}I-LDL give a single precipitin line against anti-LDL antiserum (Fig. 4A,B). In the case of either unlabeled or labeled HDL, no precipitin line can be observed even at a 100-times higher protein concentration (100 μg and 10 μg, respectively) (Fig. 4A,B). Likewise, either labeled or unlabeled HDL gives a single precipitin line when tested against rabbit anti-HDL antiserum (Fig. 5A,B), whereas either unlabeled or labeled LDL at a 100-times higher protein concentration does not give a precipitin line (Fig. 5A,B). These results demonstrate that the unlabeled and iodinated LDL preparations contain less than 1% HDL, if any at all, and vice versa.

PREPARATION OF UNILAMELLAR PC LIPOSOMES

Commercial PC preparations are either not very soluble, even after sonication, or are partly oxidized. They are therefore not very active, and it becomes important to purify PC from egg yolk by acetone precipitation, alumina chromatography, and high-performance liquid chromatography (HPLC), as described by Szoka and Papahadjopoulos [1980]. Aliquots of the PC in chloroform in vials sealed under argon gas are stored at $-20°C$ until use [Mayhew et al., 1980]. To prepare unilamellar liposomes of PC, the chloroform is evaporated under nitrogen gas and the PC is suspended in Dulbecco's modified Eagle's medium (DMEM) at a concentration of 10 mg/ml in a tube sealed under nitrogen. Liposomes are formed during 10–15 min sonication in a water bath sonicator (Laboratory Supplies, Hicksville, NY).

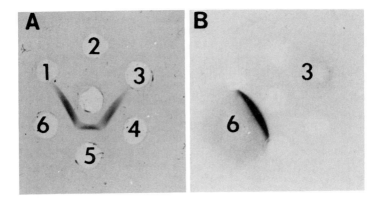

Fig. 4. Immunodiffusion of labeled and unlabeled human HDL and LDL against rabbit anti-human apoprotein (anti-LDL). Native and iodinated HDL and LDL were incubated (18 h, 24°C) in Ouchterlony plates (0.6% agarose in PBS) against rabbit anti-human β-apoprotein (anti-LDL; Cappel Laboratories, Cochranville, PA) in the center well. Plates were then washed extensively with PBS (6 changes over a 96-h period), fixed with 7% acetic acid (25/volume), and stained with 0.1% amido black (A). Plates were then dried and subjected to autoradiography on Kodak NS-2T X-ray film (Eastman Kodak Co., Rochester, NY) for a 3-h exposure time (B). The anti-β-lipoprotein is in the center well against 100 μg and 10 μg of unlabeled HDL, 1 μg of ^{125}I-HDL, 1 μg and 4 μg of unlabeled LDL, and 1 μg of ^{125}I-LDL placed in wells Nos. 1,2,3,4,5, and 6, respectively.

This results, as seen by transmission electron microscopy of negatively stained samples, in the formation of small (30-70 nm in diameter) unilamellar PC liposomes. The clear but slightly opalescent suspension is then filtered through a 0.45 μm Millipore filter. Complete recovery of PC after filtration can be confirmed by studies with radiolabeled PC. PC liposomes are used for experiments within 1 day following their preparation [Fujii et al., 1983].

USE OF LIPOPROTEINS AND PC LIPOSOMES IN CELL CULTURE

The presence of human lipoproteins has been shown to be required for the proliferation in vitro of a number of normal diploid cell types and established cell lines or tumor cell lines (reviewed by Gospodarowicz et al. [1982a, 1983a]). While LDL is in general a poor mitogen, active at low concentration and toxic at concentrations higher than 100 μg protein per 1 ml, HDLs have been shown to be extremely active over the range of 10-1,000 μg protein per 1 ml and is not cytotoxic even at the highest concentrations. Among the normal diploid cell types with a finite life span that require only HDL to grow when exposed to transferrin-supplemented medium are bovine vascular en-

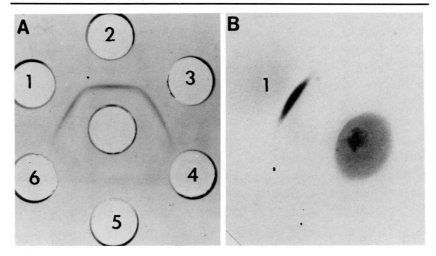

Fig. 5. Immunodiffusion of labeled and unlabeled HDL and LDL against rabbit anti-human α_1-lipoprotein (anti-HDL). Native and iodinated HDL and LDL were incubated (18 h, 24°C) in Ouchterlony plates against rabbit anti-human α_1-lipoproteins (anti-HDL) (Behring Diagnostics, American Hoechst Co., Somerville, NJ) in the center well. Plates were processed as described in Figure 4 for staining (A) or autoradiography (B). Anti-α_1-lipoprotein in the center well against 0.1 μg ^{125}I-HDL, 4 μg and 1 μg of unlabeled HDL, 10 μg of ^{125}I-LDL, and 100 μg and 10 mg of unlabeled LDL placed in wells Nos. 1,2,3,4,5, and 6, respectively.

dothelial cells [Tauber et al., 1981a] and kidney tubule epithelial cells [Gospodarowicz et al., 1983b]. Other cell types such as vascular smooth muscle cells [Gospodarowicz et al., 1981], corneal endothelial cells [Giguere et al., 1982], lens epithelial cells [Gospodarowicz and Massoglia, 1982], and rat-1 fibroblasts [Giguere and Gospodarowicz, 1983] require insulin and epidermal growth factor (EGF) or fibroblast growth factor (FGF) in addition to HDL in order to proliferate at an optimal rate when exposed to transferrin-supplemented medium. HDL is nevertheless a decisive factor, since in its absence these cells will not proliferate actively in response to a combination of transferrin, insulin, and FGF or EGF. Only in the case of cells with active steroidogenic pathways does HDL have a minor effect on cell growth. This reflects the ability of these cells to synthesize high amounts of mevalonate, from which one can derive both sterol and nonsterol products. Such is the case for adrenal cortex [Ill and Gospodarowicz, 1982] and granulosa cells [Savion et al., 1981; Savion and Gospodarowicz, 1982], which have a lower requirement for HDL than other cell types and for which insulin is the main mitogen.

In the case of tumor cells and established cell lines, HDL has been shown to be mitogenic for those studied to date [Gospodarowicz et al., 1982a,b, 1983c; Giguere and Gospodarowicz 1983; Gospodarowicz, 1983] (Table I).

TABLE I. Factors Required for the Proliferation of Normal Diploid or Tumoral and Established Cell Lines Maintained in Serum-Free Medium

	HDL (µg protein/ml)	Insulin or somato C (ng/ml)	FGF or EGF (ng/ml)		Transferrin (µg/ml)	Reference
Normal diploid cells						
Vascular endothelial cells	500	—	—	—	10	[Tauber et al., 1981b]
Corneal endothelial cells	250	2,500	100	50	10	[Giguere et al., 1982]
Vascular smooth muscle cells	250	2,500	100	50	10	[Gospodarowicz et al., 1981]
Granulosa cells	30	1,000	100	50	10	[Savion et al., 1981; Savion and Gospodarowicz, 1982]
Adrenal cortex cells	30	50	100	—	5	[Ill and Gospodarowicz, 1982]
Lens epithelial cells	250	2,500	100	—	10	[Gospodarowicz and Massoglia, 1982]
Kidney tubule cells	750	—	—	—	50	[Gospodarowicz et al., 1983b]
Embryo fibroblasts (rat-1)	1,000	5,000	—	25	25	[Giguere and Gospodarowicz, 1983]
Transformed cells (tumor)						
A-431 carcinoma	500	—	—	Toxic	10	[Gospodarowicz et al., 1982a,b]
Colon carcinoma cells	500	—	—	—	10	[Gospodarowicz et al., 1982a,b]
Ewing sarcoma cells	500	—	—	—	10	[Gospodarowicz et al., 1982a,b]
Rhabdomyosarcoma	500	—	—	—	10	[Gospodarowicz et al., 1982a,b]
MDCK (kidney-derived)	500	—	—	—	10	[Gospodarowicz et al., 1983c]
B-31 cell line	500	—	—	—	25	[Giguere and Gospodarowicz, 1983]

The mitogenic effect of HDL could be that of a nutrient or that of a mitogen. Both effects are likely. Its action may also be potentiated by LDL, which delivers cholesterol to the cells [Cohen et al., 1982a,b]. When the cells are exposed to high concentrations of LDL, it becomes toxic [Cohen et al., 1982a,b; Gospodarowicz et al., 1983c]. This could reflect cholesterol over-load. Exposure of the cells to high concentrations of HDL counteracts this effect, probably by depleting the cholesterol content of the cells or by entering into competition with low-affinity LDL-binding sites, thereby limiting the cellular uptake of cholesterol-derived LDL.

One effect of HDL on the cellular activity that correlates well with its ability to stimulate cell proliferation is its ability to stimulate 3-hydroxy-3-methylglutaryl Coenzyme A reductase (HMG CoA reductase) [Cohen et al., 1982a,b; Gospodarowicz et al., 1983c]. This enzyme is responsible for the synthesis of mevalonic acid, which is later utilized for the synthesis of nonsterol products such as dolichol, ubiquinone, isopentenyl adenyl adenine, or of sterols such as cholesterol (reviewed in Brown and Goldstein [1980]). Of these compounds, the nonsterol products, and in particular isopentenyl adenyl adenine, are essential for initiation of DNA synthesis [Quesney-Huneeus et al., 1980]. Since vascular endothelial cells or Madin-Darby canine kidney (MDCK) cell cultures exposed to compactin, a competitive inhibitor of HMG CoA reductase, will not proliferate even when provided with cholesterol derived from LDL, while they resume proliferation when provided with either mevalonate or HDL which superinduced the activity of HMG CoA reductase [Cohen et al., 1982a; Gospodarowicz et al., 1983c], further evidence is provided that inadequate synthesis of nonsterol products is a more limiting factor of cell proliferation than is cholesterol. In recent studies, the active moiety of HDL has been shown to be composed in part of phosphatidyl choline, since its substitution for HDL induced cells to proliferate. However, these cultures have a longer average doubling time during their exponential growth phase than when exposed to LDL, and they also show much earlier signs of senescence than their counterparts exposed to HDL [Gospodarowicz et al., 1982a; Fujii et al., 1983].

The substrate upon which cells are maintained is an important factor to consider if cells are to respond to HDL. For example, while bovine vascular endothelial cells proliferate in response to HDL when exposed to medium supplemented with transferrin (Fig. 6), they do so best when maintained on BCE-ECM and do so poorly when maintained on HR-9-ECM. The reverse is true of kidney epithelial cells, which do not respond to HDL when maintained on BCE-ECM but are highly responsive to it when maintained on HR-9-ECM. Similar observations have been made in the case of tumor cells. Ewing

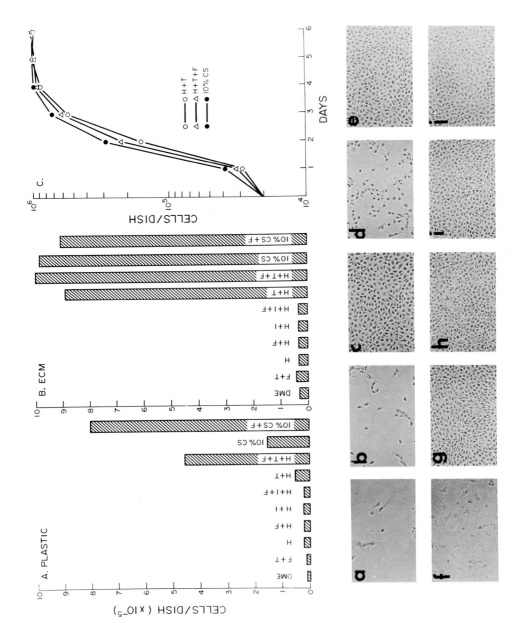

sarcoma cells exposed to serum-supplemented medium (which by definition contains HDL) proliferate in the floating configuration but not when attached loosely to plastic [Vlodavsky et al., 1980; Gospodarowicz, 1983]. Yet, when maintained on BCE-ECM, the attached populations proliferate equally well when exposed to either serum- or HDL-supplemented medium.

Although there is perfect receptor cross-reactivity between LDL and HDL of human or bovine origin, one should be aware that because of the complexity of the apoproteins making up the HDL particle [Innerarity et al., 1980] and of the structure of a given apoprotein, there could be strong species specificity so far as receptor cross-reactivity is concerned. While HDL in humans is composed mostly of particles containing ApoAI and AII and has a minor fraction of HDL_2 and HDL_1 particles containing an ApoE capable of interacting with LDL high-affinity receptor sites, rat HDL, for example, contains an appreciable fraction of HDL_1 particles containing ApoE capable of binding to LDL receptors and of delivering cholesterol to the cells [Innerarity et al., 1980]. In the case of human and rat LDL, which contain a single species of apoprotein (ApoB), human LDL is unable to interact with the cell surface receptors of rat fibroblasts [Innerarity et al., 1980]. Rat LDL do, however, react quite well with human fibroblasts. Human LDL also cross-reacts poorly with porcine and canine fibroblasts and smooth muscle cells,

Fig. 6. Comparison of the proliferation and morphological appearance of vascular endothelial cells seeded in total absence of serum on plastic versus ECM-coated dishes and exposed to serum-free medium supplemented with HDL, transferrin, insulin, and FGF added either singly or in combination versus those of cultures exposed to 10% calf serum with or without FGF. A,B. Bovine vascular endothelial cells (2×10^4 cells) were seeded in absence of serum on either 35-mm plastic dishes (A) or ECM-coated dishes (B) and exposed to DME (H-16) alone or supplemented with FGF-transferrin (F+T); HDL alone (H); HDL-FGF (H+F); HDL-insulin (H+I); HDL-insulin-FGF (H+I+F); HDL-transferrin (H+T); HDL-transferrin-FGF (H+T+F); 10% calf serum (CS); 10% calf serum-FGF (10% CS+F). The concentration of HDL added was 500 μg protein per ml, that of transferrin was 10 μg/ml, and those of insulin and FGF were 2.5 μg/ml and 100 ng/ml respectively. HDL and transferrin were added only once at day 0, whereas FGF was added every other day and insulin every 4 days. After 6 days in culture, triplicate dishes representing each condition were trypsinized and counted. C. Growth rate of low-density vascular endothelial cells seeded in absence of serum on ECM-coated dishes (2×10^4 cells per 35-mm dish) and exposed to HDL-transferrin ($\bigcirc$, H+T); HDL-transferrin-FGF ($\triangle$, H+T+F); or 10% calf serum ($\bullet$, 10% CS). Schedule of addition and concentrations of transferrin, HDL, and FGF were as described in A,B. Every day, triplicate plates representing each condition were trypsinized and counted. The standard deviation in the different determinations did not exceed 10% of the mean. a–j. Morphological appearance of vascular endothelial cell cultures maintained on plastic (a–e) or ECM-coated (f–j) dishes and exposed to serum-free medium supplemented with HDL-insulin-FGF (a,f); HDL-transferrin (b,g); HDL-transferrin-FGF (c,h); 10% calf serum (d,i); or 10% calf serum + FGF (e,j). Pictures were taken once the cultures reached confluence on day 6 (phase contrast, ×70).

while porcine and canine LDL bind with high affinity to the receptors on human cells. There may therefore be significant evolutionary differences in either the apoprotein or receptor structure among various species that result in quantitative differences in the affinities of heterologous lipoprotein-receptor systems. This species specificity of lipoproteins and their poor cross-reactivity with receptors of heterologous species could in part explain the difficulty of growing cells from a given species in heterologous systems. This is a serious factor to consider when one uses heterologous lipoproteins to trigger the proliferation of animal cells.

REFERENCES

Brown M, Goldstein J (1978): The low density lipoprotein pathway and its relation to atherosclerosis. Annu Rev Biochem 46:897–930.

Brown M, Goldstein J (1980): Multivalent feedback regulation of HMG CoA reductase, a control mechanism coordinating isoprenoid synthesis on cell growth. J Lipid Res 21:505–517.

Cohen DC, Massoglia SL, Gospodarowicz D (1982a): Correlation between two effects of high density lipoproteins on vascular endothelial cells: The induction of 3-hydroxy-3-methylglutaryl Coenzyme A reductase activity and the support of cellular proliferation. J Biol Chem 257:9429–9437.

Cohen DC, Massoglia S, Gospodarowicz D (1982b): Feedback regulation of 3-hydroxy-3-methylglutaryl Coenzyme A reductase in vascular endothelial cells: Separate sterol and nonsterol components. J Biol Chem 257:11106–11112.

Ewing AM, Freeman K, Lindgren FT (1965): The analysis of human serum lipoproteins. Adv Lipid Res 3:25–36.

Fielding P, Vlodavsky I, Gospodarowicz D, Fielding CJ (1979): Effect of contact inhibition on the regulation of cholesterol metabolism in cultured vascular endothelial cells. J Biol Chem 244:749–755.

Fujii DK, Cheng J, Gospodarowicz D (1983): Phosphatidyl choline and the growth in serum-free medium of vascular endothelial and smooth muscle cells and corneal endothelial cells. J Cell Physiol 114:267–278.

Giguere L, Gospodarowicz D (1983): Effect of RSV transformation of rat-1 fibroblasts upon their growth factor and anchorage requirements in serum-free medium. Cancer Res 43:2121–2130.

Giguere L, Cheng J, Gospodarowicz D (1982): Factors involved in the control of proliferation of bovine corneal endothelial cells maintained in serum-free medium. J Cell Physiol 110:72–80.

Gospodarowicz D (1983): The control of mammalian cell proliferation by growth factors, extracellular matrix, and lipoproteins. J Inv Dermatology 81:41–50.

Gospodarowicz D, Massoglia SL (1982): Plasma factors involved in the in vitro control of proliferation of bovine lens cells grown in defined medium. Effect of fibroblast growth factor on cell longevity. Exp Eye Res 35:259–270.

Gospodarowicz D, Hirabayashi K, Giguere L, Tauber JP (1981): Factors controlling the proliferative rate, final cell density, and life span of bovine vascular smooth muscle cells in culture. J Cell Biol 89:568-578.

Gospodarowicz D, Cohen DC, Fujii DK (1982a): Regulation of cell growth by the basal lamina and plasma factors: Relevance to embryonic control of cell proliferation and differentiation. In "Cold Spring Harbor Conferences on Cell Proliferation. Growth of Cells in Hormonally Defined Media," Vol 9: "Hormones and Cell Culture." Cold Spring Harbor, New York: Cold Spring Harbor Laboratory, pp 95-124.

Gospodarowicz D, Lui GM, Gonzalez R (1982b): High density lipoproteins and the proliferation of the human tumor cells maintained on extracellular matrix-coated dishes and exposed to defined medium. Cancer Res 42:3704-3713.

Gospodarowicz D, Lepine J, Massoglia S (1983a): Control of cell proliferation and differentiation by extracellular matrices. J Natl Cancer Inst (in press).

Gospodarowicz D, Lepine J, Massoglia S (1983b): Ability of various basement membranes to support differentiation in vitro of normal diploid bovine kidney tubule cells. J Cell Biol (submitted for publication).

Gospodarowicz D, Cohen DC, Massoglia S (1983c): Effect of high density lipoprotein on the proliferative ability of MDCK cells and HMG CoA reductase activity. J Cell Physiol 117:76-90.

Havel R, Eder HS, Bragdon JA (1955): The distribution and chemical composition of ultracentrifugally separated lipoproteins in human serum. J Clin Invest 34:1345-1353.

Ill CR, Gospodarowicz D (1982): Plasma factors involved in supporting the growth and steroidogenic functions of bovine adrenal cortex cells maintained on an extracellular matrix and exposed to a defined medium. J Cell Physiol 113:373-384.

Innerarity TL, Pitas RE, Mahley RW (1980): Disparities in the interaction of rat and human lipoproteins with cultured rat fibroblasts and smooth muscle cells. J Biol Chem 255:11163-11172.

Jonas A (1972): Physicochemical properties of bovine serum high density lipoprotein. J Biol Chem 247:7767-7772.

Lewis B (1976): "The Hyperlipidoemias. Clinical and Laboratory Practice." Oxford: Blackwell Scientific, pp 31-55.

Lindgren FT, Nicholls AV, Hayes TL, Freeman NK, Gofman JW (1959): Structure and homogeneity of the low density serum lipoproteins. Ann NY Acad Sci 72:826-832.

Lowry OH, Rosebrough NH, Farr AL, Randall J (1951): Protein measurement with the Folin phenol reagent. J Biol Chem 193:263-275.

MacFarlane AS (1958): Nature (Lond) 182:53-54.

Maxwell MAK, Hars SM, Bieber LL, Tolbert NE (1978): A modification of the Lowry procedure to simplify protein determination in membrane and lipoprotein samples. Anal Biochem 87:206-210.

Mayhew E, Rustun Y, Szoka FC, Papahadjopoulos D (1980): Cancer Treat Rep 63:1923-1928.

Quesney-Huneeus V, Wiley HH, Siperstein HD (1980): Isopentenyladenine as a mediator of mevalonate-regulated DNA replication. Proc Natl Acad Sci USA 77:5842-5846.

Savion N, Gospodarowicz D (1982): Role of hormones, growth factors, and lipoproteins in the control of proliferation and differentation of cultured bovine granulosa cells. In "Cold Spring Harbor Conferences on Cell Proliferation: Growth of Cells in Hormonally Defined Media," Vol 9: "Hormones and Cell Culture." Cold Spring Harbor, New York: Cold Spring Harbor Laboratory, pp 1141-1169.

Savion N, Lui GM, Laherty R, Gospodarowicz D (1981): Factors controlling proliferation and progesterone production by bovine granulosa cells in serum-free medium. Endocrinology 109:409-421.

Szoka F, Jr, Papahadjopoulos D (1980): Annu Rev Biophys Bioeng 9:467-508.

Tauber JP, Cheng J, Massoglia S, Gospodarowicz D (1981a): High density lipoproteins and the growth of vascular endothelial cells in serum-free medium. In Vitro 17:519-530.

Tauber JP, Goldminz D, Gospodarowicz D (1981b): Up regulation in vascular endothelial cells of high density lipoprotein receptor sites induced by 25-hydroxycholesterol. Eur J Biochem 119:327-339.

Tauber JP, Goldminz D, Vlodavsky I, Gospodarowicz D (1981c): The interactions of the high density lipoproteins with cultured vascular endothelial cells. Eur J Biochem 119:317-325.

Vlodavsky I, Fielding P, Fielding C, Gospodarowicz D (1978): Role of contact inhibition in the regulation of receptor mediated uptake of low density lipoprotein in cultured vascular endothelial cells. Proc Natl Acad Sci USA 75:356-360.

Vlodavsky I, Lui GM, Gospodarowicz D (1980): Morphological appearance, growth behavior and migratory activity of human tumor cells maintained on extracellular matrix versus plastic. Cell 19:607-616.

Methods for Preparation of Mitogenic Peptides

Methods for Preparation of Media, Supplements, and Substrata
for Serum-Free Animal Cell Culture, pages 89–109

4
Preparation of Human Platelet-Derived Growth Factor

Elaine W. Raines and Russell Ross

Serum supports the growth of most diploid cells in culture and is conventionally used as a supplement for cell culture media. In the absence of serum, cells become growth-arrested in the G_0/G_1 phase of the cell cycle. Upon readdition of serum, DNA synthesis and cell growth are stimulated [Todaro et al., 1965; Temin, 1966; Yoshikura et al., 1967] and the final cell density will be dependent on the serum concentration used [Temin, 1966, 1968; Todaro et al., 1967; Holley and Kiernan, 1971]. Serum as a growth supplement is a complex mixture of nutrients, cofactors, inhibitors, and macromolecular growth factors. Platelet-derived growth factor (PDGF), a 30,000 molecular weight protein released by the platelet during coagulation, is one of the principal macromolecules in human and primate serum capable of stimulating DNA synthesis and cell growth in connective tissue cells [Kohler and Lipton, 1974; Ross et al., 1974; Westermark and Wasteson, 1976]. Although its physiologic role is unknown, PDGF may be involved in wound healing and tissue repair [Ross and Vogel, 1978; Scher et al., 1979] and in the pathogenesis of atherosclerosis [Ross and Glomset, 1976; Ross, 1981]. In addition to possible physiologic roles, the study of PDGF-stimulated cell growth in culture is important as an aid in understanding the regulation of cell proliferation.

This chapter describes a method for purification of PDGF from outdated human platelet-rich plasma [Raines and Ross, 1982] that provides an overall yield of 18.4% as determined by thymidine incorporation (26.8% by radioreceptor assay) and employs readily available reagents. Large-scale purification

Department of Pathology (E.W.R., R.R.) and Department of Biochemistry (R.R.), University of Washington School of Medicine, Seattle, Washington 98195

procedures have been reported by several laboratories [Antoniades, 1981; Deuel et al., 1981; Heldin et al., 1981a; Raines and Ross, 1982]. These were difficult to develop because of the extremely cationic and hydrophobic properties of PDGF [Heldin et al., 1979; Ross et al., 1979] and the low concentrations of PDGF (17.5 ng/ml) normally present in whole-blood serum [Singh et al., 1982; Bowen-Pope et al., submitted for publication]. The properties of PDGF as they affect the various purification steps and different methods to assay for PDGF during its purification are discussed.

ASSAYS FOR PDGF

Measurement of ^{3}H-thymidine incorporation into trichloroacetic acid-precipitable material in quiescent cultures of mouse 3T3 cells [Antoniades, 1981; Deuel et al., 1981; Raines and Ross, 1982] and cultures of glial cells [Heldin et al., 1981a] was used to follow the mitogenic activity in the purification of PDGF. A typical dose-response curve using purified PDGF is shown in Figure 1A. Increasing amounts of platelet-rich plasma (PRP), a CM-Sephadex fraction, and purified PDGF were added directly to confluent cultures of Swiss 3T3 cells that were preincubated for 1–2 days in media containing 2% calf CMS I (calf serum depleted of PDGF by filtration on CM-Sephadex; see Vogel et al., [1978]). Twenty h after addition of test sample the medium was removed from the wells and replaced with fresh medium containing ^{3}H-thymidine. After an additional 2-h incubation, the cells were harvested and ^{3}H-thymidine incorporation into trichloroacetic acid-precipitable material was determined.

Although it is widely used, there are a number of limitations to the thymidine incorporation assay that must be kept in mind. First of all, since the ability of cells to synthesize DNA is affected by a number of different growth factors and cofactors, the assay is not specific for PDGF. This is demonstrated by comparing ^{3}H-thymidine incorporation data (Fig. 1A) with data obtained with the radioreceptor assay (Fig. 1B), in which the ability of test substances to compete with ^{125}I-PDGF for binding to the PDGF cell surface receptor is determined in a 4-h incubation at 4°C [Bowen-Pope and Ross, 1982]. In matched cultures, the amount of purified PDGF required to induce half-maximal stimulation of thymidine incorporation (0.9 ng) and the quantity required to induce half-maximal competition for binding in the radioreceptor assay (0.85 ng) are comparable. However, for platelet-rich plasma, half-maximal stimulation of thymidine incorporation is induced by 0.8 mg, but to obtain half-maximal competition for binding, 1.8 mg of PRP is required. Only 44% of the mitogenic activity in platelet-rich plasma can be accounted for by

PDGF. This is in good agreement with data obtained by Heldin et al. [1981c] and our laboratory (unpublished observations), in which 50% of the biological activity of platelet lysates could be removed by anti-PDGF antibody bound to Sepharose. Thus, the thymidine incorporation assay tends to underestimate the purification and yields achieved, particularly at early stages in the purification (see Table I). Thymidine incorporation can also reflect changes in the media unrelated to DNA synthesis such as fluctuations in pool size or transport of thymidine [Lindberg et al., 1969; Nordenskjöld et al., 1970] and the presence of thymidine-degrading activity [Heldin et al., 1977]. The chances of thymidine incorporation reflecting these changes rather than DNA synthesis are greatly reduced by performing the assay as described. The sample medium is removed after 20 h and fresh medium containing ^{3}H-thymidine (2.6 μM) and 5% calf serum is added for the 2-h labeling period. Alternatively, the cells can be plated in the presence of 10^{-5} M unlabeled thymidine as described previously [Raines and Ross, 1982]. It has also been observed that thymidine incorporation in cells varies with serum lots (unpublished observations). Therefore, if serum is used as a reference in comparing test samples from assay to assay, it is important that the reference serum lot be the same. Finally, because the test samples are incubated at 37°C for 20 h with the cells, any chemicals present in the test samples can affect the health of the cells and therefore their ability to incorporate ^{3}H-thymidine. For example, purified PDGF, shown in Figure 1, was eluted from a Phenyl-Sepharose column in 50% ethylene glycol, 0.1 N ammonium acetate, pH 7. If this material is assayed for thymidine incorporation without dialysis or even after lyophilization, no mitogenic activity is detected because of toxic effects of the buffer. Dialysis eliminates this toxic effect. The toxicity can be shown to be specifically due to the buffer by addition of the buffer to a sample of known mitogenicity (data not shown).

The radioreceptor assay [Bowen-Pope and Ross, 1982] shown in Figure 1B is specific for molecules binding with high affinity (10^{-11} M) to the PDGF cell surface receptor. It is also less sensitive to toxic effects of buffer as described above. For example, Phenyl-Sepharose-purified PDGF in 50% ethylene glycol, 0.1 N ammonium acetate (pH 7), lyophilized Phenyl-Sepharose PDGF, and dialyzed Phenyl-Sepharose PDGF are indistinguishable in the radioreceptor assay (data not shown). Because of the specificity of the PDGF radioreceptor assay, the reduced sensitivity of the radioreceptor assay to toxic buffer effects, the shorter time (4 h) of the assay, the fact that the assay does not require the cells to be quiescent (only that cell number per well be constant), and the simplicity of counting gamma-emitters, the PDGF radioreceptor assay provides a specific, sensitive, and convenient assay to follow the purification of PDGF.

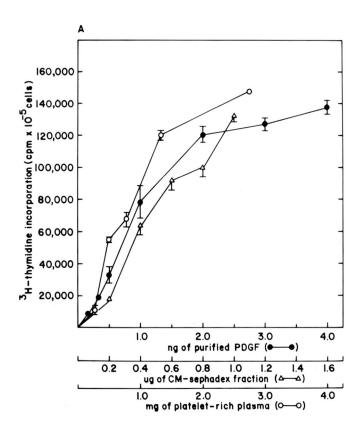

Fig. 1. Comparison between the ^{3}H-thymidine incorporation assay and the radioreceptor assay for PDGF. Platelet-rich plasma (O—O), a CM-Sephadex fraction (△—△), and purified PDGF (●—●) were assayed on confluent cultures of 3T3 cells for ^{3}H-thymidine incorporation into trichloroacetic acid precipitable material (A), and competition for ^{125}I-PDGF binding in the radioreceptor assay (B). For ^{3}H-thymidine incorporation (A), confluent cultures of Swiss 3T3 cells were prepared in 2-cm^2 Costar 24-well culture dishes (final cell density approximately 8 × 10^4 cells per well) in 1 ml Dulbecco's modified Eagle's medium with L-glutamine and D-glucose (Gibco) supplemented to give the final concentrations of the following: 10% (vol./vol.) calf serum (Gibco); 0.225% sodium bicarbonate, 100 units/ml of penicillin G sodium, and 100 μg/ml of streptomycin sulfate (Gibco); and 1 mM sodium pyruvate (Gibco). After 3–5 days the cells were switched to 1 ml of the same media containing 2% calf CMS I (calf serum depleted of PDGF by filtration on CM-Sephadex; see Vogel et al. [1978]). Test samples were added directly to the wells in 100 μl of 10 mM acetic acid and incubated for an additional 20 h at 37°C. The medium was removed from the wells after 20 h and replaced with 0.5 ml of fresh medium containing 2 μCi/ml ^{3}H-thymidine (76 Ci/mmole) and 5% calf serum. After an additional 2-h incubation, the cells were harvested by removing the medium, washing twice with 1 ml of ice-

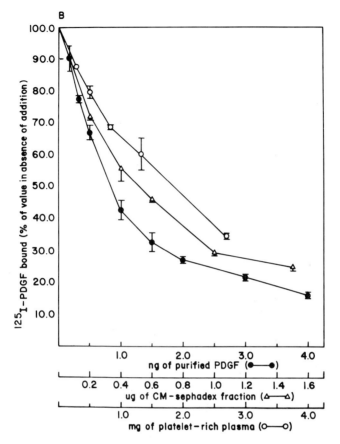

cold 5% trichloroacetic acid, and dissolving trichloroacetic acid-insoluble material in 0.8 ml of 0.25 N NaOH with mixing. Of this, 0.6 ml was added to 5 ml of Aquasol and CPM were determined with a Packard Tricarb liquid scintillation counter. For the radioreceptor assay (B), matched cultures were prepared as described for the thymidine incorporation, assay and levels of PDGF were determined by a modification of the radioreceptor assay described by Bowen-Pope and Ross [1982]. The cultures were aspirated, washed once with cold phosphate-buffered saline containing 0.1% BSA, and replaces with 1 ml per well of ice-cold binding medium (HEPES-buffered Ham's medium F-12 [Gibco] adjusted to pH 7.4) containing 2% (vol./vol.) calf CMS I and the test substances. The cultures were incubated at 4°C on an oscillating table for 3 h. After 3 h the test solutions were removed, and the wells were rinsed once with phosphate-buffered saline containing 0.1% BSA followed by an additional 1-h incubation at 4°C with 1 ml per well of binding medium containing 0.16 ng/ml ^{125}I-PDGF. Binding was terminated by rinsing three times with phosphate-buffered saline containing 0.1% BSA, and cell-associated ^{125}I-PDGF was extracted with 1.0 ml of 1% Triton X-100 in 0.1% BSA. Binding was plotted without correction for nonspecific binding, which was determined to be 3% (data not shown). For both assays, the mean ± range of duplicate determinations is plotted.

TABLE I. Purification of Platelet-Derived Growth Factor From 4 Liters of Outdated Platelet-Rich Plasma

Fraction	Total protein[a] (mg)	^{3}H-Thymidine incorporation			PDGF radioreceptor assay		
		ED_{50}[b] (ng)	Purification factor (-fold)	Yield (% of initial)	ED_{50}[c] (ng)	Purification (-fold)	Yield (% of initial)
Defibrinogenated platelet-rich plasma	195,000	878,000	—	100.0	1,600,000	—	100.0
CM-Sephadex	60.1	560	1,568	55.0	555	2,883	88.8
Sephacryl S-200	4.89	100	8,780	25.0	115	13,913	34.9
Heparin-Sepharose	0.93	20	43,900	23.8	24	66,667	31.8
Phenyl-Sepharose	0.036	1.0	878,000	18.4	1.10	1,454,545	26.8

[a]Protein was determined by absorption employing E_{280} 1% = 10.0 for platelet-rich plasma and subsequent steps, except the final step (Phenyl-Sepharose), where protein was determined by the method of Lowry et al. [1951] with bovine serum albumin as a standard. The protein values represent the mean of 49 preparations.

[b]ED_{50} is the amount of a given fraction required to give half-maximal stimulation of thymidine incorporation in the 3T3 cell assay described in the text and illustrated in Figure 1A. Values represent the mean of at least five preparations.

[c]ED_{50} is the amount of a given fraction required to give half-maximal competition for binding in the PDGF radioreceptor assay in matched cultures of 3T3 cells as described in the text and illustrated in Figure 1B. Values represent the mean of at least five preparations.

Another radioreceptor assay that could be used to follow the purification of PDGF is the ^{125}I-epidermal growth factor (EGF) radioreceptor assay. It has been shown the preincubation of PDGF or fibroblast-derived growth factor (a basic heat- and acid-stable polypeptide isolated from serum-free conditioned medium of SV40-transformed BHK cells) at 4°C [Rozengurt et al., 1982; Bowen-Pope et al., 1983b] or 37°C [Wrann et al., 1980; Heldin et al., 1982b; Rozengurt et al., 1982; Shupnik et al., 1982; Wharton et al., 1982] inhibits the subsequent binding of ^{125}I-EGF. The inhibition demonstrated by PDGF at 4°C does not result from competition by PDGF for binding to the EGF receptor, but seems to result from an increase in K_d for ^{125}I-EGF binding [Bowen-Pope et al., 1983b]. Thus, the inhibition of ^{125}I-EGF binding is detected at very low concentrations of PDGF (0.05–0.25 ng/ml). Like ^{3}H-thymidine incorporation, the inhibition of ^{125}I-EGF binding is not specific (FDGF and EGF compete, as well as potentially other, as yet undefined, growth factors), and in fact it has been recently reported [Bowen-Pope and Ross, 1983a] that the original values reported for EGF in serum as determined by EGF radioreceptor assay [Ladda et al., 1979; Nexø et al., 1980] can be almost entirely attributed to PDGF concentrations in serum. The ^{125}I-EGF radioreceptor assay, although less specific than the PDGF radioreceptor assay, offers temporal and practical advantages, particularly for a laboratory in which this assay is routinely employed.

PURIFICATION PROCEDURE

Starting Material

The method described here (Table I) employs outdated human platelet-rich plasma. A procedure has been reported for preparation of PDGF from fresh platelets [Heldin et al., 1981a] and offers the advantage that plasma proteins are excluded from the starting material. However, fresh platelets are not generally available to most laboratories in large amounts. Outdated platelets are processed with the plasma because approximately 90% of the growth-promoting activity is already present in the plasma when the outdated platelet-rich plasma is received (24–72 h after collection) from the blood bank [Raines and Ross, 1982]. No differences have been observed in activity or mobility after polyacrylamide-SDS gel electrophoresis between PDGF prepared from outdated platelets and PDGF purified from fresh platelets (Heldin and Raines, unpublished observations).

Units of outdated platelets are frozen upon receipt from the blood bank and frozen and thawed three times before use. Eighty units are processed at one time and all glassware and plasticware used in the purification is siliconized with either Aquasil or Surfasil (Pierce Chemical). The procedures, unless otherwise noted, are carried out at 4°C to prevent bacterial contamination.

Heat Defibrinogenation of Frozen-Thawed Platelet-Rich Plasma

The frozen-thawed platelet-rich plasma is heat-defibrinogenated at 55–57° for 8 min in approximately 800-ml batches, transferred to 250-ml centrifuge bottles, and left at 4°C overnight. The platelet membranes and precipitated fibrinogen are removed by centrifugation at 27,000g for 30 min at 4°C. The supernatant is collected by filtration over glass wool and kept at 4°C while the pellets are processed. Although the supernatant contains the bulk of the mitogenic activity, the membrane/precipitate pellet contains a significant amount of growth-promoting activity that can be eluted by washing the pellet. The pellets are combined into two bottles and first stirred with a total of 200 ml of 0.01 M Tris, 0.39 M NaCl, pH 7.4, for 1 h at room temperature, then centrifuged at 27,000g for 30 min at 4°C, and the supernatant is added to the original supernatant. The pellets are then resuspended in a total of 200 ml of 0.01 M Tris, 1.09 M NaCl, pH 7.4 for 1 h at room temperature, centrifuged, and pooled as in the first wash. The pH of the pooled supernatants is adjusted to pH 7.4 by the addition of 1.0 M Tris-base and the conductivity is determined. If the conductivity is greater than 13.0 mmho (room temperature), it is adjusted to this value by the addition of 0.01 M Tris, pH 7.4.

Carboxymethyl Sephadex Chromatography

The supernatant from the defibrinogenated platelet-rich plasma (approximately 4 liters) is stirred overnight at 4°C with 1 liter of swollen CM-Sephadex that has been equilibrated in 0.01 M Tris, 0.09 M NaCl, pH 7.4. The supernatant/gel slurry is loaded at a flow rate of 150 ml/h on a 1,500-ml CM-Sephadex, C-50 column (7.5 cm × 35 cm) equilibrated in the same buffer. Following addition of sample the column is washed with 0.19 M NaCl, 0.01 M Tris, pH 7.4 (4 liters), and PDGF is eluted with 0.5 M NaCl, 0.01 M Tris, pH 7.4 (4 liters). The elution profile is shown in Figure 2. Only the fractions indicated by the solid bar are pooled and further purified. The normal yield from 80 units of platelet-rich plasma is 89 mg (n = 49). Mitogenic activity (10–40% of the defibrinogenated PRP) was also detected in the fraction not adsorbed to CM-Sephadex. This nonadsorbed mitogenic fraction, the anionic fraction previously described by Heldin et al. [1977], is not recognized by anti-PDGF antibody nor does it compete for binding in the PDGF radioreceptor assay (data not shown). The column is emptied, washed with 1.0 M NaCl, 0.5 N NaOH, distilled water, and reequilibrated for use.

The mitogenic fractions eluted with 0.5 M NaCl from CM-Sephadex (Fig. 2) are pooled and concentrated in an Amicon concentrator with PM-10 filter

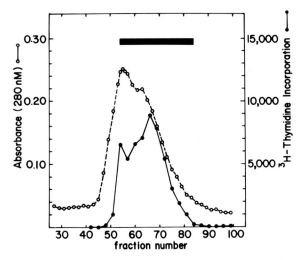

Fig. 2. CM-Sephadex chromatography of defibrinogenated, human platelet-rich plasma— elution profile. As described in the text, the platelet-rich plasma was loaded on a CM-Sephadex-C-50 column equilibrated in 0.01 M Tris, 0.09 M NaCl, pH 7.4, followed by stepwise elution with 0.01 M Tris, 0.19 M NaCl (pH 7.4), and 0.01 M Tris, 0.5 M NaCl (pH 7.4). The elution of PDGF with 0.01 M Tris, 0.5 M NaCl (pH 7.4) is shown. Fractions (25 ml) were collected and protein elution was monitored by absorbance at 280 nM (O—O) and mitogenic activity was monitored by ^{3}H-thymidine incorporation (●—●), as described in the text. The solid bar indicates the fractions pooled for further purification.

(molecular weight cutoff of approximately 10,000) to a volume of 20-50 ml (3-4 mg/ml). The concentrated fraction is dialyzed against 1.0 N acetic acid, pH 2.2 (with Spectrapor 1 dialysis tubing).

Sephacryl S-200 Molecular Sieving

The concentrated and dialyzed CM-Sephadex fraction (average yield of 60.1 mg, n = 49) is spun at 27,000g for 30 min to remove particulate matter and the pH is adjusted to 3.5 by the addition of concentrated ammonium hydroxide. The fraction is loaded at a flow rate of 20 ml/h on a Sephacryl S-200 column (5 cm × 92 cm, 1,806 ml packed gel volume) equilibrated in 1.0 N acetic acid adjusted to pH 3.5 by the addition of concentrated ammonium hydroxide. As demonstrated in Figure 3, the mitogenic activity elutes with a peak molecular weight of 43,000 and a molecular weight range of 29,000-45,000. The levels of two other platelet alpha-granule proteins (platelet factor 4 and beta-thromboglobulin) are also illustrated in Figure 3. The fractions indicated by the solid bar are pooled for further purification. This pool contains approximately 65% of the total activity recovered from the

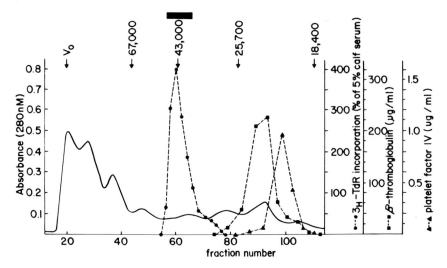

Fig. 3. Sephacryl S-200 gel filtration of CM-Sephadex-purified PDGF. CM-Sephadex purified PDGF (117 mg) was prepared as described in the text and loaded at a flow rate of 20 ml/h on a Sephacryl S-200 column (5 cm by 92 cm) equilibrated in 1.0 N acetic acid with pH adjusted to 3.5 with ammonium hydroxide. Fractions (7.8 ml) were collected and monitored for absorbance at 280 nM (—); for mitogenic activity by measurement of ^{3}H-thymidine incorporation (●--●) as described in text; for beta-thromboglobulin distribution (■--■) by radioimmunoassay with a New England Nuclear kit; and for levels of platelet factor 4 (▲--▲) by radioimmunoassay by the method of Files et al. [1981]. The molecular weight standards used to calibrate the column were: bovine serum albumin (67,000), ovalbumin (43,000), chymotrypsinogen (25,700), beta-lactalbumin (18,400), and ^{3}H$_2$O (Total column volume). The solid bar indicates the fractions pooled for further purification. Adapted from Raines, E.W. and Ross, R. (1982): J. Biol. Chem. 257:5154–5160.

column (70%). If a wider pool is taken to include all of the active fractions, the contaminating proteins are not readily removed in subsequent purification steps. The active side fractions can be recycled on Sephacryl S-200 and then further purified, or we routinely use the active side fractions to determine the nonspecific binding in the radioreceptor assay.

The presence of a dissociating agent (acetic acid, guanidine, or sodium dodecyl sulfate) during gel filtration on any matrix was found to be essential to completely separate the mitogenic activity from the void volume fraction. We now know that these higher-molecular-weight complexes detected under non-dissociating conditions are PDGF associated with plasma-binding proteins [Raines et al., submitted for publication].

Heparin-Sepharose Chromatography

PDGF has an isoelectric point of 9.5–10.3 [Heldin et al., 1977; Ross et al., 1979; Deuel et al., 1981] and therefore readily adsorbs to cation exchangers.

Two matrices, SP-Sephadex and Heparin-Sepharose, were investigated for optimal use in the purification of PDGF. Figure 4 demonstrates the results of a series of experiments at different pH values in which the recovery of mitogenic activity and purification factor were monitored for the two cation exchangers. At a lower pH, where presumably more of the proteins are positively charged (and therefore adsorb to the columns), the purification factor was not impressive. At the higher pH, the recovery dropped dramatically. This is probably due to the hydrophobic character of PDGF [Ross et al., 1979] (see Phenyl-Sepharose Chromatography, below). Of the conditions investigated, Heparin-Sepharose run at pH 7 gave the optimal purification and recovery and is routinely employed in the purification procedure.

Sephacryl S-200-purified material from 3 to 5 preparations (240–400 units of PRP) are chromatographed at one time on a Heparin-Sepharose column (1.5 cm × 13 cm, 23 ml packed gel volume). The Sephacryl S-200 fractions are used directly without dialysis by adjusting the pH to 7 by the addition of concentrated ammonium hydroxide and diluting the pool to a conductivity of 28.3 mmho (determined at 20°C) with distilled water. Buffers for the column elution are prepared in the same manner: 1.0 N acetic acid adjusted to pH 7 with concentrated ammonium hydroxide and adjusted to 28.3 mmho with

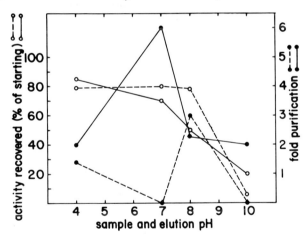

Fig. 4. Comparison between chromatography of S-200-purified PDGF on SP-Sephadex and Heparin-Sepharose. Fractions (1 mg) of S-200-purified were adjusted to indicated pH values as described in the text by addition of different amounts of ammonium hydroxide and then diluted to give an acetate concentration of 0.1 M. The samples were loaded on 1 ml packed gel volume of SP-Sephadex (----) or Heparin-Sepharose (—) equilibrated in buffer with the same pH and conductivity of the samples. The columns were eluted stepwise with 0.2 M, 0.4 M, 0.6 M, and 1.0 M ammonium acetate with the same pH. Activity recovered (○) and purification factor (-fold) (●) were determined by measuring ^{3}H-thymidine incorporation in the starting S-200 fraction and adsorbed and eluted samples as described in the text.

distilled water, for the beginning of the gradient; and for the end of the gradient, 2.0 N acetic acid adjusted to pH 7 and diluted to 80.5 mmho with distilled water. All conductivities before and after chromatography are determined at 20°C. The column is equilibrated and run at 4°C.

The Heparin-Sepharose column is loaded at a flow rate of 20 ml/h with a maximum protein load of 1 mg/ml of swollen packed gel. The column is washed with the loading buffer (5 times the column volume) and eluted with a linear gradient of ammonium acetate (conductivity increasing from 28.3 mmho to 80.5 mmho). Figure 5 shows a typical gradient elution protein profile and the mean distribution of mitogenic activity of five preparations. The majority of PDGF is eluted between 39.0 mmho and 53.0 mmho, and these fractions (indicated by solid bar) are pooled for further purification. The column is washed with the 2 M guanidine HCl (5 times the column volume) before reequilibration.

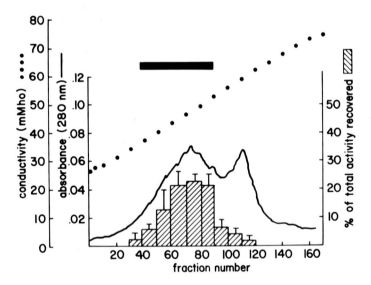

Fig. 5. Heparin-Sepharose gradient elution of Sephacryl S-200-purified PDGF. Sephacryl-S-200 purified PDGF was chromatographed at 4°C on Heparin-Sepharose as described in the text. The linear gradient elution (ammonium acetate, pH 7, increasing from 28.3 mmho to 80.5 mmho) of the adsorbed fraction is shown with absorbance at 280 nm indicated by the solid line, the conductivity measured at 20°C indicated by the solid dots (●), and the percentage of total mitogenic activity recovered shown by slashed bar graphs. The mitogenic activity was determined by measuring ^{3}H-thymidine incorporation as described in the text, and the bar graphs represent the means and standard deviations for five preparations. The solid bar indicates the fractions pooled for further purification. Adapted from Raines, E.W. and Ross, R. (1982): J. Biol. Chem. 257:5154–5160.

Phenyl-Sepharose Chromatography

The final step in the purification takes advantage of the hydrophobic character of PDGF. The pooled Heparin-Sepharose fractions are diluted with distilled water to a conductivity of 34.0 mmho (determined at 20°C) and loaded on a Phenyl-Sepharose column (0.9 cm × 12 cm) equilibrated at 4°C in 1.0 N acetic acid in which the pH is adjusted to 7.0 by the addition of concentrated ammonium hydroxide and then diluted with distilled water to the same conductivity (34.0 mmho at 20°C). The sample is loaded at a flow rate of 15 ml/h and the maximum protein load is 1 mg/ml of swollen gel. After loading, the column is eluted stepwise as follows: a) 10% (v/v) 1.0 N acetic acid adjusted to pH 7 with concentrated ammonium hydroxide, 60% (v/v) water, and 30% (v/v) ethylene glycol; b) 10% (v/v) 1.0 N acetic acid adjusted to pH 7 with concentrated ammonium hydroxide, 40% (v/v) water, and 50% (v/v) ethylene glycol. The growth-promoting activity is eluted in the buffer front of the last buffer as a single peak. This purified fraction must be dialyzed in the presence of bovine serum albumin (BSA) (2 mg/ml final concentration) before it can be assayed for mitogenic activity (see Assays for PDGF, above, for detailed discussion). The Phenyl-Sepharose column is washed with 2 M guanidine-HCl and then reequilibrated in the loading buffer before reuse.

CHARACTERISTICS OF PURIFIED PDGF

Molecular Properties of PDGF

Figure 6 shows the 15% SDS-PAGE analysis of nonreduced PDGF from various steps in the purification procedure. This Coomassie-stained gel illustrates the complexity of proteins removed during the purification. Only at the final step in the purification can the molecular complex ($M_r = 30,000$) responsible for the mitogenic activity be distinguished. Three molecular weight species ($M_r = 31,000, 29,000$, and $27,000$) can be distinguished with Coomassie blue staining. These three species represent greater than 90% of the total stainable material as determined by the densitometric scans.

With the increased sensitivity of the silver stain procedure [Switzer et al., 1979] and analysis of different preprations, four molecular weight species can be identified in the nonreduced state. Figure 7 shows the four bands with molecular weights of 31,000, 29,000, 28,500, and 27,000. These four molecular weight species have all been shown to be associated with the biological activity by a) assay of mitogenic activity after separation on SDS-PAGE and b) analysis of specifically cell-bound ^{125}I-PDGF [Raines and Ross, 1982]. In addition, two-dimensional peptide mapping of the four molecular weight

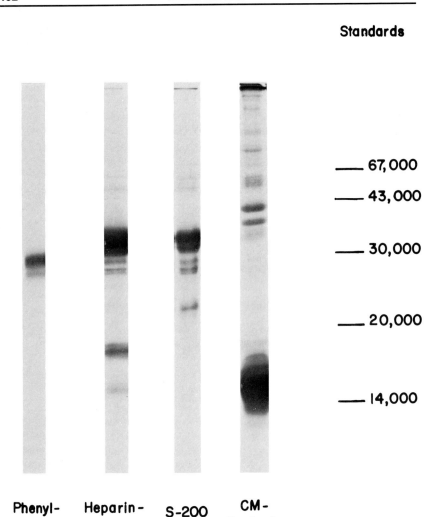

Standards

___ 67,000

___ 43,000

___ 30,000

___ 20,000

___ 14,000

Phenyl- **Heparin-** **S-200** **CM-**
Sepharose **Sepharose** **Sephadex**

Fig. 6. SDS-PAGE of mitogenically active fractions from various stages in the purification of PDGF. SDS-slab gel electrophoresis was performed [Laemmli, 1970] with a 15% separating gel and a 4.5% stacking gel of 1.5-mm thickness. Samples were boiled for 5 min in SDS sample buffer without reducing agent. Sample load varied with the purity of the fractions: Phenyl-Sepharose, 7.5 μg; Heparin-Sepharose, 50 μg; S-200, 100 μg; and CM-Sephadex, 100 μg. The gels were stained with Coomassie blue. Calibration standards were phosphorylase b (94,000), bovine serum albumin (67,000), ovalbumin (43,000), carbonic anhydrase (30,000), soybean trypsin inhibitor (20,100), and alpha-lactalbumin (14,400) from Pharmacia. Adapted from Raines, E.W. and Ross, R. (1982): J. Biol. Chem. 257:5154–5160.

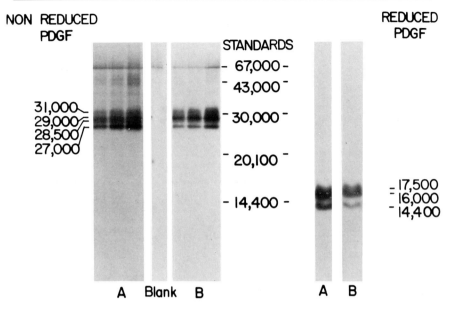

Fig. 7. Two preparations of PDGF varying in their proportion of four molecular weight components characterized by silver-stained SDS-PAGE. SDS gels (15%) were run as described in Figure 6 and stained by the sensitive silver stain procedure of Switzer et al. [1979]. A and B represent two different preparations of Phenyl-Sepharose-purified PDGF that varied in their proportion of four molecular weight components. The three lanes of the nonreduced PDGF for each preparation represent increasing amounts loaded on the gel (left to right): 350, 525, and 700 ng. The reduced gel lanes were run with a protein load of 1 μg of each preparation. The molecular weights indicated in the margins for the nonreduced and reduced PDGF were determined from a linear log plot of molecular weights of the standards indicated in the middle of the figure and their determined R_f values. The blank gel is shown to illustrate the nonspecific staining seen around $M_r = 67,000$ when no protein is loaded in the well. Adapted from Raines, E.W. and Ross, R. (1982): J. Biol. Chem. 257:5154-5160.

species demonstrated extensive peptide homology [Raines and Ross, 1982]. Deuel et al. [1981] analyzed two molecular weight species, PDGF I (M_r = 31,000) and PDGF II (M_r = 28,000), and found them to vary only slightly in their content of five amino acids, each protein containing three tyrosine residues and 18 half-cysteine residues. However, significant differences were found in the carbohydrate content of the two forms—approximately 7% of PDGF I and 3.8% of PDGF II was estimated to be neutral and amino sugars [Deuel et al., 1981]. The similarity in the molecular properties of the different forms and the small amounts of purified PDGF available have made separation and analysis of the different molecular forms difficult.

Reduction of PDGF, under various conditions, destroys its biological activity. SDS-PAGE analysis after reduction indicates the presence of at least three molecular weight bands: 14,400, 16,000, and 17,500 [Antoniades, 1981; Heldin et al., 1981a; Raines and Ross, 1982]. A two-chain structure joined by disulfide bonds has been proposed for PDGF [Heldin et al., 1979, 1981a; Antoniades, 1981; Johnsson et al., 1982; Raines and Ross, 1982]. The data further suggest that the multiplicity observed in nonreduced PDGF [Antoniades, 1981; Deuel et al., 1981; Johnsson et al., 1982; Raines and Ross, 1982] is the result of modification of one chain [Johnsson et al., 1982; Raines and Ross, 1982]. Isolation of two different forms of nonreduced PDGF and analysis by SDS-PAGE after reduction [Antoniades, 1981] and peptide maps of the reduced chains [Raines and Ross, 1982] would support a molecular model for native PDGF in which the reduced 14,400 chain is combined with either the 16,000- or 17,500-dalton chain. Based on separation of the reduced chains by HPLC and SDS-PAGE analysis of the separated peaks, Johnsson et al. [1982] proposed that the 16,000 molecular weight chain is combined with a chain with a molecular weight of 18,000, 15,000, 14,000, or 11,000. The multiple forms of PDGF may be due to differential glycosylation, genetic polymorphism, or proteolytic cleavage of one or more chains, or all four molecular forms may be proteolytic cleavage products of a higher-molecular-weight precursor.

Stability of PDGF

PDGF is a very stable molecule. Its biological activity is not affected by incubation in 2% sodium dodecyl sulfate, in 6 M guanidine, or in 6 M urea, nor over a pH range of 2-11 [Ross et al., 1979]. Reduction of PDGF, under a number of conditions, and treatment with protein degrading enzymes like trypsin are the only reported treatments that destroy the biological activity of PDGF. The Phenyl-Sepharose-purified PDGF appears stable in the elution buffer (longest time tested is 2 years at 4°C).

The strongly cationic and hydrophobic character of PDGF makes it more difficult to handle in the purified state. Dialysis of purified PDGF, chromatography of purified PDGF on a gel matrix such as Sephacryl S-200 which contains unreacted carboxyl groups, or incubation with a bare tissue culture well results in significant loss of PDGF due to adsorption [Bowen-Pope and Ross, 1982; Raines and Ross, 1982; Smith et al., 1982]. This can be minimized by addition of exogenous protein (2 mg/ml Pentex BSA or gelatin is routinely used), which presumably blocks the binding sites. For any procedure involving a number of transfers and manipulation, trace levels of ^{125}I-PDGF are added and used to determine losses during manipulation.

Purification of PDGF From Other Species

A phylogenetic analysis of PDGF by radioreceptor assay recently found that all tested specimens of clotted whole blood from phylum Chordata contain a mitogen that competes with human PDGF for specific receptor binding, but sera from tunicates down on the chordate line were negative [Singh et al., 1982]. In our laboratory polyclonal antibodies produced in a goat to purified human PDGF recognize a PDGF-like molecule in all chordates tested (except those closely related to the goat) which further suggests that the PDGF molecule has been highly conserved (Raines and Ross, unpublished observations). There have been no reports of purification of PDGF or PDGF-like molecules from the serum of other species. However, preliminary observations from our laboratory, made in collaboration with Dr. Hisao Kato, Kyushu University, Japan, suggest that PDGF from the bovine species fractionates in a manner very similar to that of human PDGF.

Biological Properties of PDGF

PDGF was purified and defined by its ability to induce ^{3}H-thymidine incorporation and cell growth in connective tissue cells in culture. These processes are temporally detected at least 20 h after addition of PDGF to cultured cells. However, it is known that cells respond very rapidly to PDGF. Within minutes after addition, the following changes are observed: saturable and specific binding of PDGF to responsive cells [Heldin et al., 1981b, 1982a,b; Bowen-Pope and Ross, 1982; Huang et al., 1982; Williams et al., 1982] to what appears to be a 170,000 MW specific membrane receptor [Glenn et al., 1982; Heldin et al., 1982a]; tyrosine phosphorylation of a 170,000 MW membrane protein [Ek and Heldin, 1982; Nishimura et al., 1982]; phosphorylation of a 33,000 MW cell protein [Nishimura and Deuel, 1981]; tyrosine phosphorylation of several specific cellular proteins [Cooper et al., 1982]; increased phosphatidylinositol turnover [Habenicht et al., 1981]; the appearance of membrane ruffles and microvilli [Schmidt et al., 1982]; inhibition of ^{125}I-EGF binding [Wrann et al., 1980; Heldin et al., 1982b; Shupnik et al., 1982; Wharton et al., 1982; Bowen-Pope et al., 1983b]; increased amino acid uptake [Owen et al., 1982]; stimulation of the Na-K pump [Mendoza et al., 1980]. Over a period of hours, PDGF increases the number of LDL receptor [Chait et al., 1980; Witte et al., 1981] and somatomedin C receptors [Clemmons et al., 1980], induces chemotaxis in fibroblasts [Seppä et al., 1982] and smooth muscle cells [Grotendorst et al., 1982], and increases the rate of fluid pinocytosis in smooth muscle cells [Davies and Ross, 1978]. The use of purified PDGF in experiments to further define these and other early changes in cells in culture will be important in the attempt to understand growth regulation. It may also be important in defining a much broader family of growth factors. PDGF-like growth factors have been reported to be produced by endothelial cells [Dicorleto and Bowen-Pope, 1983], a line of osteosarcoma cells [Heldin et al., 1980], and SV-40-transformed BHK cells [Dicker et al., 1981]. Purified PDGF and antibodies to purified PDGF will be necessary to determine whether PDGF plays a role in wound healing and tissue repair and in the pathogenesis of atherosclerosis.

ACKNOWLEDGMENTS

This work was supported by National Institutes of Health grants HL18645 and AM13970, and a grant from R.J. Reynolds, Inc.

REFERENCES

Antoniades HN (1981): Human platelet-derived growth factor (PDGF): Purification of PDGF-I and PDGF-II and separation of their reduced subunits. Proc Natl Acad Sci USA 78: 7314-7317.

Bowen-Pope DF, Ross R (1982): Platelet-derived growth factor. II. Specific binding to cultured cells. J Biol Chem 257:5161-5171.

Bowen-Pope DF, Ross R (1983a): Is epidermal growth factor present in human blood? Interference with the radioreceptor assay for epidermal growth factor. Biochem Biophys Res Comm 114:1036-1041.

Bowen-Pope DF, DiCorleto PE, Ross R (1983b): Interactions between the receptors for platelet-derived growth factor and epidermal growth factor. J Cell Biol 96:679-683.

Bowen-Pope DF, Malpass TW, Foster DM, Ross R (submitted for publication): Platelet-derived growth factor in vivo: Levels, activity, and rate of clearance.

Chait A, Ross R, Albers J, Bierman EL (1980): Platelet-derived growth factor stimulates activity of low density lipoprotein receptors. Proc Natl Acad Sci USA 77:4084-4088.

Clemmons DR, Van Wyk JJ, Pledger WJ (1980): Sequential addition of platelet factor and plasma to BALB/c 3T3 fibroblast cultures stimulates somatomedin-C binding early in cell cycle. Proc Natl Acad Sci USA 77:6644-6648.

Cooper JA, Bowen-Pope DF, Raines E, Ross R, Hunter T (1982): Similar effects of platelet-derived growth factor and epidermal growth factor on the phosphorylation of tyrosine in cellular proteins. Cell 31:263-273.

Davies PF, Ross R (1978): Mediation of pinocytosis in cultured arterial smooth muscle and endothelial cells by platelet-derived growth factor. J Cell Biol 79:663-671.

Deuel TF, Huang JS, Proffitt RT, Baenziger JU, Chang D, Kennedy BB (1981): Human platelet-derived growth factor: Purification and resolution into two active protein fractions. J Biol Chem 256:8896-8899.

Dicker P, Pohjanpelto P, Pettican P, Rozengurt E (1981): Similarities between fibroblast-derived growth factor and platelet-derived growth factor. Exp Cell Res 135:221-227.

DiCorleto PE, Bowen-Pope DF (1983): Endothelial cell-derived protein that binds to the platelet-derived growth factor receptor. Proc Natl Acad Sci USA 80:1919-1923.

Ek B, Heldin C-H (1982): Characterization of a tyrosine-specific kinase activity in human fibroblast membranes stimulated by platelet-derived growth factor. J Biol Chem 257:10486-10492.

Files JC, Malpass TW, Yee EK, Ritchie JL, Harker LA (1981): Studies of human platelet α-granule release in vivo. Blood 58:607-618.

Glenn K, Bowen-Pope DF, Ross R (1982): Platelet-derived growth factor: III. Identification of a platelet-derived growth factor receptor by affinity labeling. J Biol Chem 257:5172-5176.

Grotendorst GR, Chang T, Seppä HEJ, Kleinman HK, Martin GR (1982): Platelet-derived growth factor is a chemoattractant for vascular smooth muscle cells. J Cell Physiol 113:261-266.

Habenicht AJR, Glomset JA, King WC, Nist C, Mitchell CD, Ross R (1981): Early changes in phosphatidylinositol and arachidonic acid metabolism in quiescent Swiss 3T3 cells stimulated to divide by platelet-derived growth factor. J Biol Chem 256:12329-12335.

Heldin C-H, Wasteson Å, Westermark B (1977): Partial purification and characterization of platelet factors stimulating the multiplication of normal human glial cells. Exp Cell Res 109:429-437.

Heldin C-H, Westermark B, Wasteson Å (1979): Platelet-derived growth factor: Purification and partial characterization. Proc Natl Acad Sci USA 76:3722-3726.

Heldin C-H, Westermark B, Wasteson Å (1980): Chemical and biological properties of a growth factor from human-cultures osteosarcoma cells: Resemblance with platelet-derived growth factor. J Cell Physiol 105:235-246.

Heldin C-H, Westermark B, Wasteson Å (1981a): Platelet-derived growth factor: Isolation by large-scale procedure and analysis of subunit composition. Biochem J 193:907-913.

Heldin C-H, Westermark B, Wasteson Å (1981b): Specific receptors for platelet-derived growth factor on cells derived from connective tissue and glia. Proc Natl Acad Sci USA 78:3664-3668.

Heldin C-H, Westermark B, Wasteson Å (1981c): Demonstration of an antibody against platelet-derived growth factor. Exp Cell Res 136:255-261.

Heldin C-H, Ek B, Wasteson Å (1982a): Interaction between platelet-derived growth factor and its fibroblast receptor. J Cell Biochem Suppl 6:146.

Heldin C-H, Wasteson Å, Westermark B (1982b): Interaction of platelet-derived growth factor with its fibroblast receptor: Demonstration of ligand degradation and receptor modulation. J Biol Chem 257:4216-4221.

Holley RW, Kiernan JA (1971): Studies of serum factors required by 3T3 and SV3T3 cells. In Wolstenholme GE, Knight J (eds): "Growth Control in Cultures." London: Churchill Livingstone, pp 3-10.

Huang JS, Huang SS, Kennedy B, Deuel TF (1982): Platelet-derived growth factor: Specific binding to target cells. J Biol Chem 257:8130-8136.

Johnsson A, Heldin C-H, Westermark B, Wasteson Å (1982): Platelet-derived growth factor: Identification of constituent polypeptide chains. Biochem Biophys Res Commun 104:66-74.

Kohler N, Lipton A (1974): Platelets as a source of fibroblast growth-promoting activity. Exp Cell Res 87:297-301.

Ladda RL, Bullock LP, Gianopoulous T, McCormick L (1979): Radioreceptor assay for epidermal growth factor. Anal Biochem 93:286-294.

Laemmli UK (1970): Cleavage of structural proteins during the assembly of the head of Bacteriophage T4. Nature (Lond) 227:680-685.

Lindberg U, Nordenskjöld BA, Reichard P, Skoog L (1969): Thymidine phosphate pools and DNA synthesis after polyoma infection of mouse embryo cells. Cancer Res 29:1498-1506.

Lowry OH, Rosebrough NJ, Farr AL, Randall RJ (1951): Protein measurement with the Folin phenol reagent. J Biol Chem 193:265-275.

Mendoza SA, Wigglesworth NM, Pohjanpelto P, Rozengurt E (1980): Na entry and Na-K pump activity in murine, hamster, and human cells—Effect of monensin, serum, platelet extract, and viral transformation. J Cell Physiol 103:17-27.

Nexø E, Hollenberg MD. Figueroa A, Pratt RM (1980): Detection of epidermal growth factor-urogastrone and its receptor during fetal mouse development. Proc Natl Acad Sci USA 77:2782-2785.

Nishimura J, Deuel TF (1981): Stimulation of protein phosphorylation in Swiss mouse 3T3 cells by human platelet-derived growth factor. Biochem Biophys Res Commun 103:355-361.

Nishimura J, Huang JS, Deuel TF (1982): Platelet-derived growth factor stimulates tyrosine-specific protein kinase activity in Swiss mouse 3T3 cell membranes. Proc Natl Acad Sci USA 79:4303-4307.

Nordenskjöld BA, Skoog L, Brown NC, Reichard P (1970): Deoxyribonucleotide pools and deoxyribonucleic acid synthesis in culture mouse embryo cells. J Biol Chem 245: 5360-5368.

Owen AJ III, Geyer RP, Antoniades HN (1982): Human platelet-derived growth factor stimulates amino acid transport and protein synthesis by human diploid fibroblasts in plasma-free media. Proc Natl Acad Sci USA 79:3203-3207.

Raines EW, Ross R (1982): Platelet-derived growth factor. I. High yield purification and evidence for multiple forms. J Biol Chem 257:5154-5160.

Raines EW, Bowen-Pope DF, Ross R (submitted for publication): Plasma binding proteins for platelet-derived growth factor that inhibit its binding to cell-surface receptors.

Ross R (1981): Atherosclerosis: A problem of the biology of arterial wall cells and their interactions with blood components. Arteriosclerosis 1:293-311.

Ross R, Glomset JA (1976): The pathogenesis of atherosclerosis. N Engl J Med 295:369-377, 420-425.

Ross R, Vogel A (1978): The platelet-derived growth factor. Cell 14:203-210.

Ross R, Glomset J, Kariya B, Harker L (1974): A platelet-dependent serum factor that stimulates the proliferation of arterial smooth muscle cells in vitro. Proc Natl Acad Sci USA 71(4): 1207-1210.

Ross R, Vogel A, Davies P, Raines E, Kariya B, Rivest MJ, Gustafson C, Glomset J (1979): The platelet-derived growth factor and plasma control cell proliferation. In Sato GH, Ross R (eds):"Cold Spring Harbor Conferences on Cell Proliferation," Vol 6: "Hormones and Cell Culture." Cold Spring Harbor, New York: Cold Spring Harbor Laboratory, pp 3-16.

Rozengurt E, Collins M, Brown KD, Pettican P (1982): Inhibition of epidermal growth factor factor binding to mouse cultured cells by fibroblast-derived growth factor: Evidence for an indirect mechanism. J Biol Chem 257:3680-3686.

Scher CD, Shepard RC, Antoniades HN, Stiles CD (1979): Platelet-derived growth factor and the regulation of the mammalian fibroblast cell cycle. Biochim Biophys Acta 560:217-241.

Schmidt RA, Glomset JA, Wight TN, Habenicht AJR, Ross R (1982): A study of the influence of mevalonic acid and its metabolites on the morphology of Swiss 3T3 cells. J Cell Biol 95:144-153.

Seppä H, Grotendorst G, Seppä S, Schiffmann E, Martin GR (1982): Platelet-derived growth factor is chemotactic for fibroblasts. J Cell Biol 92:584-588.

Shupnik MA, Antoniades HN, Tashjiian AH (1982): Platelet-derived growth factor increases prostaglandin production and decreases epidermal growth factor receptors in human osteosarcoma cells. Life Sci 30:347-353.

Singh JP, Chaikin MA, Stiles CD (1982): Phylogenetic analysis of platelet-derived growth factor by radioreceptor assay. J Cell Biol 95:667-671.

Smith JC, Sing JP, Lillquist JS, Goon DS, Stiles CD (1982): Growth factors adherent to cell substrate are mitogenically active in situ. Nature 296:154-156.

Switzer RC, Merril CR, Shifrin S (1979): A highly sensitive silver stain for detecting proteins and peptides in polyacrylamide gels. Anal Biochem 98:231-237.

Temin HM (1966): Studies on carcinogenesis by avian sarcoma viruses. III. The differential effect of serum and polyanions on multiplication of uninfected and converted cells. J Natl Cancer Inst 37:167-175.

Temin HM (1968): Studies on carcinogenesis by avian sarcoma viruses. VIII. Glycolysis and cell multiplication. Int J Cancer 3:273-282.

Todaro GJ, Lazar GK, Green H (1965): The initiation of cell division in a contact-inhibited mammalian cell line. J Cell Comp Physiol 66:325-333.

Todaro GJ, Matsuya Y, Bloom S, Robbins A, Green H (1967): Stimulation of RNA synthesis and cell division in resting cells by a factor present in serum. In Defendi V, Stoker M (eds): "Growth Regulating Substances for Animal Cells in Culture." Wistar Institute Symposium. Philadelphia: Wistar Institute Press, Monograph No. 7, pp 87-98.

Vogel A, Raines E, Kariya B, Rivest M-J, Ross R (1978): Coordinate control of 3T3 cell proliferation by platelet-derived growth factor and plasma components. Proc Natl Acad Sci USA 75:2810-2814.

Westermark B, Wasteson Å (1976): A platelet factor stimulating human normal glial cells. Exp Cell Res 98:170-174.

Wharton W, Leof E, Pledger WJ, O'Keefe EJ (1982): Modulation of the epidermal growth factor receptor by platelet-derived factor and choleragen: Effects on mitogenesis. Proc Natl Acad Sci USA 79:5567-5571.

Williams LT, Tremble P, Antoniades HN (1982): Platelet-derived growth factor binds specifically to receptors on vascular smooth muscle cells and the binding becomes nondissociable. Proc Natl Acad Sci USA 79:5867-5870.

Witte LD, Cornicelli JA, Miller RW, Goodman DS (1982): Effects of platelet-derived and endothelial cell-derived growth factors on the low density lipoprotein receptor pathway in cultured human fibroblasts. J Biol Chem 257:5392-5401.

Wrann M, Fox CF, Ross R (1980): Modulation of epidermal growth factor receptors on 3T3 cells by platelet-derived growth factor. Science 210:1363-1365.

Yoshikura H, Hirokawa Y, Yamada M (1967): Synchronized cell division induced by medium change. Exp Cell Res 48:226-228.

Methods for Preparation of Media, Supplements, and Substrata
for Serum-Free Animal Cell Culture, pages 111–138

5
Purification of Multiplication-Stimulating Activity

Lawrence A. Greenstein, S. Peter Nissley, Alan C. Moses, Patricia A. Short, Yvonne W.-H. Yang, Lilly Lee, and Matthew M. Rechler

BACKGROUND

Dulak and Temin [1973a,b] first described the purification of a family of polypeptides designated multiplication-stimulating activity (MSA) from medium conditioned by the BRL-3A rat liver cell line. Although the term MSA had been used earlier by Temin [1971] to designate the factor(s) in serum that support(s) the multiplication of cells in culture, and Pierson and Temin [1972] had partially purified this activity from calf serum, MSA has usually referred to the polypeptides produced by the BRL-3A cell line.

Relationship to Human Somatomedins or Insulin-Like Growth Factors

Dulak and Temin [1973a] showed that MSA from the BRL-3A cells was related to a class of polypeptides called somatomedins or nonsuppressible insulin-like activity (NSILA) that had been purified from human plasma [Zapf et al., 1981]. This relationship was strengthened by the demonstration that MSA and the human somatomedins or NSILA (later renamed insulin-like growth factor [IGF]) bound to the same cell surface receptors and interacted

Endocrine Section, Metabolism Branch, National Cancer Institute (L.A.G., S.P.N., A.C.M., P.A.S., L.L.) and Section on Biochemistry of Cell Regulation, Laboratory of Biochemical Pharmacology, National Institute of Arthritis, Diabetes, and Digestive and Kidney Diseases (Y.W.-H.Y., M.M.R.), National Institutes of Health, Bethesda, Maryland 20205

A.C. Moses is now at the Endocrinology Diabetes Unit, Beth Israel Hospital, Boston, Massachusetts 02215.

with the same serum-binding proteins [Rechler et al., 1980, 1981]. The close relationship of MSA to the human IGFs was firmly established by Marquardt et al. [1981], who showed that the primary structure of one of the MSA species was homologous with the primary structure of human IGF-II with the exception of only five amino acid residues.

Terminology in the somatomedin/IGF field has been confusing but has been recently clarified by the demonstration that IGF-I (formerly NSILA-I), somatomedin C, and basic somatomedin probably are identical [Zapf et al., 1981; Svoboda et al., 1980; Van Wyk et al., 1980; Bala and Bhaumick, 1979]. In addition, somatomedin preparations designated somatomedin A [Fryklund et al., 1974] and purified on the basis of a neutral isoelectric point may contain IGF-I and IGF-II [Hall and Enberg, 1982; Spencer et al., 1983; Hintz et al., 1983]. The mitogenic activity of somatomedin B has been shown to be due to contamination with epidermal growth factor (EGF) [Heldin et al., 1981]. Thus, our current state of knowledge suggests that well-characterized prototypes of this family of growth factors can be reduced to two: IGF-I and IGF-II. MSA is the rat homolog of human IGF-II.

Insulin-Like Growth Factors and Cells in Culture

Following the initial demonstration that MSA was a mitogen for chick embryo fibroblasts in culture [Dulak and Temin, 1973a], there have been other reports of IGFs serving as growth factors for cells in culture [Rechler et al., 1978; Stiles et al., 1979; Weidman and Bala, 1980; Ewton and Florini, 1980; Zapf et al., 1981; Clemmons and Van Wyk, 1981; Nagarajan and Anderson, 1982]. In most instances, a mixture of factors is required for optimal growth. This is the case for 3T3 cells and human skin fibroblasts, for example [Stiles et al., 1979; Clemmons and Van Wyk, 1981]. Platelet-derived growth factor (PDGF) or fibroblast growth factor (FGF) serve as so-called competence factors, whereas a combination of somatomedin C and other factors present in platelet-poor plasma are required for these cells to progress through S phase.

Recent results from competitive binding and affinity cross-linking studies point to two types of IGF receptors: One (type I) prefers IGF-I over IGF-II and also recognizes insulin; a second type (type II) prefers IGF-II over IGF-I and does not recognize insulin [Rechler et al., 1980; Kasuga et al., 1981; Massague and Czech, 1982; Rechler and Nissley, 1983]. Except for cells such as chick embryo fibroblasts, which have only the type I receptor [Kasuga et al., 1982], it is not clear which receptor type is mediating the biologic response to IGF-I and IGF-II. Since MSA (rat IGF-II) interacts with both types of receptors but with a lower potency for the type I receptor, it should be a suitable substitute for IGF-I if used at a sufficiently high concentration. This has practical implications since the purification of IGF-I from human or rat serum is a laborious procedure [Rinderknecht and Humbel, 1976; Rubin et al., 1982]. Since MSA is relatively easily purified from the serum-free medium conditioned by the BRL-3A cell line, MSA serves as a convenient substitute not only for human IGF-II but also for IGF-I.

High concentrations of insulin have been used in many formulations of chemically defined tissue culture medium [Barnes and Sato, 1980]. Since insulin interacts with the type I IGF receptor at high concentrations, it is likely that insulin is functioning as an IGF analog for some cell types. In some instances, however, insulin clearly stimulates cell multiplication by acting through its own receptor [Koontz and Iwahashi, 1981; Nagarajan and Anderson, 1982].

MSA Is a Family of Polypeptides

Dulak and Temin [1973b] showed that MSA from the BRL-3A cell line is a family of polypeptides. Moses et al. [1980a] described at least seven different MSA species: 1 of MW = 16,300 (MSA I), 4 of MW = 8,700 (MSA II), and 2 of MW = 7,100 (MSA III). On the basis of similar behavior in a bioassay, a competitive protein binding assay, radioreceptor assays, and a radioimmunoassay, it was proposed that the smaller MSA species were derived from the larger ones [Moses et al., 1980a,b]. Recently, Acquaviva et al. [1982] have identified a presumed pre-pro-MSA of MW = 22,000 in a reticulocyte lysate translation system that utilized BRL-3A mRNA, and Yang et al. [1982] have identified an MW = 20,000 MSA species (possibly pro-MSA) in whole-cell labeling experiments. Both studies identified these larger MSA species by immunoprecipitation with antisera that had been raised against the MW = 8,700 (MSA II) species. Thus it seems probable that all of the MSA species are products of a single pre-pro-MSA.

Moses et al. [1980a] showed that the MW = 7,100 (MSA III-2) species is more active than the larger MSA species in a bioassay and in radioreceptor assays. Although the MW = 8,700 (MSA II) species usually predominates in BRL-3A-conditioned medium, it is difficult to resolve the different MSA II species from each other. The more active MSA III-2 species is usually present in lower amounts than MSA II.

In this chapter we will describe the purification of MSA from BRL-3A-conditioned medium by two different methods: method A [Moses et al., 1980a], using ion exchange chromatography, Sephadex G-75 gel filtration, and preparative polyacrylamide gel electrophoresis (PAGE); and method B [Marquardt et al., 1981], which utilizes Bio-Gel P-10 gel filtration and high-pressure liquid chromatography (HPLC). It should be noted that in the BRL-3A-conditioned medium MSA is found associated with a binding protein [Moses et al., 1979]; both methods A and B utilize gel permeation chromatography in 1 M acetic acid to dissociate and separate MSA from the binding protein.

PURIFICATION OF MSA
Method A

Method A is as described by Moses et al. [1980a] and was modified from the original purification scheme described by Dulak and Temin [1973a]. By following this procedure through the Sephadex G-75 gel filtration step, it is possible to obtain relatively large quantities of a mixture of the MSA II (MW = 8,700) polypeptides as well as MSA III (MW = 7,100) species. For laboratories not having access to HPLC, preparative PAGE can then be used to prepare highly purified MSA II-1 (MW = 8,700) or MSA III-2 (MW = 7,100) from the MSA II and MSA III pools obtained from Sephadex G-75 gel filtration. The entire purification scheme (method A) is carried out at room temperature.

Ion exchange chromatography. A 35-cm × 10-cm column containing 2 liters of AG 50W-X8, 100–200 mesh (Bio-Rad Laboratories, Richmond, CA) is washed successively with 8 liters each of H_2O, 1.2 N HCl, H_2O, 1.2 N NaOH, H_2O, and 0.15 M NaCl. Four liters of conditioned medium (see Appendix) are thawed and warmed to room temperature, and the pH is adjusted to 7.1 with 1 M HCl. The conditioned medium is applied to the resin, now in the Na^+ form, and the column is washed with 3 liters of 0.15 M NaCl at a flow rate of 300 ml/min. The column is developed by the addition of 3 liters of 0.1 M $NaHCO_3$, pH 9.0, and 3 liters of 0.1 M NH_4OH. During elution with the NH_4OH (pH 11) the eluate is continually neutralized (the residual phenol red being used as an indicator) by addition of 2 N HCl. The NH_4OH eluate (containing MSA) is dialyzed overnight against 16 liters of 2% acetic acid (one change) using Spectrapor 3 membrane tubing, MW cutoff: 3,500 (Spectrum Medical Industries, Inc., Los Angeles, CA). The dialyzate is shell-frozen in three 2-liter freeze-drying flasks and lyophilized. The flasks are washed with 60 ml of 1 M acetic acid, and the dissolved material is dialyzed overnight against 4 liters of 1 M acetic acid. Insoluble material is removed by centrifugation in a #40 rotor at 27,000 rpm for 1 h at 4°C in a Beckman L-50 ultracentrifuge. This preparation is referred to as Dowex MSA and contains many protein components in addition to MSA (Fig. 7). Because MSA species have neutral or acidic isoelectric points and MSA is not eluted from Dowex with pH 9.0 $NaHCO_3$, Dulak and Shing [1976] have proposed that it is a MSA-binding protein complex that is binding to the negatively charged ion exchange column.

Sephadex G-75 gel filtration. Most of the MSA in the BRL-3A-conditioned medium is complexed to a binding protein. This binding protein has been purified from BRL-3A-conditioned medium by Knauer et al. [1981] and shown to have a MW = 60,000 and to be made up of two subunits. The MSA can be dissociated from the binding protein at low pH. Gel filtration on Sephadex G-75 (Pharmacia Fine Chemicals, Inc., Piscataway, NJ) in 1 M acetic acid separates the binding protein from the MSA species. The binding protein elutes just after the void volume and is resolved from MSA I (MW = 16,300).

Lyophilized Dowex MSA from two 4-liter batches of conditioned medium is dissolved in 7 ml of 1 M acetic acid and is applied to a Sephadex G-75 column (2.6 × 88 cm) (Fig. 1). The sample is eluted with 1 M acetic acid at 30 cm pressure (H_2O) at room temperature. While most of the protein is found in the exclusion volume of the column, MSA elutes in a broad region from Kd = 0.16 to Kd = 0.66. The MSA elution profile has been divided into three regions designated I, II, and III on the basis of the Sephadex G-75 elution position and the protein staining pattern on analytical polyacrylamide gel electrophoresis (see Appendix). The analytical PAGE pattern for fractions from each of the three regions in shown in Figure 2.

The major brand in region I of the Sephadex G-75 elution profile has an R_f of 0.36 by analytical PAGE. This is an MSA species of MW = 16,300 that we have designated MSA I [Moses et al., 1980a]. This species has not been further purified, but presumably it represents a precursor of the smaller MSA species.

Four closely migrating MSA species found in the region II (MSA II-1,2,3,4) have R_f values of 0.49–0.41 by PAGE and indistinguishable MW values by gel filtration on Sepharose 6B in 6 M guanidine HCl under nonreducing conditions (MW = 8,700) [Moses et al., 1980a]. It should be noted that a protein with an R_f value identical to the R_f of MSA II-1 (0.49) is present in fractions 49–53 (Fig. 2). This protein has been isolated by preparative PAGE and shown not to be MSA (inactive in the [^{3}H]-thymidine incorporation assay in chick embryo fibroblasts).

Region III from the Sephadex G-75 column contains two MSA species, designated III-1 and III-2, with R_f values by PAGE of 0.59 and 0.56, respectively, and the same molecular weight (7,100) [Moses et al., 1980a].

By selecting fractions from the Sephadex G-75 region II on the basis of the protein staining pattern on PAGE, it is possible to obtain a preparation of MSA

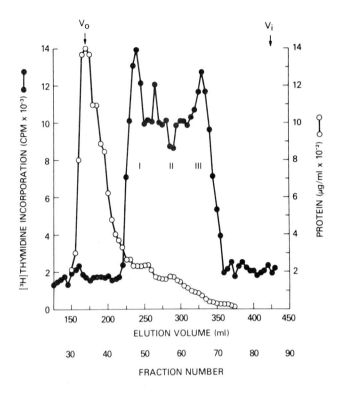

Fig. 1. Sephadex G-75 gel filtration of Dowex MSA. Dowex MSA from 8 liters of BRL-3A-conditioned medium was applied to a Sephadex G-75 column (2.6 × 88 cm) in 1 M acetic acid. Five-milliliter fractions were collected and 5-μl aliquots were added directly to wells in the chick embryo fibroblast ^{3}H-thymidine incorporation assay (see Appendix) to measure MSA (●—●). Regions of the MSA elution profile are designated I, II, and III on the basis of presence of MSA I (MW = 16,300), MSA II (MW = 8,700), and MSA III (MW = 7,100) species determined by PAGE (Fig. 2). Protein was measured by the Bradford method [Bradford, 1976] on aliquots from each fraction (○—○). The excluded volume (Vo) and included volume (Vi) are shown.

II with only minor contaminating bands. There is evidence that the four MSA II species represent microheterogeneity [Moses et al., 1980a,b]. MSA II-1 is equipotent with a mixture of MSA II-2, II-3, and II-4 by bioassay and radioimmunoassay. The mixture of MSA II species obtained from Sephadex G-75 gel filtration is almost equipotent with MSA II-1 prepared by preparative PAGE. Although MSA II is less potent than MSA III-2, the mixture of MSA II polypeptides should be satisfactory for use in chemically defined tissue culture medium—for example, where relatively large amounts of growth factor are required.

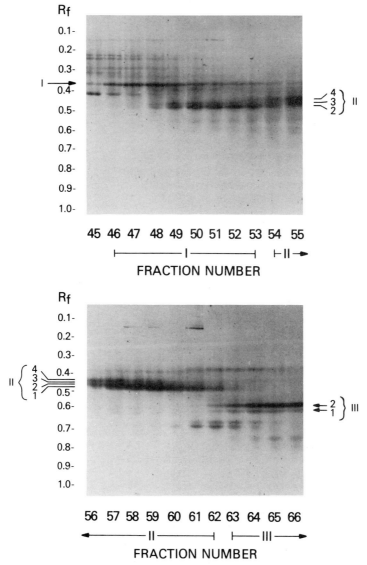

Fig. 2. Analytical polyacrylamide gel electrophoresis of Sephadex G-75 column fractions. Aliquots (10 µg protein) of Sephadex G-75 column fractions (#45-55, upper panel; #56-66, lower panel) from Figure 1 were analyzed by acid-urea PAGE (see Appendix). The gels were stained for protein with Coomassie blue. The mobility (R_f) is shown on the left. The locations of individual MSA I, MSA II, and MSA III species are indicated in the left and right margins. Horizontal lines drawn beneath the fraction numbers indicate the fractions containing MSA I, II, and III species, respectively.

The relative amounts of MSA I, II, and III species (Table I) vary for different lots of conditioned medium, possibly depending upon the extent of proteolysis prior to collection of the medium.

Preparative polyacrylamide gel electrophoresis. To prepare highly purified MSA II-1 (MW = 8,700) by preparative PAGE, Sephadex G-75 column fractions are pooled from the II region on the basis of the analytical PAGE patterns (Fig. 2). This pool is lyophilized and fractionated by electrophoresis on a preparative PAGE system (Tables II and III) (Fig. 3). The preparative PAGE system differs from the analytical system in several respects

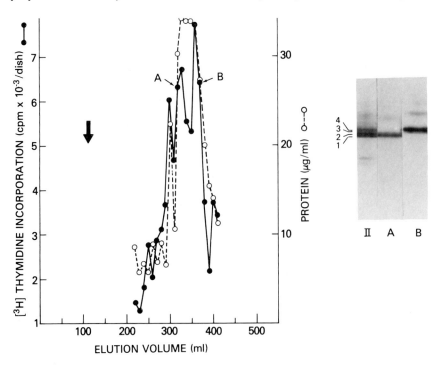

Fig. 3. Preparative acrylamide gel electrophoresis of MSA II (MW = 8,700) purified by Sephadex G-75 gel filtration. MSA II pooled from four Sephadex G-75 gel filtrations (16 liters conditioned medium) was further purified by preparative acid-urea PAGE (see Appendix). Ten-milliliter fractions were collected and 5-μl aliquots were tested in the ^{3}H-thymidine incorporation assay (see Appendix) ($\bullet$—$\bullet$). Protein ($\bigcirc$—$\bigcirc$) was measured by the Lowry procedure [Lowry et al., 1951]. In the left (graph) panel the vertical arrow indicates the elution position of the methylene blue tracking dye. In the right (gel) panel, analytical acid-urea PAGE of aliquots of the Sephadex G-75 MSA II pool (left lane) and fractions A and B (as indicated in the left panel) (middle and right lanes) are shown. Fraction A contains mainly MSA II-1 with some MSA II-2; fraction B contains MSA II-3 and II-4. Modified from Moses et al. [1980].

TABLE I. Purification of Multiplication-Stimulating Activity From 12 Liters of BRL-3A-Conditioned Medium [Moses et al., 1980a]

Fraction	Total protein (mg)	Specific activity (ng insulin equiv./ng)	Recovery (%)
Conditioned medium	2,008	0.02	(100)
Dowex 50	225	0.10	56
Sephadex G-75			
Peak I	9.8	0.20[a]	5
Peak II	6.7	1.0	17
Peak III	9.0	2.0	45
Preparative electrophoresis (Sephadex G-75 peak II)			
Polypeptide II-1	0.5	2.0	2.5[b]

Protein content of conditioned medium was determined on the basis of protein excluded by Sephadex G-25 in 1 M acetic acid. Protein was measured by the procedure of Lowry et al. [1951]. The specific activity was measured as stimulation of ^{3}H-thymidine incorporation into chick embryo fibroblast DNA compared to the same lot of porcine insulin used by Smith and Temin [1974].

[a]At concentrations greater than 500 ng/ml the biological activity exceeded that of peaks II and III. Peak I MSA yielded a dose-response curve in thymidine incorporation studies in chick embryo fibroblasts that was parallel to and two times more potent than Dowex 50 MSA.

[b]The 2.5% recovery of polypeptide II-1 from preparative disc acrylamide electrophoresis represents MSA biological activity as a percentage of total conditioned medium activity. Recovery of biological activity and protein after preparative acrylamide gel electrophoresis ranged from 40% to 70% and was partially dependent on protein load.

(Table III): The lower buffer contains 5 M urea and an elution buffer is utilized. The running gel is 35 ml of 12.5% acrylamide, and the stacking gel is 6 ml of 3.5% acrylamide.

We use a Poly-Prep 100 (Haake-Buchler, Saddle Brook, NJ) apparatus. (The Poly-Prep 100 has been discontinued, but the larger-capacity Poly-Prep 200 is available.) The gel is polymerized on the day prior to applying the sample and is maintained overnight at room temperature with a circulating water bath. The sample is dissolved in 2 ml of upper buffer containing 10% sucrose and 0.1% methylene blue and applied to the upper gel. Electrophoresis is at 20 mA for the stacking gel and 40 mA for the running gel. The elution buffer is pumped from a reservoir at 1 ml/min with a peristaltic pump. Beginning shortly before elution of the methylene blue marking dye, 5-ml fractions are collected. Fractions are dialyzed against 1 M acetic acid in Spectropor 3 dialysis membranes or desalted on a Sephadex G-25 column.

Preparative PAGE of MSA II is shown in Figure 3. The protein profile shows a broad peak without apparent resolution of individual MSA II components. Only by performing analytical PAGE is it possible to identify fractions containing individual MSA II species such as MSA II-1.

Table I summarizes the purification of MSA II-1 from 12 liters of conditioned medium according to method A. It is frequently difficult to obtain a true

**TABLE II. Stock Solutions for Acid-Urea
Polyacrylamide Gel Electrophoresis**

A. 4.0 N KOH	1.5 ml	
Glacial acetic acid	26.6 ml	
TEMED[a]	0.6 ml	
10 M Urea	TM[b] 50.0 ml	
		pH 2.9[c]
B. 4.0 N KOH	6.0 ml	
Glacial acetic acid	1.5 ml	
10 M Urea	40.0 ml	
H_2O	TM 50.0 ml	
		pH 5.9[c]
C. Acrylamide	12.5 g	
bis-Acrylamide	0.2 g	
10 M Urea	TM 25.0 ml	
D. Acrylamide	2.5 g	
bis-Acrylamide	0.2 g	
10 M Urea	20.0 ml	
H_2O	TM 25.0 ml	
E. Riboflavin	1 mg	
10 M Urea	25 ml	
Upper buffer stock		
Glycine	28.0 gm	
Glacial acetic acid	3.0 ml	
H_2O	TM 1,000 ml	
		pH 4.0[c]
Lower and elution buffer stock		
4.0 N KOH	60.0 ml	
Glacial acetic acid	1,064.0 ml	
H_2O	T M 2,000 ml	

[a]TEMED: N,N,N′,N′-tetramethylethylenediamine.
[b]TM: to make.
[c]These pH measurements were made on solutions in which H_2O replaced 10 M urea.

**TABLE III. Formulations for Polyacrylamide Gel
Electrophoresis Gels and Buffers**

Running gel		Stacking gel	
Stock	Volume (ml)	Stock	Volume (ml)
A	2.0	B	2.0
C	8.0	D	4.0
10 M Urea	18.0	E	2.0
E	2.0	10 M Urea	4.0
AP	2.0	H_2O	3.0
		AP^a	1.0
Upper buffer			
Upper buffer stock		75.0	
10 M Urea		125.0	
H_2O		50.0	
Lower buffer			
Lower and elution buffer stock		600	
H_2O		600	
Lower buffer (preparative)			
Lower and elution buffer stock		25.0	
10 M Urea		200.0	
H_2O		TM 400.0	
Elution buffer (preparative)			
Lower and elution buffer stock		75.0	
10 M Urea		1,080	
H_2O		TM 1,200	

[a]AP: ammonium persulfate (freshly dissolved) 70 mg/10 ml 10 M urea.

activity measurement on the conditioned medium because of the presence of inhibitors.

Similarly, MSA III-2 (MW = 7,100) can be purified by preparative PAGE starting with MSA III pooled from Sephadex G-75 chromatography fractions (Figs. 1 and 2). Figures 4 and 5 show an example of the purification of MSA III-2 by preparative PAGE. (In this example, MSA has been measured by a radioimmunoassay in which MSA III-2 is considerably less potent than MSA II [Moses et al., 1980b]; consequently MSA II appears to be present in a relatively greater amount than MSA III.) By performing analytical PAGE on individual fractions from the MSA III region (Fig. 5), it is possible to select fractions that contain only MSA III-2 (fractions 24 and 25). Thus, in this example, 780 μg of MSA III-2 has been purified from 24 liters of conditioned medium.

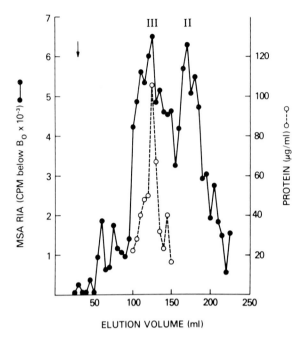

Fig. 4. Preparative acrylamide gel electrophoresis of MSA III (MW = 7,100) purified by Sephadex gel filtration. MSA III (containing some MSA II) pooled from six Sephadex G-75 gel filtrations (24 liters conditioned medium) was further purified by acid-urea preparative PAGE (see Appendix and text). Five-milliliter fractions were collected and MSA was measured by a radioimmunoassay [Moses et al., 1980b]. The radioimmunoassay data are expressed as inhibition of ^{125}I-MSA binding to the antiserum (counts per minute below B_0). (This radioimmunoassay utilizing ^{125}I-MSA-II-1 preferentially measures MSA II compared to MSA III; consequently the immunoreactivity in the second peak (MSA II) is high relative to the immunoreactivity in the first peak (MSA III) even though the extent of contamination of this MSA III preparation with MSA II is relatively low.) Protein was measured by the method of Lowry et al. [1951]. The vertical arrow on the left indicates the elution position of the methylene blue tracking dye.

In Figure 6 preparations of MSA II-1 and III-2 are analyzed by analytical PAGE; a preparation of Dowex MSA is shown for comparison. The MSA II-1 preparation has been shown to also migrate as a single component in an alkaline PAGE system [Moses et al., 1980a]. End-group analysis of MSA II-1 shows a carboxy-terminal glycine and no amino terminal amino acid, a finding that is consistent with a blocked amino terminus [Moses et al., 1980a].

Method B

Marquardt and his colleagues [Marquardt et al., 1981] described a simplified scheme for the purification of MSA from medium conditioned by the

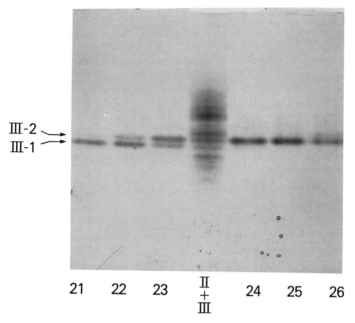

III-2 →
III-1 →

21   22   23   II   24   25   26
                        +
                        III

Fig. 5. Analytical polyacrylamide gel electrophoresis of fractions from preparative gel electrophoresis of MSA III (Fig. 4). Aliquots (10 µg protein) of fractions corresponding to elution volume 105-130 ml (Fig. 4) were analyzed by acid-urea PAGE. An impure mixture of MSA II and MSA III from Sephadex G-75 is shown for reference in the center lane.

BRL-3A cell line. They used this procedure to purify an MSA species (calculated MW = 7,484) that was shown to have a primary structure 93% homologous with human IGF-II. This two-step purification scheme uses gel filtration on Bio-Gel P-10, which provides better resolution of MSA II from MSA III than does Sephadex G-75. MSA III-2 is then purified by HPLC, which has the advantages of speed, ease of operation, and greater resolving power compared with preparative PAGE.

Marquardt employed a radioreceptor assay to measure MSA during the purification, whereas in the example of method B shown here we use an MSA radioimmunoassay. As discussed in the Appendix, similar results could be achieved with the ^{3}H-thymidine incorporation assay. Although method B described here is for the purification of MSA III-2, our preliminary experiments show that the MSA II species can also be resolved from each other by reverse-phase HPLC. We have not determined which MSA II preparation (Sephadex G-75 gel filtration in method A versus Bio-Gel P-10 filtration in method B) is

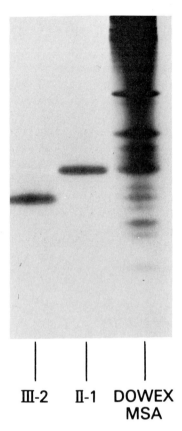

Ⅲ-2   Ⅱ-1   DOWEX
MSA

Fig. 6. Analytical polyacrylamide gel electrophoresis of MSA II-1 and MSA III-2 purified according to method A. Ten micrograms of MSA II-1 and MSA III-2 and 50 μg of Dowex MSA were analyzed by acid-urea PAGE. The gels were stained with Coomassie blue. Modified from Moses et al. [1980].

the best starting material to use for HPLC purification of individual MSA II species.

Gel filtration on Bio-Gel P-10. Two liters of conditioned medium (see Appendix) are thawed and then dialyzed for 60 h against 8–10 volumes of 0.1 M acetic acid (several changes) at 4°C in Spectropor 3 dialysis tubing (3,500 MW cutoff). The dialyzate is then lyophilized, dissolved in 50 ml 1 M acetic acid, lyophilized, and dissolved in 10 ml 1 M acetic acid. After centrifugation at 40,000 rpm for 60 min in a 60 Ti rotor in a model L3-50 Beckman ultracentrifuge, the supernatant is applied to a 2.5 × 90-cm column containing Bio-Gel P-10, 200-400 mesh (Bio-Rad, Richmond CA) in 1 M acetic acid (Fig. 7). The column is run at a flow rate of 0.5-0.7 ml/min at room temperature, and 5-ml fractions are collected. Aliquots of fractions can be tested directly in the ^{3}H-thymidine incorporation assay (the MSA radioimmunoassay is shown here) and analyzed by PAGE. The two peaks of activity correspond to MSA II and III. The binding protein is found in the exclusion volume of the column; and as we note in the Appendix, we have not detected MSA I (MW = 16,300) when method B is used to purify MSA from Waymouth medium conditioned by the BRL-3A cell line. Figure 8 shows the results of analytical PAGE of aliquots of fractions constituting the MSA III peak (fractions 61-68). The dark-staining band indicated by the arrow has an R_f value of 0.56 corresponding to MSA III-2. Fractions 61-69 were pooled for further purification by reverse-phase HPLC.

Purification on reverse-phase HPLC. MSA III from Bio-Gel P-10 is further purified by reverse-phase HPLC (Fig. 9). We use a system from Waters

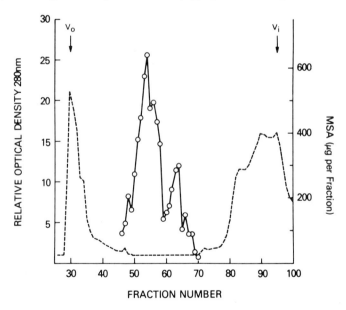

Fig. 7. Bio-Gel P-10 gel filtration of MSA according to method B. Two liters of conditioned medium were dialyzed against 0.1 M acetic acid for 60 h and lyophilized. The lyophilized material was dissolved in 1 M acetic acid and applied to a Bio-Gel P-10 column (2.5 × 90 cm). Five-milliliter fractions were collected. MSA was measured by radioimmunoassay. The amount of MSA (O—O) in each fraction was calculated based on MSA II standard (fraction 46-58) or MSA III-2 standard (fraction 59-70). Relative optical density at 280 nm (----) is shown on an arbitrary scale. The excluded volume (Vo) and included volume (Vi) are shown.

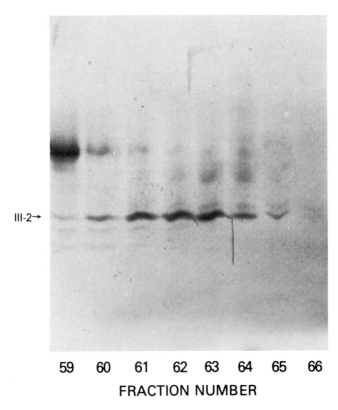

III-2→

59   60   61   62   63   64   65   66

FRACTION NUMBER

Fig. 8. Analytical polyacrylamide gel electrophoresis of fractions from Bio-Gel P-10 gel filtration of MSA (Fig. 7). Aliquots (2 μg of MSA by radioimmunoassay) from individual fractions in the second peak of immunoreactivity in Figure 7 were analyzed by acid-urea PAGE. The gel was stained by the silver method (see Appendix).

Associates (Milford, MA) that includes two M6000A solvent delivery pumps; an M710A WISP as the injector; an M720 systems controller; an M730A data module; and two detectors, the M440 single channel detector (280 nm) and the M450 variable wavelength detector (212 nm). We also use an LKB 2111 Multirack fraction collector (Bromma, Sweden), which we integrate with the systems controller. All separations are performed on a Synchropak RP-P column (250 × 4.1 mm) (Synchrome, Linden, IN).

The solvents used are 0.1% trifluoroacetic acid (TFA) (Pierce Chemical Co., Rockford, IL) in water (pH 1.9-2.1) as the aqueous solvent and 0.075% TFA in acetonitrile (Burdick and Jackson, Muskegon, MI) as the organic solvent. Water is filtered through a Milli-Q Reagent Grade Water System

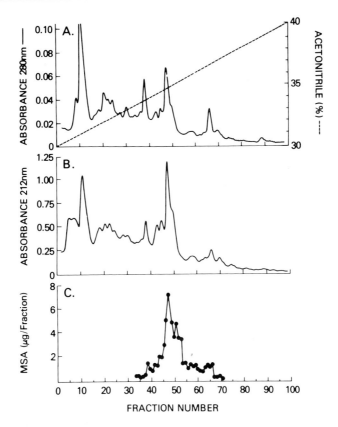

Fig. 9. High-pressure liquid chromatography of MSA III purified by Bio-Gel P-10 filtration. Fractions (60-69) from the second peak of immunoreactivity in Figure 7 (MSA III) and a similar pool from a second Bio-Gel P-10 gel filtration were combined, lyophilized, and dissolved in 2.0 ml 1 M acetic acid, and 0.6 ml (240 µg protein by the Bradford method) was injected onto the HPLC column (see text). The absorbance at 280 nm (panel A) and 212 nm (panel B) are shown. The acetonitrile concentration gradient is indicated by the dotted line in panel A. MSA was measured by radioimmunoassay (panel C). Other than the solvent front there were no other peaks of absorbance noted for parts of the program not shown here.

(Millipore Corp., Bedford, MA). Solvents are filtered with a solvent clarification kit (Waters) through HAWP and FHUP filters (Millipore) for aqueous and organic solvents, respectively. The solvents are degassed under vacuum prior to use.

The column is operated at a flow rate of 1 ml/min at room temperature. The column is initially equilibrated with 20% acetonitrile. After injection, the acetonitrile concentration is kept at 20% for 10 min. A linear gradient (20-

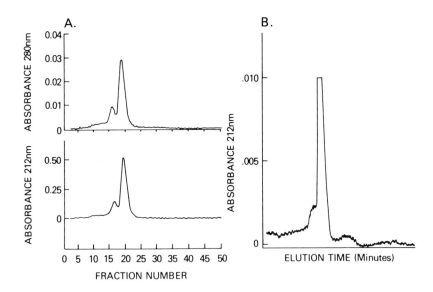

Fig. 10. Repeat high-pressure liquid chromatography of MSA III. Fractions 46–49 (42 μg protein based on 212 nm absorbance) from Figure 9 were pooled, and 34 μg was rechromatographed by the same program (panel A). (The fraction numbers do not coincide with those in Figure 9 because collection was begun later in the program.) An aliquot of fractions 18–20 was analyzed by HPLC in panel B. (The top of the peak is not shown because the absorbance was off-scale.)

40% acetonitrile) is programmed from minutes 10–50. This is followed by a second linear gradient to 80% acetonitrile from minutes 50–60. The column is maintained at 80% acetonitrile for an additional 10 min before being reequilibrated to initial conditions (20% acetonitrile) for 30 min. The column is rinsed and stored in 50:50 water:methanol (Burdick and Jackson).

Fractions (0.2 min) are collected into 12 × 75-mm polypropylene tubes during the 20–40% acetonitrile gradient. All fractions are taken to dryness with a Speed Vac Concentrator (Savant Instruments, Hicksville, NY). The dried fractions are dissolved in 200 μl 1 M acetic acid and aliquots are assayed for MSA.

Figure 9 shows the chromatography of MSA III on reverse-phase HPLC. The absorbance profile of the effluent is shown at 280 nm and 212 nm, as well as the immunoreactivity of individual fractions. Most of the activity is confined to two peak regions over a 10-fraction range. The major MSA peak elutes at 35% acetonitrile. Fractions 46–49 are pooled and rechromatographed as shown in Figure 10A.

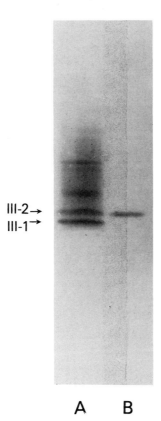

Ⅲ-2→
Ⅲ-1→

A   B

Fig. 11. Analytical polyacrylamide gel electrophoresis of MSA III purified by method B. MSA III pooled from Sephadex G-75 column fractions (10 μg MSA by radioimmunoassay) (left lane) was analyzed in parallel with MSA purified by method B (5 μg protein based on 212 nm absorbance) (right lane). The gel was stained by the silver method [Switzer et al., 1979]. HPLC-purified MSA (right lane) comigrated with MSA III-2 (left lane).

Fractions 18–20 constituting the major peak in Figure 10A are pooled, and a small aliquot is rechromatographed in Figure 10B and appears to be greater than 90% pure. To confirm that this MSA species was MSA III-2, an aliquot was compared to an MSA III preparation from Sephadex G-75 gel filtration (method A) in the acid-urea PAGE system (Fig. 11). The mobility of the HPLC-purified MSA corresponds to MSA III-2. (The MW of MSA III-2 [7,100] was determined by gel filtration on Sepharose 6B in 6 M guanidine HCl [Moses et al., 1980a], since the error of this method is at least 10%, the value of 7,100 is not significantly different from the MW of 7,484 calculated from the amino acid sequence data by Marquardt et al. [1981].) By method B it is possible to purify approximately 180 μg of MSA III-2 from 6 liters of BRL-3A-conditioned medium.

REFERENCES

Acquaviva AM, Bruni CB, Nissley SP, Rechler MM (1982): Cell-free synthesis of rat insulin-like growth factor II. Diabetes 31:656-658.

Bala RM, Bhaumick B (1979): Purification of a basic somatomedin from human plasma Cohn fraction IV-1 with physiochemical and radioimmunoassay similarity to somatomedin-C and insulin-like growth factor. Can J Biochem 57:1289-1298.

Barnes D, Sato G (1980): Methods for growth of cultured cells in serum-free medium. Anal Biochem 102:255-270.

Bradford MM (1976): A rapid and sensitive method for the quantitation of microgram quantities of protein utilizing the principle of protein-dye binding. Anal Biochem 72:248-254.

Clemmons DR, Van Wyk JJ (1981): Somatomedin C and platelet derived growth factor stimulate human fibroblast replication. J Cell Physiol 106:361-367.

Daughaday WH, Parker KA, Borowsky S, Trivedi B, Kapadia M (1982): Measurement of somatomedin-related peptides in fetal, neonatal, and maternal rat serum by insulin-like growth factor (IGF) I radioimmunoassay, IGF-II radioreceptor assay (RRA) and multiplication-stimulating activity RRA after acid-ethanol extraction. Endocrinology 110:575-581.

Dulak NC, Shing YW (1976): Large scale purification and further characterization of a rat liver cell conditioned medium multiplication stimulating activity. J Cell Physiol 90:127-138.

Dulak NC, Temin HM (1973a): A partially purified polypeptide fraction from rat liver cell conditioned medium with multiplication-stimulating activity for embryo fibroblasts. J Cell Physiol 81:153-160.

Dulak NC, Temin HM (1973b): Multiplication-stimulating activity for chick embryo fibroblasts from rat liver cell conditioned medium: A family of small polypeptides. J Cell Physiol 81:161-170.

Ewton DZ, Florini JR (1980): Relative effects of the somatomedin, multiplication-stimulating activity and growth hormone on myoblasts and myotubes in culture. Endocrinology 106:577-583.

Fryklund L, Uthne K, Sievertsson H (1974): Identification of two somatomedin A active polypeptides and in vivo effects of a somatomedin A concentrate. Biochem Biophys Res Commun 61:957-962.

Hall K, Enberg G (1982): Isolation of somatomedin A from whole serum by affinity chromatography. Symposium on Insulin-like Growth Factors/Somatomedins, Nairobi, Kenya, p 80 (abstract).

Heldin CH, Wasteson A, Fryklund L, Westermark B (1981): Somatomedin B: Mitogenic activity derived from contaminant epidermal growth factor. Science 2134:1122-1123.

Hintz RL, Liu F, Chang D (1983): The use of synthetic peptides for the development of radioimmunoassays for the insulin-like growth factors. In Spencer EM (ed): "Insulin-like Growth Factors/Somatomedins. Basic Chemistry, Biology and Clinical Importance." New York: Walter de Gruyter & Co (in press).

Kasuga M, Van Obberghen E, Nissley SP, Rechler MM (1981): Demonstration of two subtypes of insulin-like growth factor receptors by affinity crosslinking. J Biol Chem 256:5305-5308.

Kasuga M, Van Obberghen E, Nissley SP, Rechler MM (1982): Structures of the insulin-like growth factor receptor in chick embryo fibroblasts. Proc Natl Acad Sci USA 79:1864-1868.

Knauer DJ, Wagner FW, Smith GL (1981): Purification and characterization of multiplication-stimulating activity (MSA) carrier protein. J Supramol Struct Cell Biochem 15:177-191.

Koontz JW, Iwahashi M (1981): Insulin as a potent, specific growth factor in a rat hepatoma cell line. Science 211:947-949.

Lowry OH, Rosebrough NJ, Farr AL, Randall RJ (1951): Protein measurements with Folin phenol reagent. J Biol Chem 193:265-275.

Marquardt H, Todaro GJ, Henderson LE, Oroszlan S (1981): Purification and primary structure of a polypeptide with multiplication stimulating activity from rat liver cell cultures; homology with human insulin-like growth factor II. J Biol Chem 256:6859-6865.

Massague J, Czech MP (1982): The subunit structures of two distinct receptors for insulin-like growth factors I and II and their relationship to the insulin receptor. J Biol Chem 257:5038-5045.

Moses AC, Nissley SP, Passamani J, White RM, Rechler MM (1979): Further characterization of growth hormone dependent somatomedin-binding proteins in rat serum and demonstration of somatomedin-binding proteins produced by rat liver cells in culture. Endocrinology 104:536-546.

Moses AC, Nissley SP, Short PA, Rechler MM, Podskalny JM (1980a): Purification and characterization of multiplication-stimulating activity, insulin-like growth factors purified from rat-liver-cell-conditioned medium. Eur J Biochem 103:387-400.

Moses AC, Nissley SP, Short PA, Rechler MM (1980b): Immunological cross-reactivity of multiplication-stimulating activity polypeptides. Eur J Biochem 103:401-408.

Nagarajan L, Anderson WB (1982): Insulin promotes the growth of F9 embryonal carcinoma cells apparently by acting through its own receptor. Biochem Biophys Res Commun 106:974-980.

Neville DM (1967): Fractionation of cell membrane protein by disc electrophoresis. Biochim Biophys Acta 133:168-170.

Nissley SP, Short PA, Rechler MM, Podskalny JM, Coon HG (1977): Proliferation of Buffalo rat liver cells in serum-free medium does not depend upon multiplication stimulating activity (MSA). Cell 11:441-446.

Nissley SP, Rechler MM, Moses AC, Eisen HJ, Higa OZ, Short PA, Fennoy I, Bruni CB, White RM (1979): Evidence that multiplication stimulating activity, purified from the BRL-3A rat liver cell line, is found in rat serum and fetal liver organ culture. In "Cold Spring Harbor Conferences on Cell Proliferation," Vol. 6: "Hormones and Cell Culture." Cold Spring Harbor, New York: Cold Spring Harbor Laboratory, pp 79-94.

Pierson RW, Temin HM (1972): The partial purification from calf serum of a fraction with multiplication-stimulating activity for chick embryo fibroblasts in cell culture and with non-suppressible insulin-like activity. J Cell Physiol 79:319-330.

Rechler MM, Nissley SP (1983): Receptors for insulin-like growth factors. In Posner BI (ed): "Receptors for Polypeptide Hormones." New York: Marcel Dekker (in press).

Rechler MM, Fryklund L, Nissley SP, Hall K, Podskalny J, Skottner A, Moses AC (1978): Purified human somatomedin A and rat multiplication-stimulating activity (MSA): Mitogens for cultured fibroblasts that cross-react with the same growth peptide receptors. Eur J Biochem 82:5-12.

Rechler MM, Zapf J, Nissley SP, Froesch ER, Moses AC, Podskalny JM, Schilling EE, Humbel RE (1980): Interactions of insulin-like growth factors I and II and multiplication stimulating activity with receptors and serum carrier proteins. Endocrinology 107:1451-1459.

Rechler MM, Nissley SP, King GL, Moses AC, Van Obberghen-Schilling EE, Romanus JA, Knight AB, Short PA, White RM (1981): Multiplication-stimulating activity (MSA) from the BRL-3A rat liver cell line: Relation to human somatomedins and insulin. J Supramol Struct Cell Biochem 15:411-444.

Rinderknecht E, Humbel RE (1976): Polypeptides with nonsuppressible insulin-like and cell-growth promotion activities in human serum: Isolation, chemical characterization, and some biological properties of forms I and II. Proc Natl Acad Sci USA 73:2365-2369.

Rubin JS, Mariz I, Jacobs JW, Daughaday WH, Bradshaw RA (1982): Isolation and partial sequence analysis of rat basic somatomedin. Endocrinology 110:734-740.

Smith GL, Temin H (1974): Purified multiplication-stimulating activity from rat liver cell conditioned medium: Comparison of biological activities with calf serum, insulin, and somatomedin. J Cell Physiol 84:181-192.

Spencer EM, Ross M, Smith B (1983): The identity of human insulin-like growth factors I and II with somatomedins A and C and homology with rat IGF-I and II. In Spencer EM (ed): "Insulin-like Growth Factors/Somatomedins. Basic Chemistry, Biology, and Clinical Importance." New York: Walter de Gruyter & Co (in press).

Stiles CD, Capone GT, Sher CD, Antoniades HN, Van Wyk JJ, Pledger WJ (1979): Dual control of cell growth by somatomedins and platelet-derived growth factor. Proc Natl Acad Sci USA 76:1279-1283.

Svoboda ME, Van Wyk JJ, Klapper DBN, Fellows RE, Grissom FE, Schlueter RJ (1980): Purification of somatomedin-C from human plasma: Chemical and biological properties, partial sequence analysis, and relationship to other somatomedins. Biochemistry 19:790-797.

Switzer RC, Merrill CR, Shifrin S (1979): A highly sensitive silver stain for detecting proteins and peptides in polyacrylamide gels. Anal Biochem 98:231-237.

Temin HM (1971): Stimulation by serum of multiplication of stationary chicken cells. J Cell Physiol 78:161-170.

Van Wyk JJ, Svoboda ME, Underwood LE (1980): Evidence from radioligand assays that somatomedin-C and insulin-like growth factor-I are similar to each other and different from other somatomedins. J Clin Endocrinol Metab 50:206-208.

Weidman ER, Bala RM (1980): Direct mitogenic effects of human somatomedin on human embryonic lung fibroblasts. Biochem Biophys Res Commun 92:577-585.

Yang, YW-H, Romanus JA, Acquaviva AM, Liu T-Y, Nissley SP, Rechler MM (1982): Intracellular biosynthetic precursor of multiplication stimulating activity (MSA) in a rat liver cell line. Program of the 22nd Annual Meeting, American Society of Cell Biology, 196A.

Zapf J, Froesch ER, Humbel RE (1981): The insulin-like growth factors (IGF) of human serum: Chemical and biological characterization and aspects of their possible physiological role. Curr Top Cell Reg 19:257-309.

APPENDIX
Preparation of Conditioned Medium

The BRL-3A cell line was started by Hayden Coon in 1968 from the liver of a normal 5-week-old female Buffalo rat [Nissley et al., 1977]. Dulak and Temin [1973a] described the purification of MSA from medium conditioned by this cell line. The BRL-3A cell line (ATCC CRL 1442) is available from the American Type Culture Collection (Rockville, MD). The BRL-3A cells require serum-containing medium for cell attachment but multiply almost as well in serum-free medium as in serum-containing medium [Nissley et al., 1977]. After reaching confluence the cells continue to multiply to form ridges (Fig. 12).

Dulak and Temin [1973a] grew the BRL-3A cells in Temin's modified Eagle medium with 20% (vol./vol.) tryptose phosphate broth: however, Dulak and Shing [1976] eliminated the tryptose phosphate broth and used Dulbecco's modified Eagle medium (4.5 g/liter glucose). Marquardt et al. [1981] (method B) used a modified Waymouth medium (SR-1027, Meloy Laboratories, Springfield, VA). The cells are maintained at 37°C in an atmosphere of 95% air–5% CO_2. The buffering capacity of Waymouth medium is lower than for Temin's or Dulbecco's modified Eagle medium, requiring a medium change every 2 days to avoid sloughing of the cell layer. In our experience the MW = 16,300 species is low in amount or absent in MSA preparations from Waymouth-conditioned medium when method B is used, raising the possibil-

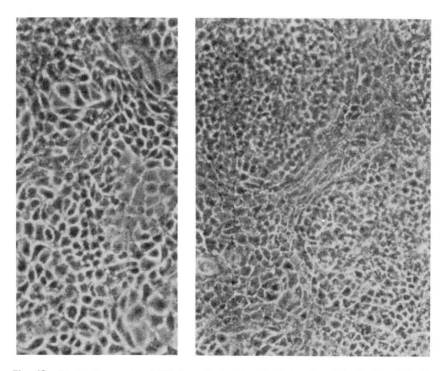

Fig. 12. Photomicrographs of BRL-3A cells at different cell densities. After the BRL-3A cells grow to confluence, the cells become arranged in ridges.

ity that the more acidic culture conditions promote the conversion of MSA I to the smaller MSA species.

The BRL-3A cells can be grown in dishes, flasks, or roller bottles, the latter being more convenient for large-scale collections [Dulak and Shing, 1976]. The cells are plated in medium containing 1-10% calf serum. On the following day the medium is changed to serum-free medium. Medium collections are begun after the cells have reached confluence. It is important to discard the first serum-free medium collection to avoid the presence of serum proteins in the medium to be used for MSA purification. Collections are made twice a week for cells maintained in Temin's or Dulbecco's modified Eagle medium and every 2 days for cells maintained in Waymouth medium. Since there is a lag in the time of appearance of biologically active MSA in the medium following medium change, and the amount of MSA in the medium increases linearly with time [Nissley et al., 1979], it is not advantageous to make more frequent medium collections.

We do not attempt to passage cells from these dense cultures from which media collections have been made. Instead, cultures are started from frozen cell stocks or from small-scale cultures that have been maintained in serum-free Ham's modified F-12 or Dulbecco's modified Eagle media and passaged when the cells are confluent. Cells can be detached from the plastic surface by exposure to Coon's CTC formulation (2% chicken serum, 0.075% trypsin, 7 U/ml collagenase) or 0.125% trypsin in TD buffer (NaCl, 8 g/liter; KCl, 0.038 g/liter; Na_2HPO_4, 0.1 g/liter; dextrose, 1 g/liter; Tris, 3 g/liter, pH 7.4). The cell monolayer usually detaches in a sheet, and a cell suspension is produced by pipetting.

Measurement of MSA; Stimulation of ^{3}H-Thymidine Incorporation Into DNA in Chick Embryo Fibroblasts

Preparation of cell cultures. All of the surgical instruments and glassware are autoclaved. The egg (12-day) is placed pointed end down in a holder. The top surface is swabbed with 70% ethanol and allowed to dry. The top of the egg is carefully broken open with the tip of a blunt-end forceps and the shell is removed in pieces. The underlying membrane is peeled away with a curved-end forceps. With another curved-end forceps, the embryo is raised by the neck and placed in a 100-mm plastic tissue culture dish. The embryo is decapitated and eviscerated. The carcass is transferred to a second tissue culture dish and is minced with two scalpels. The minced tissue is washed with 10 ml of TD buffer, the liquid is aspirated, and the tissue is washed a second time. Trypsin (0.125%) in 10 ml TD buffer is added, and the minced tissue is transferred to a trypsinization flask (Bellco #1995-00125, modified by placing a perforated glass disk in the side-arm) with a 5-ml plastic pipette from which the tip has been broken off. The tissue is stirred for 3-4 min on a magnetic stirrer at room temperature, the large pieces of tissue are allowed to settle, and the supernatant is decanted from the side-arm and discarded. Five milliliters of trypsin is added through the side-arm by forcing the liquid through each of the perforations in the side-arm disk. After stirring for 3-4 min, the supernatant is decanted into a 50-ml polypropylene conical tube containing 10 ml of Temin's modified Eagle medium with 20% (vol./vol.) tryptose phosphate broth and 10% fetal calf serum (growth medium). (Tryptose phosphate broth is prepared as for bacteriologic use, which includes autoclaving; a 20% [vol./vol.] solution is then prepared with Temin's or Dulbecco's modified Eagle medium.)

The digestion is repeated with two 5-ml aliquots of trypsin, and the supernatants are added to the same 50-ml tube. The tube is centrifuged for 10 min at 2,000 rpm, the supernatant is discarded, and the pellet is resuspended in 8 ml of growth medium. The suspension is forced through a metal Swinny filter to break up cell clumps. To each of eight 100-mm tissue culture dishes is added 15 ml of growth medium and 1 ml of the cell suspension. The cultures are incubated at 37°C in an atmosphere of 95% air-5% CO_2. (Dulbecco's modified Eagle medium [4.5 g/liter glucose] can be substituted for Temin's modified Eagle medium, but 20% [vol./vol.] tryptose phosphate broth is required for optimal growth.)

Two days later the primary cultures are trypsinized by first washing the monolayer with TD buffer and then adding 5 ml of 0.125% trypsin in TD buffer. After approximately 15 min at 37°C the cell suspensions are transferred to 50-ml tubes containing growth medium. After low-speed centrifugation, the pellets are resuspended in growth medium and distributed to 20 100-mm dishes.

The cells are passaged for a second time 2 days later in a similar fashion except that the cells are plated at a density of 1×10^6 cells per 2 ml per well in 35-mm Linbro cluster dishes (Flow Laboratories) in growth medium containing 0.25% fetal calf serum. The cells are conveniently dispersed by attaching a 5-ml glass Cornwall syringe by rubber tubing to a glass tube introduced through one of the side-arms of a cell suspension culture bottle (Bellco #1769). The entire unit can be autoclaved. Sufficient cells for at least 100 wells are usually obtained from a single embryo. These tertiary cultures become serum-starved and synchronized in G_1 of the cell cycle, and they can be used for ^{3}H-thymidine incorporation assay 3-5 days later.

Measurement of ^{3}H-thymidine incorporation into DNA [Temin and Dulak, 1973a]. The medium is aspirated from the wells, and samples are added in 1 ml of serum-free medium without tryptose phosphate broth. The cultures are incubated for 11-13 h. At the end of the incubation, the medium is aspirated and 1 ml of minimal essential Eagle medium containing 0.2 μCi/ml of ^{3}H-methyl-thymidine, 10-21 Ci/mmole (Schwarz-Mann, Orangeburg, NJ) is added to each well. After 1 h incubation, the medium is aspirated and the cell monolayer is washed twice with 2 ml of cold phosphate-buffered saline, twice with 2 ml of cold 10% trichloroacetic acid, and once with ethanol-ether (3:1). The wells are air-dried and the precipitated cellular residue (including DNA) is solubilized with 1.5 ml of 0.2 N NaOH. After 15 min, a 1-ml aliquot is removed from each well and added to a counting vial containing 10 ml of Aquasol (New England Nuclear, Boston, MA) and 100 μl of glacial acetic acid. Radioactivity is measured in a scintillation counter.

Other Assays for MSA

MSA can also be measured during the purification procedure by radiore-ceptor assays [Rechler et al., 1980; Daughaday et al., 1982], a competitive protein-binding assay that utilizes IGF-binding proteins [Moses et al., 1979; Rechler et al., 1980], and radioimmunoassay [Moses et al., 1980b]. The radioreceptor assays utilize purified membrane preparations or intact cells as sources of receptor, and the competitive protein-binding assay uses IGF-binding proteins found in serum. The membrane preparations (rat liver, rat placenta) can be stored frozen for months, and rat serum can be used directly as source of binding protein. These assays require a highly purified MSA preparation for use as radioligand, and in the case of the radioimmunoassay antiserum directed against MSA is also essential.

Both method A and method B utilize gel filtration in 1 M acetic acid to dissociate and separate MSA from the binding protein present in the BRL-3A-conditioned medium. When the radioreceptor assays and the radioimmu-noassay are used to measure MSA in column fractions, it should be recognized that the stripped binding protein will behave as apparent activity in the assay because the stripped binding protein competes with the receptor or antibody for binding to radioligand.

For the laboratory with a continuing requirement for large quantities of MSA, one of these alternative assays may be less time-consuming than the chick embryo fibroblast ^{3}H-thymidine incorporation assay described above. The reader is referred to the literature for detailed descriptions of these assays [Rechler et al., 1980; Daughaday et al., 1982; Moses et al., 1979, 1980b].

Polyacrylamide Gel Electrophoresis (PAGE)

An acid-urea disc acrylamide gel electorphoresis system described by Neville [1967] is used to assess the purity and relative amounts of the different

MSA species during the purification. The PAGE system is run at pH 2.7 in the presence of 9 M urea in a 12.5% acrylamide gel.

A slab gel unit such as a Hoefer SE 500 is more convenient than a system in which the gel is formed in tubes. The stock solutions are listed in Table II. The 10 M urea is deionized on a mixed bed ion exchange resin (Bio-Rad AG501-X8) just before use. The stock solutions A–E can be stored at 4°C for at least a month. The formulations for the upper and lower buffers and the stacking and running gels are given in Table III. The volumes given are suitable for two 10-cm × 14-cm gels in the Hoefer SE 500 slab gel system. The ammonium persulfate (AP) is freshly made (70 mg/10 ml 10 M urea).

The running gel (approximately 8 cm in height) is poured with a pipette or syringe. A fluorescent lamp is used to aid the polymerization. A layer of isobutanol is applied to the top of the gel to form a sharp boundary. After the running gel has hardened, the butanol layer is removed with a syringe and the top of the gel is rinsed with water.

Before pouring the stacking gel, TEMED (N,N,N′,N′-tetramethylethylene-diamine) is added to the stacking gel mix by transferring 95 μl of the mix into a separate tube, adding 5 μl of TEMED, and then returning 56 μl of the diluted TEMED to the stacking gel mix. The 1.5-mm comb is inserted between the glass plates, and the stacking gel (1.5–2.0 cm in height) is added with a syringe. Again the UV lamp is used to aid in polymerizing the gel. Fresh upper and lower buffers are prepared, the top of the hardened stacking gel is rinsed with upper buffer, and the wells are filled with upper buffer. The samples are dissolved in 25 μl of tracking dye solution. (Because of the small volume, it is important to have the dried sample concentrated in the tip of the tube; this can be accomplished conveniently with a Savant spin-concentrator in which the liquid is evaporated under vacuum while the sample is centrifuged to keep the remaining sample in the tip of the tube.) Electrophoresis is carried out at 10–12 mA per gel through the stacking gel and then at 15–17 mA through the running gel.

Coomassie blue staining method. Protein bands are precipitated in the gel with 12.5% trichloroacetic acid (20–30 min), and the gels are stained overnight with Coomassie blue G-250 (Bio-Rad) solution prepared as follows: 2 ml of 0.25% solution in water is dissolved in 40 ml of 12.5% trichloroacetic acid. The gels are destained and stored in 1 M acetic acid.

Silver staining method. The silver staining method is generally more sensitive than the Coomassie blue method except for MSA III polypeptides, which stain equally well by either method. The procedure we employ is a modification of a method described by Switzer et al. [1979]. The water used in this method is filtered through the Milli-Q Water Filtration System (Millipore

Corp.). Gel protein bands are precipitated with 15% trichloroacetic acid overnight, and the gels are washed three times in water for 15 min with agitation. The gels are fixed in 10% glutaraldehyde (American Scientific Products, McGaw Park, IL) for 30 min and are washed three times in water for 20 min. The gels are stained for 10–15 min in a diamine solution prepared as follows: mix 46 ml 0.36% NaOH and 3.1 ml fresh concentrated NH_4OH; add 9 ml 20% $AgNO_3$ (fresh) and mix until the precipitate dissolves; then add 165 ml 20% ethanol. The gels are then rinsed three times with water for 15 min with gentle agitation and developed in a solution prepared as follows: 2.5 ml 1% citric acid (fresh) and 2.5 ml 3.7% formaldehyde are added to 50 ml absolute ethanol, then diluted with water to make 500 ml. The gels are left in this solution until adequately developed (approximately 1–5 min), the solution is drained, and the gels are destained in 1 M acetic acid.

Methods for Preparation of Media, Supplements, and Substrata
for Serum-Free Animal Cell Culture, pages 139–145
© 1984 Alan R. Liss, Inc., 150 Fifth Avenue, New York, NY 10011

6
Preparation of Guinea Pig Prostate Epidermal Growth Factor

Jeffrey S. Rubin and Ralph A. Bradshaw

Epidermal growth factor (EGF) has been identified in and isolated from the guinea pig prostate gland, following the discovery of nerve growth factor (NGF) in the same tissue [Harper et al., 1979], and a preparative scheme that provides for the simultaneous purification of both EGF and NGF has been devised [Rubin, 1983]. The procedure makes use of gel filtration chromatography at an early stage to separate the factors, adopted for the isolation of EGF and NGF from the mouse submandibular gland [Rose et al., 1975]. The homogeneity of the guinea pig prostate EGF has been established by gel electrophoresis and partial amino acid sequence analysis [Rubin, 1983].

MATERIALS AND METHODS
Material

Prostate glands from sexually mature guinea pigs were obtained from Pel-Freez. The tissue is stable for at least 3 months when stored at approximately −20°C. Occasionally preparations have been made from a combination of prostate and coagulating glands without an appreciable effect on overall yield or purity. Sephadex gel filtration resins, G-100-120 and G-50-40, were obtained from Sigma Chemical Co., and the ion exchange resin, DE-52, was

Department of Biological Chemistry, Washington University School of Medicine, St. Louis, Missouri 63110

R.A. Bradshaw is now at the Department of Biological Chemistry, California College of Medicine, University of California at Irvine, Irvine, California 92717.

from either Sigma or Whatman. All chemical reagents were analytical grade. Spectropor dialysis tubing No. 3, molecular weight cutoff 3,500 (Spectrum Medical Industries, Inc.) was used throughout the protocol wherever dialysis is indicated.

Bioassays

Mitogenesis bioassays were performed according to the methods of Lieberman et al. [1980] or Thomas et al. [1980]. The protocols for the assays are similar: 1) Mouse 3T3 fibroblasts are plated at a sparse density and allowed to grow in the presence of serum for a few days; 2) the cells are washed and maintained for approximately 2 days longer in media containing low levels of serum; 3) the samples to be assayed are added to the media and several hours later ^{3}H-thymidine is introduced; 4) after a specified period (4 h in the first assay, 25 h in the second), the cells are harvested and the amount of ^{3}H incorporated into trichloroacetic acid-precipitable material is measured. The bioassays are calibrated by measuring the stimulation of ^{3}H incorporation at varying concentrations of mouse EGF, with a maximal effect attained at 1–10 ng/ml. Because of a plateau effect at higher concentrations and sensitivity at very low concentrations, an order of magnitude difference in concentration of two samples does not always show a substantial difference in biological activity unless serial dilutions are assayed. Therefore serial dilutions (10^3-fold to 10^5-fold) of selected samples are assayed to determine the position of maximal activity.

ISOLATION PROCEDURE
Preliminary Process

Purification is routinely performed with about 100 g of frozen mature guinea pig prostate tissue corresponding to 70–80 glands. Unless stated otherwise, all procedures are carried out at 4°C. The tissue is thawed and minced with scissors prior to the addition of 2–3 ml distilled water per gram gland and homogenization in a Waring blender. The blender is run for 15- to 20-sec bursts at increasing speeds until a smooth consistency is achieved, which usually requires about 2 min of operating time. Excessive vigor in blending, including use of a Polytron apparatus, can result in markedly reduced yields. Cellular debris is removed by centrifugation at 12,000g for 30 min. The pellet is resuspended in a similar volume of distilled water and the centrifugation is repeated. The supernatant is diluted with an additional 1 ml distilled water per gram gland and transferred to ice. Streptomycin sulfate (56 mg/g gland) at a concentration of 115 mg/ml in 0.1 M Tris, pH 7.5, is slowly added to the supernatant with constant stirring. After additional stirring on ice for 40 min, the contents are again centrifuged at 12,000g for 30 min. The supernatant is filtered through two layers of cheesecloth and lyophilized.

Sephadex G-100-120 Chromatography

The lyophilized powder is redissolved in 50-60 ml distilled water and centrifuged at 12,000g for 15 min to remove any undissolved material, and the supernatant is loaded on a Sephadex G-100-120 column (7.5 × 90 cm), equilibrated in 50 mM Tris, pH 7.5. The column is developed at a flow rate of 95 ml/h. The fractions containing EGF are pooled on the basis of their position in the elution profile as judged by the absorbance at 280 nm (see Fig. 1). (Radioimmunoassay analysis has established the position of EGF in the profile.) The pool is dialyzed for several hours against distilled water and lyophilized.

DE-52 Chromatography

The lyophilized powder is redissolved in 20 mM ammonium acetate, pH 5.6, and dialyzed against the same buffer. The dialysis bath is changed every few hours to expedite equilibration and minimize proteolysis. When the sample ionic strength is within 10% of the buffer ionic strength (note that the

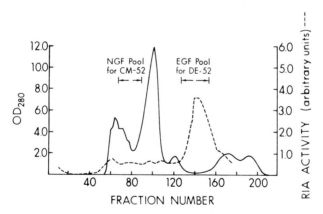

Fig. 1. Elution profile of the gel filtration of guinea pig prostate EGF on a column of Sephadex G-100-120. The processed homogenate was applied to the column (7.5 × 90 cm), equilibrated in 50 mM Tris, pH 7.5, and developed at 95 ml/h at 4°C. The fractions containing EGF were initially identified on the basis of immunocrossreactivity with a radioimmunoassay (RIA) using anti-mouse submandibular EGF and mouse ^{125}I-EGF as tracer (as indicated); subsequently fractions were pooled according to their position in the OD_{280} profile. The location of NGF is indicated to demonstrate the ease of separating the growth factors by this technique [Rubin and Bradshaw, 1981]. Fraction size was 20 ml.

sample will continue to have a higher ionic strength than the bath because of its high protein concentration) and the pH values are in agreement, the sample is applied to a column (1.6 × 20 cm) of DE-52, equilibrated in the dialysis buffer, and washed extensively at 12 ml/h with 20 mM ammonium acetate buffer until the A_{280} is near baseline (see Fig. 2). The EGF is eluted with a linear gradient of increasing ionic strength, generated by a double-chamber gradient, one chamber containing 200 ml of 20 mM and the other 200 ml of 200 mM ammonium acetate, pH 5.6. EGF is detected by its activity in a mitogenesis bioassay. While some mitogenic activity is routinely recovered in

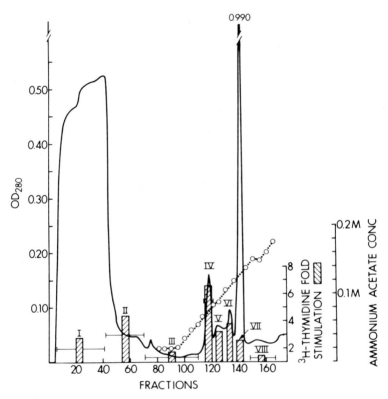

Fig. 2. Elution profile of guinea pig prostate EGF, partially purified by gel filtration, on a column (1.6 × 20 cm) of DE-52. Following addition of the sample to the column and extensive washing with the equilibration buffer, 20 mM ammonium acetate, pH 5.6, the sample was eluted with a linear gradient of increasing salt concentration prepared with 200 ml each of 0.02 and 0.2 M ammonium acetate, pH 5.6. The flow rate was 12 ml/h and the fraction size was 5.0 ml. While some mitogenic activity appeared in the breakthrough and wash, the largest distinct peak of activity (pool IV) emerged during the gradient elution.

the column breakthrough along with the bulk of protein, the largest peak of activity (Fig. 2, fraction IV) emerges from the resin approximately midway through the gradient.

Sephadex G-50-40 Chromatography

Typically the active pool contains varying amounts of higher molecular weight contaminants. This material is removed by gel filtration on a column of Sephadex G-50-40 resin (1.0 × 180 cm) equilibrated in 0.1 M ammonium bicarbonate, pH 7.9, operated at a rate of 4 ml/h at room temperature (see Fig. 3).

DISCUSSION

The protocol outlined above can be completed in two weeks and typically yields 2–3 mg EGF from 100 g prostate tissue. Owing to the reproducibility of the Sephadex G-100-120 chromatography, EGF-containing fractions can be pooled reliably on the basis of the A_{280} profile without requiring bioassays. However, variation in the A_{280} profile of the DE-52 column necessitated the use of the bioassay to determine the location of the EGF. The final product from the Sephadex G-50-40 gel filtration was also routinely tested in the bioassay to demonstrate its activity, although its position in the elution profile was highly reproducible.

The homogeneity of the preparation was suggested by a single band seen migrating near the dye front in basic gel electrophoresis. Amino-terminal

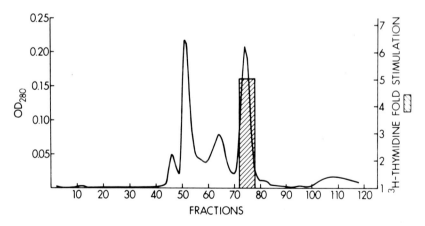

Fig. 3. Elution profile of the gel filtration of DE-52 cellulose fraction IV (Fig. 2) on a column (1.0 × 180 cm) of Sephadex G-50-40. The column was equilibrated in 0.1 M ammonium bicarbonate, pH 7.9, and developed at 4.0 ml/h at room temperature. The fraction size was 1.3 ml. The single active pool was composed of homogeneous guinea pig EGF.

sequence analysis yielded negative results, consistent with the subsequent identification of pyrollidone carboxylic acid at the amino terminus of the EGF molecule, presumably formed from the cyclization of an amino-terminal glutamine residue. The failure to detect any free N-terminal amino acids can be viewed as a measure of purity in its own right. Determination of most of the amino acid sequence of fragments generated by specific proteolysis rigorously established the homogeneity. These data also revealed an estimated 70% homology to the sequences of mouse and human EGF [Savage et al., 1972; Gregory, 1975]. Guinea pig and mouse EGF were also shown to have comparable potencies in a 3T3 fibroblast radioreceptor assay [Rubin, 1983].

The guinea pig EGF isolated by this procedure was consistently found to lack a carboxy-terminal arginine, a prominent feature of both the mouse and human EGF molecules. This residue is thought to be important for biosynthetic processing [Frey et al., 1979] as well as a site of interaction with the binding protein, an arginine esteropeptidase of the serine family [Server et al., 1976]. A similar situation occurs in mouse submandibular NGF [Angeletti and Bradshaw, 1971; Berger and Shooter, 1978]. However, in that case, recent evidence [Rubin and Bradshaw, 1981; Isackson, P.I., Rubin, J.S., Gross, K.W., Howles, P.N., and Bradshaw, R.A., manuscript in preparation] has shown that the arginine esteropeptidase of that complex (γ-subunit) is absent from the guinea pig prostate complex, suggesting that its postulated role in processing a precursor of the active subunit (β-subunit) is highly suspect.

With respect to prostate EGF, it should be noted that the EGF-binding protein has also not been detected in that tissue. However, cDNA probes have not been used in this case to show the absence of the mRNA, and des-arginine forms of mouse EGF have been isolated that are presumably products of limited proteolysis. Thus, the simple absence of the C-terminal arginine should not be taken as evidence that it will not be found in longer sequences and that it does not provide a site for precursor processing. In fact, the mitogenic material in the breakthrough fraction of the DE-52 column may well represent such a form, since it has been shown to compete with mouse EGF in a 3T3 fibroblast radioreceptor assay [Rubin, 1983].

One attempt was made to improve the preparation of EGF from the guinea pig prostate, in a manner analogous to that devised for mouse submandibular EGF by Savage and Cohen [1972]. They observed that the elution of mouse EGF was significantly retarded on Bio-gel P10 resin, equilibrated in 0.05 N HCl-0.15 M NaCl, pH 1.5. This retention resulted in a major purification step. Unfortunately, EGF from guinea pig prostate is not retarded under these conditions, invalidating this approach.

ACKNOWLEDGMENTS

This study was supported by USPHS research grant NS10229. J.S.R. was a predoctoral trainee of USPHS training grant GM07200, Medical Scientist.

REFERENCES

Angeletti RH, Bradshaw RA (1971): Nerve growth factor from mouse submaxillary gland: Amino acid sequence. Proc Natl Acad Sci USA 68:2417-2420.

Berger EA, Shooter EM (1978): Biosynthesis of β nerve growth factor in mouse submaxillary glands. J Biol Chem 253:804-810.

Frey P, Forand R, Maciag T, Shooter EM (1979): The biosynthetic precursor of epidermal growth factor and the mechanism of its processing. Proc Natl Acad Sci USA 76:6294-6298.

Gregory H (1975): Isolation and structure of urogastrone and its relationship to epidermal growth factor. Nature 257:325-327.

Harper GP, Barde Y-A, Burnstock G, Carstairs JR, Dennison ME, Suda KL, Vernon CA (1979): Guinea pig prostate is a rich source of nerve growth factor. Nature 279:160-162.

Lieberman MA, Rothenberg P, Raben DM, Glaser L (1980): Effect of 3T3 plasma membranes on cells exposed to epidermal growth factor. Biochem Biophys Res Commun 92:696-702.

Rose SP, Pruss RM, Herschman HR (1975): Initiation of 3T3 fibroblast cell division by epidermal growth factor. J Cell Physiol 86:593-598.

Rubin JS (1983): Purification and characterization of three polypeptide growth factors from mammalian sources. Doctoral Thesis, Washington University, St. Louis, Missouri.

Rubin JS, Bradshaw RA (1981): Isolation and partial amino acid sequence analysis of nerve growth factor of guinea pig prostate. J Neurosci Res 6:451-464.

Savage CR Jr, Cohen S (1972): Epidermal growth factor and a new derivative: Rapid isolation procedures and biological and chemical characterization. J Biol Chem 247:7609-7611.

Savage CR Jr, Inagami T, Cohen S (1972): The primary structure of epidermal growth factor. J Biol Chem 247:7612-7621.

Server AC, Sutter A, Shooter EM (1976): Modification of the epidermal growth factor affecting the stability of its high molecular weight complex. J Biol Chem 251:1188-1196.

Thomas KA, Riley MC, Lemmon SK, Baglan NC, Bradshaw RA (1980): Brain fibroblast growth factor: Nonidentity with myelin basic protein fragments. J Biol Chem 255:5517-5520.

Methods for Preparation of Media, Supplements, and Substrata
for Serum-Free Animal Cell Culture, pages 147–158
© 1984 Alan R. Liss, Inc., 150 Fifth Avenue, New York, NY 10011

7
Purification of Human Epidermal Growth Factor From Urine

C. Richard Savage, Jr. and Robert A. Harper

Epidermal growth factor (EGF) is a single-chain polypeptide containing 53 amino acid residues that exhibits potent mitogenic activity for a variety of cell types both in vivo and in vitro. Much information is available concerning the mechanism of action and biological effects of EGF (for recent reviews, see Das [1982], Carpenter [1981], Gospodarowicz [1981], and Carpenter and Cohen [1979]). This molecule was first described by Cohen [1962], who isolated EGF from the submaxillary glands of adult male mice. Although EGF has been isolated from the rat [Moore, 1978] and human urine [Cohen and Carpenter, 1975; Gregory and Willshire, 1975], the most abundant source known is the submaxillary glands of the mouse, in which it is present in levels of up to 0.5% of the dry weight of the gland. Milligram quantities of EGF can be purified readily from the mouse (m-EGF); however this is not the case for human EGF (h-EGF).

Cohen and Carpenter [1975] first published a procedure for the purification of h-EGF, starting from crude concentrates of urinary proteins that were available commercially from a pharmaceutical company. Unfortunately, this material is no longer available as a source of h-EGF. That same year Gregory and Willshire [1975] published a procedure for the isolation of urogastrone from human urine. Urogastrone is a polypeptide hormone that causes inhibition of gastric acid secretion and is thought to be identical to h-EGF [Gregory and Willshire, 1975; Gregory, 1975]. However, this procedure involves the fractionation of up to 3,000 liters of human urine by a 12-step procedure that results in relatively low yields of h-EGF (3–5%).

Departments of Biochemistry (C.R.S., R.A.H.) and Dermatology (R.A.H.), Temple University School of Medicine, Philadelphia, Pennsylvania 19140

Recently Savage and Harper [1981] reported a new method for the purification of h-EGF from raw urine that incorporates only two chromatographic steps requiring detailed elution profile analysis. From 20 liters of urine, three biologically active forms of h-EGF can routinely be isolated by straightforward purification techniques. In addition, this procedure results in increased yields of h-EGF from relatively small quantities of urine that are readily manageable with common laboratory equipment. In this review we summarize the published procedure and present a modification that employs affinity chromatography using a rabbit antibody to h-EGF.

PURIFICATION OF h-EGF FROM URINE BY STANDARD TECHNIQUES
Radioreceptor Assay

This assay employs 24-multiwell tissue culture plates containing a confluent monolayer of human fibroblasts [Savage and Harper, 1981] as a source of EGF receptors. The assumption is made that h-EGF competes equally with radioiodinated m-EGF for binding to the receptors located on the plasma membrane of the cells [Carpenter and Cohen, 1976]. A typical standard curve (Fig. 1) shows that h-EGF-1, h-EGF-2, and m-EGF all compete effectively for binding of ^{125}I-m-EGF to the fibroblasts. This assay procedure permits the detection of a minimum of 0.5 ng EGF. The advantages of using multiwell plates instead of culture dishes are: 1) increased sensitivity; 2) faster, more efficient removal of unbound ^{125}I-m-EGF by washing; and 3) requirement of fewer cells for the assay. This assay was used to estimate the amounts of h-EGF present during the purification procedures.

Step 1. Adsorption to Bio-Rex 70. Twenty liters of adult human urine were acidified with 1 liter of glacial acetic acid and the pH of the mixture was adjusted to 3.0-3.3 by the addition of concentrated HCl. Twenty liters of urine contained a total of 1.82×10^6 absorbance units and 700-900 μg of h-EGF as determined by the radioreceptor assay (Table I). Bio-Rex 70 ion-exchange resin was suspended in water, and sufficient glacial acetic acid was added to lower to the pH to 3.1. The resin was then washed by decantation five times with five volumes of 5% acetic acid as 25°C. The settled resin (240 ml) was added to the 20 liters of urine and the mixture was stirred for 18 h at 4°C. The resin was allowed to settle for 2-4 h and the supernatant fraction was discarded. The resin was transferred into a 4-cm-diameter glass column and washed with 3.6 liters of 10^{-3}N HCl at 25°C. The wash was discarded and the adsorbed h-EGF was eluted with 960 ml of 1.0 M ammonium acetate that had been adjusted to pH 8.0 with concentrated NH_4OH. The eluate, containing 8,050 absorbance units and 800-900 μg h-EGF, was lyophilized.

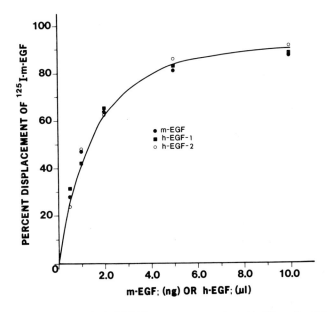

Fig. 1. Radioreceptor assay for h-EGF. The assay was conducted in 24-multiwell tissue culture plates using confluent monolayers of fibroblasts ($\sim 2 \times 10^5$ cells per well). The growth medium was removed and the binding reaction was conducted in Eagle's minimal essential medium containing 100 U/ml penicillin, 100 μg/ml streptomycin, 0.2% bovine serum albumin, and 25 mM HEPES buffer, pH 7.4. A standard curve was generated in a final volume of 0.2 ml of binding medium containing 0.5 ng of ^{125}I-labeled-m-EGF (9.6×10^4 cpm) and 0–10 ng of nonlabeled m-EGF. Also shown is the effect of increasing concentrations of purified h-EGF-1 and h-EGF-2. Since the absolute amount of protein present in such small quantities is difficult to determine, the concentrations of h-EGF are expressed in microliters. After 40 min at 37° the binding medium was removed and the cells were washed four times by flooding the entire plate with 100 ml per wash ice-cold Hanks' balanced salt solution, pH 7.4, containing 25 mM Hepes buffer. The plates were drained, the cells were dissolved in 1.0 ml of 10% NaOH for 20 min, and the amount of radioactivity bound was determined. In all cases, nonspecific binding was determined in the presence of 1,000-fold excess of nonlabeled m-EGF. This value, which was less than 6% of the total amount bound, was subtracted from each point.

Pepstatin (0.5 mg) was added and the dry residue was suspended at 25°C in 50 ml of a 2 mM aqueous solution of arginine containing 200 mg of crystalline bovine serum albumin. It should be noted that bovine serum albumin was added as a carrier protein to insure complete precipitation of the EGF by ethanol fractionation, described in the next step.

Step 2. Ethanol precipitation. To the suspension obtained from step 1, 500 ml of absolute ethanol was added with mixing. After standing for 30 min at 25°, the precipitate was collected by centrifugation and the clear brown

TABLE I. Purification of Human EGF From Urine

Step	Procedure	Total volume (ml)[a]	A_{280} (units/ml)[a]	Concentration of h-EGF (μg/ml)[b]	Total A_{280} (units)	Total h-EGF (μg)	Average specific activity (μg h-EGF per Abs. unit)[c]	Average yield (%)
	Urine	20,000	91.0	0.035–0.045	1.82×10^6	700–900	4.40×10^{-4}	100
1	Batch Bio-Rex 70	50[d]	161	16–18	8,050	800–900	0.106	>90
2	Ethanol precipitation	42[d]	96.0	19–21	4,030	800–900	0.211	>90
3	Passage over DE-52 cellulose	25[d]	50.1	28–32	1,250	700–800	0.600	>90
4	Bio-Gel, P-10 chromatography	7[d]	3.75	64–79	26.3	450–550	19.0	63
5	Adsorption to CM-52 cellulose	5[d]	0.63	70–100	3.15	350–500	135	53
6	DE-52 cellulose chromatography	h-EGF-1 1.0[d]	0.392	100–150	0.392	h-EGF-1 100–150	319	16
		h-EGF-2 1.0[d]	0.255	50–100	0.255	h-EGF-2 50–100	294	9

[a]One A_{280} unit is defined as the amount of material providing an optical density of 1.0 at 280 nm in a 1-cm light path. It should be noted that in the initial purification steps the material absorbing at 280 nm represents predominantly pigments and other nonprotein compounds.

[b]Determined by the radioreceptor assay, assuming that h-EGF and m-EGF compete equally for binding of ^{125}I-m-EGF to the membrane receptor on human skin fibroblasts.

[c]Specific activity is defined as μg h-EGF per absorbance unit at 280 nm.

[d]Indicates volumes after lyophilization and resuspension of the dry residue. This was necessary, since the presence of high concentrations of ammonium acetate interfere with the radioreceptor assay.

supernatant liquid was discarded. The pellet was suspended in 15 ml of 2 mM arginine and the pH was adjusted to 3.0 by the addition of concentrated HCl. The mixture was centrifuged and the clear dark brown supernatant fraction (42 ml) contained 800–900 μg h-EGF and 4,030 absorbance units. The pellet was discarded.

Step 3. Passage over DE-52 cellulose. A 4-cm $\times$ 14-cm column of DE-52 cellulose was prepared and equilibrated at 25°C with 0.05% formic acid, pH 3.1. The supernatant fraction from step 2 was warmed to 25°C and applied to the column. The column was washed with 530 ml of 0.05% formic acid, pH 3.1. Under these conditions h-EGF was not adsorbed to the cellulose. The column effluent was collected in a flask maintained on ice and then lyophilized. The residue was suspended in 20 ml of 0.05 N HCl at 4°C and the pH was adjusted to 1.5 by the addition of 2 N HCl. The mixture was clarified by centrifugation and the brown supernatant fraction (25 ml), containing 1,250 absorbance units and 700–800 μg h-EGF, was removed and the pellet discarded. Through this stage of purification greater than 90% of the starting amount of h-EGF was retained (Table I).

Step 4. Bio-Gel P-10 chromatography. The supernatant fraction (25 ml) obtained in step 3 was applied to a reverse-flow column of Bio-Gel P-10. Fractions were analyzed for h-EGF by the competitive binding assay, and the A_{280} was determined. A typical elution profile is shown in Figure 2. Most of the material absorbing at 280 nm was eluted in 1.5 column volumes; however, h-EGF did not begin to elute from the column before 1.7 column volumes. The fractions between the arrows were pooled, neutralized to pH 6 with concentrated ammonium hydroxide, and lyophilized. The dry residue was dissolved in 50 ml of 0.04 M ammonium acetate buffer, pH 3.9, and concentrated to 5 ml by pressure-ultrafiltration. The solution was desalted by dilution to 50 ml with the ammonium acetate buffer and concentrated again to 5 ml. This procedure was repeated again, resulting in 7 ml of a pale yellow solution that contained 26.3 absorbance units and 450–550 μg of h-EGF (Table I). This procedure is based on the observation that, at low pH, columns of polyacrylamide (Bio-Gel) are capable of selectively adsorbing m-EGF from crude homogenates [Savage and Cohen, 1972]. It is thought that the two adjacent tryptophanyl residues (res. 49–50) that are present in both mouse and human EGF are responsible for the selective adsorption to the gel.

Step 5. Adsorption to CM-52 cellulose. The solution of h-EGF from step 4 was applied to a 0.9-cm $\times$ 10-cm column of CM-52 cellulose equilibrated at 25°C with 0.04 M ammonium acetate, pH 3.9. The column was washed with 60 ml of the ammonium acetate buffer, and the adsorbed h-EGF was eluted from the column with 14 ml of 1 M ammonium acetate and

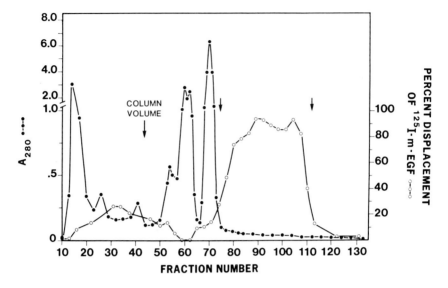

Fig. 2. Bio-Gel P-10 chromatography. A 4-cm × 90-cm column of Bio-Gel P-10 (100–200 mesh) was packed at 45 cm head pressure and equilibrated with 0.05 N HCl, pH 1.5, at 4°C. The column was operated with a 40-cm head pressure providing a flow rate of 42 ml per hour. Fractions (45 ml) were collected every 64 min.

lyophilized. The dry residue, containing 350–500 μg of h-EGF (3.15 absorbance units), was dissolved in 5 ml of 0.02 M ammonium acetate, pH 5.3, at 4°C.

Step 6. DE-52 ion exchange chromatography. The h-EGF from step 5 was applied to a 0.9-cm × 10-cm column of DE-52 cellulose equilibrated at 4° with 0.02 M ammonium acetate buffer, pH 5.3. The adsorbed h-EGF was eluted as described in the legend to Figure 3. Each fraction was analyzed for h-EGF, and A_{280} was determined. Two major peaks of h-EGF and one minor peak were observed. It is interesting that this pattern strongly resembles that obtained during the last purification step of m-EGF by DE-52 cellulose ion-exchange chromatography [Savage and Cohen, 1972]. The fractions in the first and third peaks were pooled separately and designated h-EGF-1 and h-EGF-2, respectively. The small minor peak was designated h-EGF-A. After lyophilization, the white residues were dissolved in 1.0 ml of distilled H_2O and stored at $-20°C$. As outlined in Table I, this method permits the isolation of 100–150 μg of h-EGF-1 and 50–100 μg of h-EGF-2 from 20 liters of raw urine with an overall yield of 20–30%.

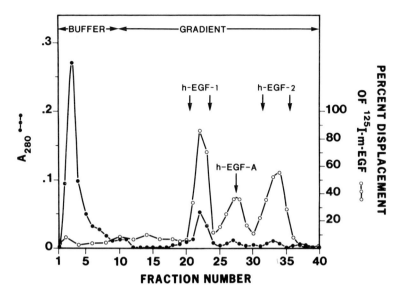

Fig. 3. DE-52 cellulose ion-exchange chromatography. A 0.9-cm × 10-cm column of DE-52 cellulose was prepared and equilibrated at 2°C with 0.02 M ammonium acetate buffer, pH 5.3. The flow rate was maintained at 8 ml per hour with a Buchler polystaltic pump. After the sample was loaded, the column was washed with 70-80 ml of the 0.02 M ammonium acetate buffer, pH 5.3. The adsorbed protein was eluted with a 0.02-0.3 M ammonium acetate gradient that was formed by allowing 0.3 M ammonium acetate, pH 5.3, to drip into a constant-volume mixing chamber containing 125 ml of 0.02 M ammonium acetate, pH 5.3. Fractions were collected every 30 min.

The electrophoretic mobilities in native polyacrylamide gels at pH 9.5 of h-EGF-1 and h-EGF-2 were compared to those of m-EGF and authentic h-EGF (Fig 4). Authentic h-EGF (h-EGF-SC) was obtained as a gift from Dr. Stanley Cohen of Vanderbilt University. Both forms of h-EGF had a greater electrophoretic mobility at pH 9.5 than m-EGF. Interestingly, the mobility of h-EGF-2 was identical to that of authentic h-EGF. We suspect that h-EGF-1 may be the native form of the hormone and h-EGF-2 may lack the COOH− terminal arg or leu-arg residue(s) [Savage and Harper, 1981]. All three forms of h-EGF isolated by the above procedure were found to be equal with respect to their ability to stimulate the growth of human skin fibroblasts in culture.

PURIFICATION OF h-EGF WITH ANTI-h-EGF AFFINITY COLUMN
Preparation of Affinity Column

Rabbit antibody to h-EGF was prepared by the injection of 200 μg of purified h-EGF with complete Freund's adjuvant into the footpads of a 90-

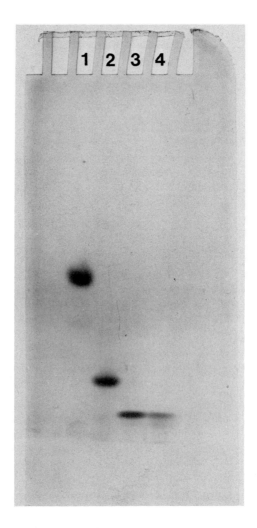

Fig. 4. Polyacrylamide gel electrophoresis of h-EGF-1 and h-EGF-2. Polyacrylamide gel electrophoresis was carried out in standard 7.5% slab gels at pH 9.5. The samples were electrophoresed at 25°C for 2.5 h at 20 mA per gel and the gels were subsequently stained for 10 min with Coomassie brilliant blue R250 (0.24% wt./vol. solution in methanol/H$_2$0/acetic acid, 5/5/1). The gel was destained by soaking in methanol/H$_2$0/ acetic acid, 4/33/3. Each lane contained 2–4 µg of protein as follows: lane 1, m-EGF; lane 2, h-EGF-1; lane 3, h-EGF-2; and lane 4, h-EGF-SC.

day-old female Albino rabbit. After 6 weeks the injection was repeated. At 4–6 weeks after the second injection, the animal was bled from the ear vein. The blood was allowed to clot at 4°C and the serum was removed and stored at −70°C. The gamma-globulin fraction containing antibody to h-EGF was precipitated at 25°C from 10 ml of antiserum by the addition of 5 ml of a solution of saturated ammonium sulfate adjusted to pH 7.4. The pellet was collected by centrifugation and dissolved in 10 ml of phosphate-buffered saline, pH 7.4, and the precipitation was repeated. The precipitate was dissolved in 5 ml of 0.1 M borate-saline buffer, pH 8.3, and exhaustively dialyzed for 2 days against same buffer with daily 1-liter changes. The dialyzed antibody fraction was clarified by centrifugation, and solid NaCl was added to provide a final concentration of 0.5 M.

Two grams of cyanogen bromide-activated Sepharose 4B were rapidly washed with ice-cold 1 mM HCl, and the gel was combined with the antibody to h-EGF. The mixture was rotated overnight at 4°C. The Sepharose was sedimented by centrifugation and treated at room temperature for 1 h with 1 M ethanolamine, pH 8.3, to block unreacted groups on the CN-Br-activated Sepharose. The Sepharose was collected by filtration and washed at 4°C with 20 ml of 0.1 M borate-saline buffer, pH 8.3. Next, the Sepharose was washed with three washing cycles (200 ml each) of buffer to remove noncovalently adsorbed protein. Each cycle consisted of a wash at pH 4.3 with 0.1 M ammonium acetate containing 1 M NaCl followed by a wash at pH 8.1 with 0.1 M borate buffer containing 1 M NaCl. After three cycles of washing, the gel was then equilibrated with 0.1 M borate-saline buffer, pH 8.3, followed by a wash with 0.01 M borate-saline buffer, pH 8.0. A column (about 7 ml) was prepared in a plastic 10-ml pipette. Prior to use, the antibody column was equilibrated at 25°C with 0.01 M borate buffer containing 0.15 M NaCl, final pH 8.0.

Step 1. Rabbit antibody affinity column. The starting material for the affinity chromatography was obtained by combining two preparations at the stage of lyophilization from step 3 (Table I). The dry residue containing about 1,500 μg h-EGF was dissolved in 60 ml of 0.01 M sodium borate buffer, pH 8.0, containing 0.15 M NaCl. The preparation at this point is extremely brown, owing to the large amount of melanin pigments present. The solution is clarified by centrifugation and loaded onto the anti-h-EGF affinity column ($\sim$0.8 $\times$ 17.5 cm) at a flow rate of 1 drop per 12 sec. The column was washed with two column volumes of the 0.01 M borate-saline buffer. During this wash all the remaining pigments were removed from the column. The column was then washed with three column volumes of 0.01 M ammonium acetate buffer, pH 8.0, followed by three column volumes of 1 M ammonium acetate buffer, pH 5.0. The 1 M ammonium acetate wash removes significant amounts of extraneous protein with little or no loss of EGF. The h-EGF was eluted from the column with 21 ml of 1 M acetic acid and lyophilized. The dry residue, containing 900-1,100 μg of h-EGF (Table II), was dissolved in 5 ml of 0.02 M ammonium acetate, pH 5.3, at 4°C.

The pattern of electrophoretic mobilities of the three forms of h-EGF, together with m-EGF, is shown in Figure 5. It can be seen that affinity chromatography yields a mixture of the three forms of h-EGF in a highly purified state.

Step 2. DE-52 cellulose chromatography. The three forms of h-EGF obtained from the above step can be separated by DEAE-cellulose chromatography (step 6, Table I) exactly as described in the legend to Figure 3. As seen in Table II, this method permits the isolation of 300-350 μg of h-EGF-1, 200-250 μg of h-EGF-2, and 150-200 μg of h-EGF-A from 40 liters of raw urine with an average yield of ~50%. All three forms of h-EGF are biologically active.

There are advantages to using affinity chromatography rather than older conventional methods. First, the large amounts of pigments and other colored compounds present in urine and urine concentrates are extremely heterogeneous mixtures that are difficult to remove through purification. The affinity chromatography efficiently removes essentially all of these and other contaminants and yields a mixture of the three active forms of h-EGF in a highly purified state. Second, since the affinity column replaces Bio-Gel P-10 chromatography (step 4) and adsorption to CM-52 cellulose (step 5, Table I), it permits easy processing of larger quantities of starting material in approximately half the time. Thus, this new procedure that employs affinity chroma-

TABLE II. Purification of h-EGF by Affinity Chromatography

Step	Procedure	Total volume (ml)	Concentration of h-EGF (μg/ml)[a]	Total h-EGF (μg)	Average yield (%)
1	Affinity chromatography[b]	30	30-37	900-1100	67
2	DE-52 cellulose chromatography	1.0[c]	300-350	h-EGF-1 300-350	22
		1.0[c]	200-250	h-EGF-2 200-250	15
		1.0[c]	150-200	h-EGF-A 150-200	12

[a]Determined by the radioreceptor assay.
[b]The starting material for the affinity column was obtained from step 3, Table I (~1,500 μg h-EGF). This amount of starting material represents two separate preparations that were combined.
[c]Indicates the volume after lyophilization and resuspension of the dry residue.

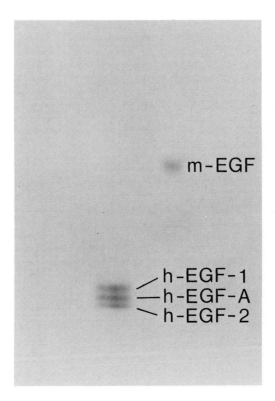

Fig. 5. Polyacrylamide gel electrophoresis of h-EGF-1, h-EGF-A, and h-EGF-2. Procedures were carried out as described in the legend to Figure 4. About 5 μg of each form of h-EGF and 3 μg of m-EGF were applied to the gel.

tography provides a rapid, convenient method for the isolation of relatively large quantities of h-EGF.

ACKNOWLEDGMENTS

The authors wish to thank Kirk Baumeister for carrying out preliminary studies concerning affinity chromatography. This research was supported by grants GM-29257, AM-25436, and AM-21545 from the National Institutes of Health.

REFERENCES

Carpenter G (1981): Epidermal growth factor. In Baserga R (ed): "Handbook of Experimental Pharmacology." New York: Springer-Verlag, Vol 57, pp 89-132.

Carpenter G, Cohen S (1976): ^{125}I-labeled human epidermal growth factor—binding, internalization, and degradation in human fibroblasts. J Cell Biol 71:159-171.

Carpenter G, Cohen S (1979): Epidermal growth factor. Annu Rev Biochem 48:193-216.

Cohen S (1962): Isolation of a mouse submaxillary gland: Protein accelerating incisor eruption and eyelid opening in the newborn animal. J Biol Chem 237: 1555-1562.

Cohen S, Carpenter G (1975): Human epidermal growth factor: Isolation and chemical and biological properties. Proc Natl Acad Sci USA 74:1317-1321.

Das M (1982): Epidermal growth factor: Mechanisms of action. Int Rev Cytol 78:233-256.

Gospodarowicz D (1981): Epidermal and nerve growth factors in mammalian development. Annu Rev Physiol 43:251-263.

Gregory H (1975): Isolation and structure of urogastrone and its relationship to epidermal growth factor. Nature 257:325-327.

Gregory H, Willshire IR (1975): The isolation of urogastrones—inhibition of gastric acid secretion—from human urine. Hoppe-Seyler's Z Physiol Chem 356:1765-1774.

Moore JB Jr (1978): Purification and partial characterization of epidermal growth factor isolated from the male rat submaxillary gland. Arch Biochem Biophys 189:1-7.

Savage CR Jr, Cohen S (1972): Epidermal growth factor and a new derivative: Rapid isolation procedures and biological and chemical characterization. J Biol Chem 247:7609-7611.

Savage CR Jr, Harper R (1981): Human epidermal growth factor/urogastrone: Rapid purification procedure and partial characterization. Anal Biochem 111:195-202.

Methods for Preparation of Media, Supplements, and Substrata
for Serum-Free Animal Cell Culture, pages 159–179

8
Isolation of Growth Factors From Human Milk

Yuen W. Shing and Michael Klagsbrun

Milk is a biological fluid that is important for the nutrition, growth, and development of the mammalian newborn. Milk, both human [Klagsbrun, 1978] and bovine [Klagsbrun and Neumann, 1979], contains mitogens that are capable of stimulating DNA synthesis and cell division in cultured cells. The growth-promoting properties of milk in vitro have been demonstrated by the ability of milk to replace serum as a supplement for the medium used for the growth of a variety of cultured cells including epithelial cells, normal and transformed fibroblasts, smooth muscle cells, and chondrocytes [Klagsbrun, 1980; Steimer and Klagsbrun, 1981; Steimer et al., 1981; Sereni and Baserga, 1981].

The biochemical characteristics of the mitogens in milk have been investigated by a number of laboratories. Klagsbrun [1978], using a bioassay that measured DNA synthesis and cell division in 3T3 cells, reported that most of the growth factor activity in human milk could be attributed to an acid- and heat-stable polypeptide(s) with molecular weight between 14,000 and 18,000 and an isoelectric point between 4.4 and 4.7. Subsequently, Carpenter [1980] demonstrated that antibodies to human urine-derived epidermal growth factor (EGF) could neutralize 70% of the mitogenic activity in human milk for 3T3 cells. This result suggests that most of the growth factor activity in human milk for 3T3 cells can be accounted for by EGF. More recently, it has been found that there are at least three growth facors in human milk that stimulate DNA synthesis in 3T3 cells [Klagsbrun and Shing, 1984]. Biochemical analysis of these three growth factors indicates that one appears to be structurally similar

Departments of Biological Chemistry and Surgery, Harvard Medical School and Children's Hospital Medical Center, Boston, Massachusetts 02115

to EGF, but the other two are not EGF-like polypeptides. In this chapter, we will describe the characterization of these three factors and a method for the purification of the EGF-like polypeptide in human milk.

CHARACTERIZATION OF GROWTH FACTOR ACTIVITY IN HUMAN MILK
Biological Assay for the Growth Factor Activity

Growth factor activity is assayed by measuring DNA synthesis in mouse BALB/c 3T3 cells maintained in Dulbecco's Modified Eagle Medium (DMEM) supplemented with 10% calf serum [Klagsbrun et al., 1977]. Aliquots of 200 μl containing about 10^4 cells are plated into 0.32-cm^2 microtiter plate wells. The 3T3 cells are incubated 5-10 days without medium change in order to deplete the serum of growth-promoting factors and thereby establish confluent monolayers of quiescent cells. Test samples of up to 50 μl are added to the cells along with 10 μl of ^{3}H-thymidine (6.7 Ci/mmole, final concentration of 4 μCi/ml). After an incubation of 30-40 h, the medium is removed and the cells are washed once in 0.15 M NaCl. Measurement of the incorporation of ^{3}H-thymidine into DNA is accomplished by the following successive steps: addition of methanol twice for periods of 5 min, 4 washes with water, addition of cold 5% trichloroacetic acid twice for periods of 10 min, and 4 washes with water. Cells are lysed in 150 μl of 0.3 N NaOH and counted in 4 ml of Insta-Gel liquid scintillation cocktail with a liquid scintillation counter. A unit of activity is defined as the amount of growth factor necessary to stimulate the half-maximum incorporation of ^{3}H-thymidine into the DNA of quiescent BALB/c 3T3 cells. Under the standard conditions described above, the maximal incorporation in a typical microtiter well that contains about 20,000 cells is about 100,000 cpm, whereas background incorporation is about 1,000 cpm. Alternatively, after stimulation with growth factor, it is possible to trypsinize the cells and count them in a Coulter counter to obtain the increase in cell number. Increases in DNA synthesis correlate directly with increases in cell number. Because of the ease in measuring DNA synthesis compared to cell number, measurement of DNA synthesis is the preferred method for measuring growth factor activity.

Sources and Storage of Human Milk

Human milk can be obtained from private donors, hospitals, or milk banks. Most of our milk samples were obtained from a milk bank directed by Dr. Cuberto Garza, Department of Pediatrics, Section on Nutrition and Gastroenterology, Baylor College of Medicine, Houston, TX 77030. General information about milk banks can be obtained from Dr. Thorsten A. Fjellstedt,

Health Scientist Administrator, Section on Nutrition and Endocrinology, Clinical Nutrition and Early Development Branch, Center for Research for Mothers and Children, National Institute of Child Health and Human Development, Room 7C-17, Landow Building, 7910 Woodmont Avenue, Bethesda, MD 20205.

The growth factor activity of human milk declines as the lactation period progresses. Although growth factor activity can still be detected 6 months after birth, it is generally higher in the first month after birth. Therefore, it is recommended that milk obtained in the first few months after birth be used as a source for the purification of milk-derived factors.

Human milk can be stored as whole milk or defatted milk at $-20°C$ for at least 6 months without appreciable loss of growth factor activity.

Preliminary Fractionation of the Growth Factor Activity in Human Milk

Human milk is centrifuged at 13,000g for 30 min at 4°C. Three fractions are obtained: fat floating on top of the centrifuge tube; cells and debris at the bottom of the tube; and the rest, which is defatted acellular milk. Most of the growth factor activity (more than 75%) is found in the defatted acellular milk and about 7% is found associated with the fat and cellular fractions. The rest of the growth factor activity cannot be accounted for. The ability of the defatted milk to stimulate DNA synthesis in 3T3 cells was measured and found to be dose-dependent (Fig. 1). Half-maximal stimulation is obtained with 0.4% (vol/vol) milk (about 0.1 mg/ml) and maximal stimulation is obtained with 1.2% (vol/vol) milk (about 0.3 mg/ml). Human defatted acellular milk contains about 250 units/ml of growth factor activity for 3T3 cells.

The defatted acellular milk is further fractionated by treatment with acid. Concentrated HCl is added to the defatted milk to obtain a pH of 4.3. The acidified milk is incubated at 37°C for 50 min and is subsequently centrifuged at 30,000g for 60 min to remove precipitate, mostly casein. After centrifugation, it is found that the supernatant fraction contains about 85% of the total growth factor activity of the defatted milk. This fraction is used for further purification and characterization of milk-derived growth factors.

Gel Filtration and Size Exclusion Chromatography of Human Milk-Derived Growth Factors

Approximately 1.5 g of protein, obtained from lyophilization of about 100 ml of acidified milk, was resuspended in 20 ml of equilibration buffer and was chromatographed at 4°C on a Sephadex G-100 column (5 cm × 90 cm) equilibrated with 0.1 M NaCl and 0.01 M sodium acetate, pH 4.3 (Fig. 2).

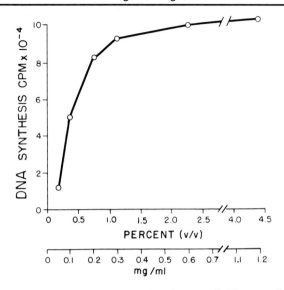

Fig. 1. Growth factor activity in defatted acellular human milk. Human milk obtained about 1 month post partum was centrifuged to remove fat and cells. Growth factor activity was measured by testing the ability of varying concentrations of human milk to stimulate DNA synthesis in 3T3 cells.

The elution was carried out at a flow rate of 40 ml/h, and fractions (18 ml) were collected and tested for the ability to stimulate DNA synthesis in 3T3 cells. Three peaks of growth factor activity with different molecular weights are found and designated as human milk growth factors (HMGF) I, II, and III [Klagsbrun and Shing, 1984]. The relative activities of HMGF I (MW greater than 100,000), HMGF II (MW 32–40,000), and HMGF III (MW less than 6,200) vary from batch to batch of milk sample. With a typical pooled milk sample as shown in Figure 2, the relative growth factor activities of HMGF I, II, and III account for about 5%, 20%, and 75% of the total growth factor activity for 3T3 cells in the acidified milk. However, with some individual samples, it has been found that HMGF III accounts for more than 95% of the total growth factor activity.

The molecular weights of HMGF I, II, and III were determined more accurately by high-performance liquid chromatography (HPLC) (Beckman Model 322 HPLC system) on TSK 2000 size exclusion columns (7.5 mm ID × 60 cm, Altex) equilibrated with 0.1 M ammonium sulfate and 0.05 M potassium phosphate, pH 7.0 (Fig. 3). HPLC is carried out at room temperature at a flow rate of 1 ml/min. Fractions of 0.85 ml are collected, dialyzed

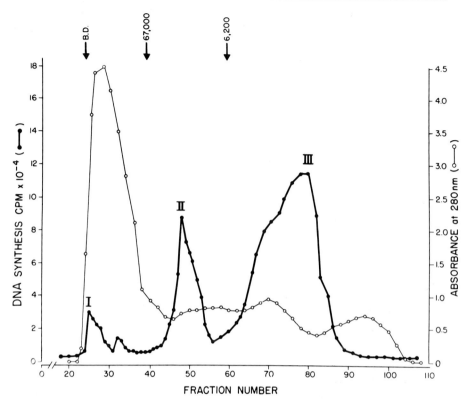

Fig. 2. Gel filtration chromatography of human milk. Defatted acellular milk (100 ml) was acidified to pH 4.3 and the precipitate formed was removed by centrifugation. The supernatant fraction was lyophilized. About 1.5 g of protein dissolved in 20 ml of equilibration buffer was applied to a Sephadex G-100 column (5 cm × 90 cm) equilibrated with 0.1 M NaCl and 0.01 M sodium acetate, pH 4.3. Fractions (18 ml) were collected, measured for absorbance at 280 nm, and tested for the ability to stimulate DNA synthesis in 3T3 cells.

against deionized distilled H_2O and tested for the ability to stimulate DNA synthesis in 3T3 cells. The columns are calibrated with the following molecular weight standards: Blue dextran (MW 2×10^6), β-galactosidase (MW 130,000), phosphorylase b (MW 94,000), ovalbumin (MW 43,000), α-chymotrypsinogen (MW 25,700), myoglobin (MW 17,800), and insulin (MW 5,800). The molecular weights of HMGF I (Fig. 3, top), II (Fig. 3, center), and III (Fig. 3, bottom) were determined to be 100,000–120,000, 30,000–34,000, and 5,000–6,000, respectively.

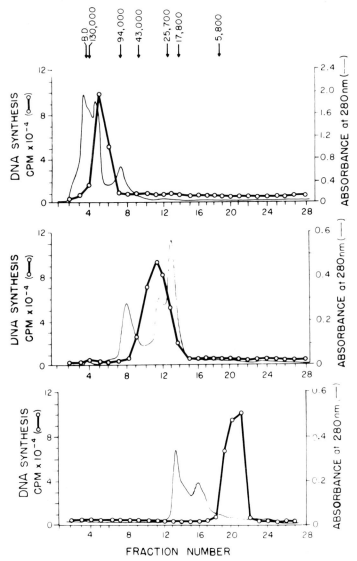

Fig. 3. Analysis of human milk growth factors by means of HPLC size exclusion columns. HMGF samples were prepared by Sephadex G-100 gel filtration as shown in Figure 2. About 8, 4, and 2 mg of HMGF I (top), II (center), and III (bottom), respectively, were resuspended in 100 μl of equilibration buffer and analyzed on TSK 2000 size exclusion columns equilibrated with 0.1 M ammonium sulfate and 0.05 M potassium phosphate, pH 7.0, at a flow rate of 1 ml/min. Fractions (0.85 ml) were collected, dialyzed against deionized, distilled H₂O, and tested for the ability to stimulate DNA synthesis in 3T3 cells.

Subunit molecular weights of HMGF I, II, and III were determined by HPLC on TSK 3000 size exclusion columns (7.5 mm ID × 50 cm, Varian) equilibrated with 6 M guanidine-HCl and 0.02 M 2-(N-morpholino)ethane sulfonic acid (MES), pH 6.5 (Fig. 4). In the presence of guanidine-HCl, the molecular weights of HMGF I (Fig. 4, top), II (Fig. 4, center), and III (Fig. 4, bottom) are 34,000-38,000, 12,000-16,000, and 5,000-6,000, respectively. HMGF I and II either form aggregates or are bound to large-molecular-weight carriers.

The sensitivity of HMGF I, II, and III to sulfhydryl reducing agent were determined by HPLC on TSK 3000 size exclusion columns equilibrated in 6 M guanidine-HCl, 5 mM dithiothreitol (DTT), and 0.02 M MES, pH 6.5 (Fig. 5). Under these conditions, HMGF I (Fig. 5, top) and II (Fig. 5, center) are inactivated irreversibly, but HMGF III (Fig. 5, bottom) remains active.

Isoelectric Point of Human Milk Growth Factors

Isoelectric points of HMGF I, II, and III isolated by Sephadex G-100 gel filtration (Fig. 2) were determined on an LKB 8100 vertical electrofocusing column with a capacity of 110 ml, as previously described [Klagsbrun, 1978; Klagsbrun and Smith, 1980] (Fig. 6). HMGF I and III are anionic polypeptides with isoelectric points between 4 and 5. HMGF II is resolved by isoelectric focusing into an anionic fraction with an isoelectric point between 3.2 and 4.8 and a more cationic fraction with an isoelectric point between 7.4 and 8.5. Whether HMGF II contains two distinct growth factors or whether there is one growth factor some of which is bound to carrier has not yet been determined.

Stability of Human Milk Growth Factors

Aliquots of HMGF I, II, and III were lyophilized and resuspended in the following solutions: 6 M guanidine HCl, 5 mM dithiothreitol, 1 M HCl (pH 1), and H_2O. After a 2-h incubation period at room temperature, all samples were dialyzed against H_2O exhaustively. The samples were lyophilized, resuspended in H_2O, and tested for the ability to stimulate DNA synthesis in 3T3 cells. None of the growth factors are inactivated by exposure to either guanidine HCl or pH 1. However, HMGF III differs from HMGF I and II by being resistant to inactivation by dithiothreitol. The stabilities of the three growth factors along with their molecular weights and isoelectric points are summarized in Table I.

PURIFICATION OF A HUMAN MILK-DERIVED EGF-LIKE POLYPEPTIDE
Isolation of Human Milk Growth Factor III

HMGF III is the major growth factor species in human milk for 3T3 cells and has been purified to homogeneity [Klagsbrun and Shing, 1984]. A flow

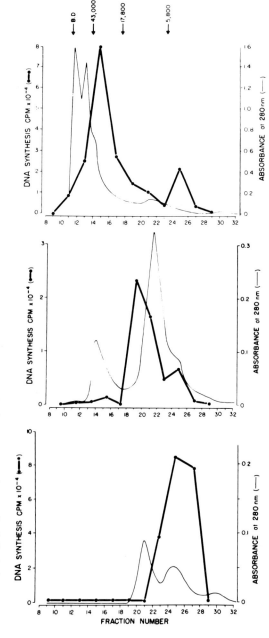

Fig. 4. Analysis of human milk growth factors by means of HPLC size exclusion columns in the presence of 6 M guanidine-HCL. HMGF samples were prepared by Sephadex G-100 gel filtration as shown in Figure 2. About 8, 4, and 2 mg of HMGF I (top), II (center), and III (bottom), respectively, were resuspended in 100 μl of equilibration buffer and analyzed on TSK 3000 size exclusion columns equilibrated with 0.02 M MES, pH 6.5, and 6 M guanidine-HCl at a flow rate of 1 ml/min. Fractions (0.85 ml) were collected, dialyzed against deionized, distilled H_2O, and tested or the ability to stimulate DNA synthesis in 3T3 cells.

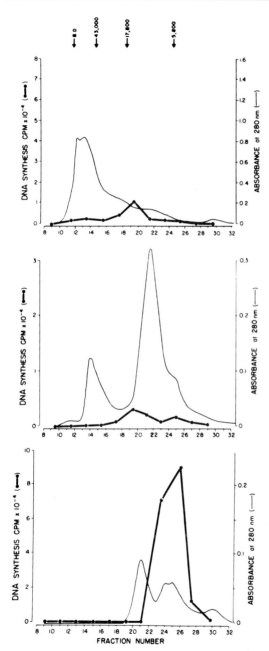

Fig. 5. Analysis of human milk growth factors by means of HPLC size exclusion columns in the presence of 6 M guanidine-HCl and 5 mM dithiothreitol. HMGF samples were prepared by Sephadex G-100 gel filtration as shown in Figure 2. About 8, 4, and 2 mg of HMGF I (top), II (center), and III (bottom), respectively, were resuspended in 100 μl of equilibration buffer and analyzed on TSK 3000 size exclusion columns equilibrated with 0.02 M MES, pH 6.5, in the presence of 6 M guanidine-HCl and 5 mM DTT at a flow rate of 1 ml/min. Fractions (0.85 ml) were collected, dialyzed against deionized, distilled H_2O and tested for the ability to stimulate DNA synthesis in 3T3 cells.

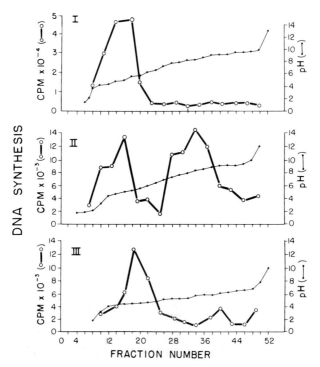

Fig. 6. Isoelectric focusing of human milk growth factors I (top), II (center), and III (bottom). The three biologically active peaks, obtained after Sephadex G-100 filtration chromatography as shown in Figure 2, were dialyzed against deionized, distilled H_2O, concentrated and analyzed by preparative isoelectric focusing with ampholytes in the 3.5–10 pH range. Fractions (2 ml) were collected, dialyzed against deionized, distilled H_2O, and tested for their ability to stimulate DNA synthesis in 3T3 cells.

chart for the purification of this growth factor is shown in Table II. About 110 ml of whole milk that had been stored at −20°C was thawed, defatted, and acidified as described on page 161. The supernatant fraction of acidified milk was chromatographed by gel filtration on Sephadex G-100 as described on pages 161–165. Three peaks of growth factor activity are obtained as shown in Figure 2. The one designated as human milk growth factor III, which contains approximately 75% of the growth factor activity for 3T3 cells recovered from the column, was further purified by anion exchange and size exclusion chromatography.

TABLE I. Characterization of Human Milk Growth Factors I, II, and III

	Growth factors		
	I	II	III
1. Percentage of total activity	5	20	75
2. Molecular weight in:			
a. Phosphate buffer, pH 7.0	100,000-120,000	30,000-34,000	5,000-6,000
b. Guanidine-HCl, pH 6.5	34,000-38,000	12,000-16,000	5,000-6,000
3. Isoelectric point	4.2-5.2	3.2-4.8 7.4-8.5	4.4-4.7
4. Inactivation by 6 M guanidine HCl	No	No	No
5. Inactivation by 5 mM DTT	Yes	Yes	No
6. Inactivation 1 M HCl, pH 1	No	No	No

HMGF I, II, and III were prepared by gel filtration chromatography on Sephadex G-100. Molecular weights were determined on HPLC TSK 2000 size exclusion columns under nondenaturing conditions (0.1 M ammonium sulfate, 0.05 M potassium phosphate, pH 7.0), and on HPLC TSK 3000 exclusion size columns under denaturing conditions (6 M guanidine hydrochloride, 0.02 M MES, pH 6.5). Isoelectric points were determined by preparative isoelectric focusing. Sensitivity to inactivation by 6 M guanidine-HCl, 5 mM DTT, and pH 1 was also tested.

Anion Exchange Chromatography

Pooled fractions of human milk growth factor III, prepared by Sephadex G-100 column chromatography as shown in Figure 2, were lyophilized and redissolved in 30 ml of water. The samples were dialyzed exhaustively against 0.01 M sodium acetate, pH 5.6, by means of Spectrapor tubing with molecular weight cutoff 3,500, and were subsequently applied to a DEAE-Sephacel column (2.5 cm × 32 cm) equilibrated at 4°C with the dialysis buffer. The column was washed with 100 ml of a NaCl gradient (0-0.1M) in 0.01 M sodium acetate, pH 5.6, at a flow rate of 20 ml/h. Growth factor activity was subsequently eluted from the column with 320 ml of a second NaCl gradient (0.1-0.8 M) and 200 ml of a third NaCl gradient (0.8-1 M) in the same buffer. Fractions (8 ml) were collected and tested for their ability to stimulate DNA synthesis in 3T3 cells. Virtually all of the growth factor activity adheres to the anionic exchanger (Fig. 7). When a gradient of NaCl (0.1-0.8 M) was applied to the column, two peaks of growth factor activity (a and b) were eluted. The major species (a) is eluted second at about 0.7 M NaCl and

TABLE II. Purification Scheme for Human Milk Growth Factor III

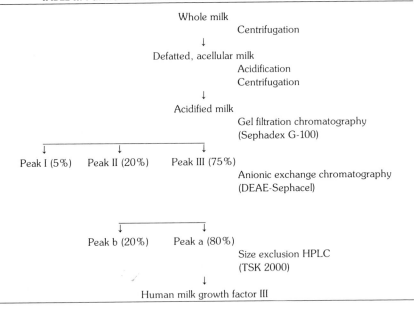

Whole milk
Centrifugation

↓

Defatted, acellular milk
Acidification
Centrifugation

↓

Acidified milk
Gel filtration chromatography
(Sephadex G-100)

Peak I (5%)   Peak II (20%)   Peak III (75%)
Anionic exchange chromatography
(DEAE-Sephacel)

Peak b (20%)   Peak a (80%)
Size exclusion HPLC
(TSK 2000)

↓

Human milk growth factor III

contains about 80% of the growth factor activity. Species b, which elutes several tubes earlier, contains about 20% of the growth factor activity.

Anion exchange HPLC can also be used for the purification of human milk growth factor. HMGF III (8 mg) in 2 ml of equilibration buffer was applied to a Synchropak AX 300 anion exchange column (4.1 mm ID × 250 mm, SynChrom) equilibrated with 0.01 M sodium acetate buffer, pH 5.6 (Fig. 8). The elution was carried out at a flow rate of 0.5 ml/min with a NaCl gradient (0–1 M) in the equilibration buffer over a period of 60 min. Fractions of 0.85 ml were collected and tested directly for the ability to stimulate DNA synthesis in 3T3 cells. Anion exchange HPLC of HMGF III is similar to that of DEAE-Sephacel chromatography in that two activity peaks are detected. Anion exchange HPLC appears to be the more efficient technique, showing better recovery (about 35%) than does DEAE-Sephacel chromatography (about 25%). However, for large-scale purification, DEAE-Sephacel chromatography is preferred because of its greater loading capapcity.

Size Exclusion Chromatography

HMGF III prepared from a peak obtained by DEAE-Sephacel chromatography as shown in Figure 7 was further purified by HPLC with the TSK 2000

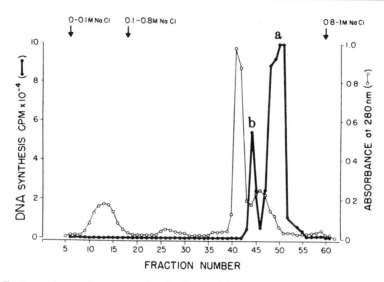

Fig. 7. Ion exchange chromatography of milk growth factor III. About 50 mg of protein in 30 ml, obtained from pooling fractions (68-84) in the peak labeled III shown in Figure 2, were applied to a DEAE-Sephacel anionic exchange column (2.5 cm × 32 cm) equilibrated with 0.01 M sodium acetate, pH 5.6. Three successive gradients of NaCl (100 ml 0.0-0.1 M, 320 ml 0.1 M-0.8 M, 220 ml 0.08 M-1.0 M) in 0.01 M sodium acetate, pH 5.6 were applied to the column. Fractions (8 ml) were collected, dialyzed against deionized, distilled H_2O, measured for absorbance at 280 nm, and tested for the ability to stimulate DNA synthesis in 3T3 cells.

size exclusion columns described on pages 161-165. HMGF IIIa migrates as a sharp peak of growth factor activity on the TSK-2000 column and has a molecular weight of about 5,000-6,000 (Fig. 9).

Homogeneity, Specific Activity, and Recovery of HMGF IIIa

HMGF III purified by DEAE-Sephacel ion exchange chromatography (Fig. 7) followed by HPLC on size exclusion columns (Fig. 9) was analyzed by SDS-polyacrylamide gel electrophoresis [Laemmli, 1970] (Fig. 10). A single sharp band with a molecular weight of about 6,000 is found after visualization with the highly sensitive technique of silver stain [Oakley et al., 1980]. HMGF IIIa (slot 1) migrates slightly slower than mouse EGF (slot 2).

Purified HMGF IIIa (Fig. 10, slot 1) shows a dose-dependent stimulation of DNA synthesis in 3T3 cells (Fig. 11). Half-maximal stimulation (1 unit of activity) occurs at a concentration of about 25 ng/ml. The purification scheme for HMGF IIIa is summarized in Table III. Starting with defatted acellular milk,

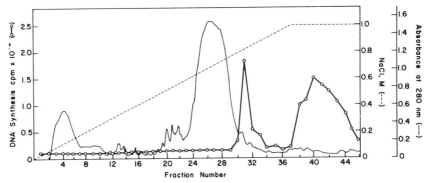

Fig. 8. Ion exchange HPLC of human milk growth factor III. About 8 mg of protein in 2 ml, obtained from pooling fractions (68–84) in the peak labeled III shown in Figure 2, were applied to a Synchropak AX 300 anionic exchange column (4.1 mm ID × 250 mm) equilibrated with 0.01 M sodium acetate, pH 5.6. A 30-ml gradient of NaCl (0–1 M) in 0.01 M sodium acetate, pH 5.6 was applied to the column over a period of 60 min. Fractions (0.85 ml) were collected at a flow rate of 0.5 ml/min, measured for absorbance at 280 nm, and tested for the ability to stimulate DNA synthesis in 3T3 cells.

TABLE III. Purification of Human Milk-Derived Growth Factor

	Total protein (mg)	Total activity (units[a])	Specific activity (units/mg)	Recovery of activity (percentage)	Fold purification
Defatted human milk, 100 ml	1,400	21,800	15.6	100	1
Acid precipitation, pH 4.3	1,100	18,000	16.4	83	1.1
Sephadex G-100, peak III	52	3,750[b]	72	17	4.6
DEAE-Sephacel, peak a	3.4	960[c]	282	4.4	18.1
HPLC, TSK-2000	0.0012 (1.2 μg)	172	143,000	0.8	9,200

[a]A unit of growth factor activity is defined as the amount of factor needed to stimulate half-maximal incorporation of methyl-^{3}H-thymidine into trichloroacetic acid-insoluble DNA. Under standard conditions of the bioassay, a microtiter well contains 20,000 confluent quiescent Balb/c 3T3 cells in a volume of 250 μl. Background incorporation is about 2,000 cpm and maximal incorporation is 100,000–120,000 cpm.
[b]An additional 800 units of activity are found associated with milk growth factors I and II.
[c]An additional 250 units of activity are found associated with peak b.

acid precipitation followed by Sephadex G-100 gel filtration chromatography, DEAE-Sephacel anion exchange chromatography, and HPLC on size exclusion TSK 2000 column results in a 9,200-fold purification. The yield of HMGF IIIa from 100 ml of defatted acellular milk, which contains about 1,400 mg of protein, is about 1.2 μg. Thus, the recovery is about 0.8%. Based on the yield and percentage of activity recovered, it can be estimated that the concentration of biologically active HMGF IIIa in human milk may be as high as 1 μg/ml.

Is HMGF III a Human EGF?

There are several lines of evidence to suggest that HMGF III is a human EGF: First, the molecular weight of HMGF III is similar to that of both mouse EGF and human EGF. HMGF III is a polypeptide with a molecular weight of about 6,000 as ascertained by measurement of biological activity on HPLC size exclusion columns (Fig. 9) and by mobility upon SDS-polyacrylamide gel electrophoresis (Fig. 10). The molecular weight of mouse EGF is 6,100

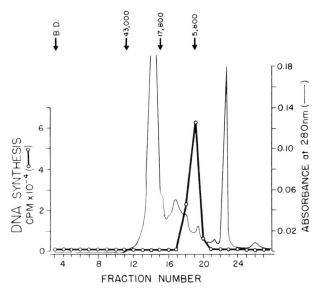

Fig. 9.   Size exclusion HPLC of human milk growth factor III. The major species of HMGF (3.4 mg of protein prepared from peak a by ion exchange chromatography as shown in Fig. 7) were applied to a HPLC TSK 2000 size exclusion column equilibrated with 0.1 M ammonium sulfate and 0.05 M potassium phosphate, pH 7.0. Fractions were collected, dialyzed against distilled H$_2$O, and tested for the ability to stimulate DNA synthesis in 3T3 cells.

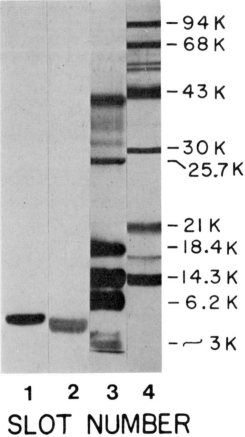

			−94K
			−68K
			−43K
			−30K
			⟍25.7K
			−21K
			−18.4K
			−14.3K
			−6.2K
			−⁓3K

1   2   3   4
SLOT NUMBER

Fig. 10. SDS-polyacrylamide gel electrophoresis of HMGF and EGF. The peak fraction of HMGF prepared by HPLC size exclusion column (Fig. 9) was analyzed by SDS-polyacrylamide gel electrophoresis. The polyacrylamide gels were devloped with a silver stain. Slot 1, human milk growth factor; slot 2, mouse EGF purified on HPLC TSK 3000 columns; slot 3, Bethesda Research Labs molecular weight standards, 400 ng/protein (Ovalbumin, α-chymotrypsinogen, β-lactoglobulin, lysozyme, bovine trypsin inhibitor, and insulin); slot 4: Bio-Rad molecular weight standards, 200 ng/protein (phosphorylase b, bovine serum albumin, ovalbumin, carbonic anhydrase, soybean trypsin inhibitor, and lysozyme).

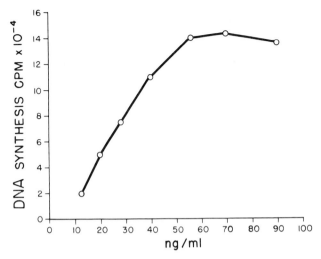

Fig. 11. Dose-dependent stimulation of DNA synthesis by purified HMGF. Various concentrations of a homogeneous preparation of human milk growth factor III (Fig. 10, slot 1) were tested for the ability to stimulate DNA synthesis in 3T3 cells. In this assay maximal incorporation, corresponding to 100% labeled nuclei, was about 150,000 cpm.

[Taylor et al., 1972], and the molecular weight of human EGF is 6,200 [Carpenter and Cohen, 1979]. The three polypeptide factors were directly compared by HPLC on TSK size exclusion columns. HMGF III comigrates with a highly purified sample of human EGF obtained from Dr. C.R. Savage, Temple University, Philadelphia, PA (Fig. 12, upper panel). The human EGF peak is broader than the HMGF III peak. Mouse EGF activity elutes from this column several fractions after HMGF III and in a region well below its known molecular weight of 6,100. Mouse EGF is apparently interacting with the TSK 2000 column and is being retarded to a great extent. However, when the HPLC TSK column is equilibrated with 6 M guanidine-HCl in order to suppress column-polypeptide interactions, HMGF III and mouse EGF comigrate, suggesting that they actually have the same molecular weight (Fig. 12, lower panel). A second line of evidence that HMGF III is a human EGF is that the isoelectric point of HMGF III is between 4.4 and 4.7 (Fig. 6), and the isoelectric point of mouse EGF and human EGF are 4.6 and 4.5, respectively [Hollenberg, 1979]. Furthermore, HMGF III can be resolved into two peaks of growth factor activity by anion exchange chromatography (Figs. 7 and 8). Both human EGF [Gregory, 1975; Savage and Harper, 1981] and mouse

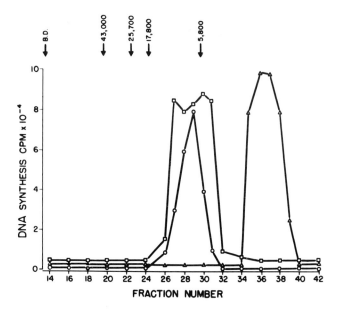

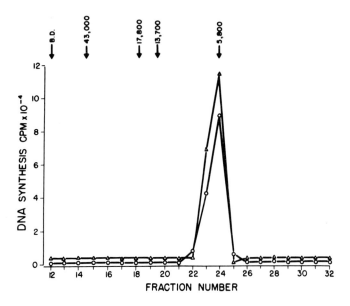

Fig. 12. Analysis of HMGF III, human EGF (urogastrone), and mouse EGF on HPLC size exclusion columns. Partially purified HMGF (4 mg, ○—○), human EGF (1 μg, □—□), and mouse EGF (20 μg, △—△) were analyzed on HPLC TSK size exclusion columns under nondenaturing and denaturing conditions. Upper panel: Composite of 3 TSK 2000 columns equilibrated in 0.1 M ammonium sulfate and 0.05 potassium phosphate, pH 7.0. Lower panel:

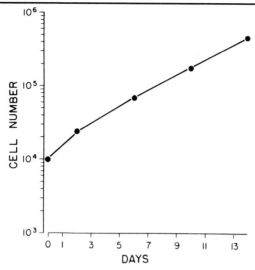

Fig. 13. Growth of Madin-Darby canine kidney (MDCK) epithelial cells in DMEM supplemented with human milk. MDCK cells were plated sparsely (10^4 cells/ml, 5×10^3 cells/cm^2) in DMEM. After 4 h, unattached cells were removed and the cultures were fed with DMEM supplemented with 1% human milk (1 month post partum) and 5 μg/ml transferrin. At 2, 6, 10, and 14 days after plating, cells were trypsinized and counted in a Coulter counter. The remaining cultures were refed.

EGF [Savage and Cohen, 1972] can be isolated by anion exchange chromatography in two forms. Finally, Carpenter [1980] has found that antibodies produced against human EGF isolated from human urine neutralize 70% of the mitogenic activity of defatted human milk for 3T3 cells. This value is consistent with our finding that about 75% of the mitogenic activity of human milk for 3T3 cells is due to HMGF III.

Thus, biological, structural, and immunological studies suggest that HMGF III is an epidermal growth factor. However, it should be pointed out that the

Composite of 2 TSK 3000 columns equilibrated with 6 M guanidine-HCl, 5 mM DTT, and 0.02 M MES, pH 6.5. The fractions (0.85 ml) for both types of columns were collected, dialyzed against deionized distilled H$_2$O, and tested for the ability to stimulate DNA synthesis in 3T3 cells. The volumes of fractions tested were adjusted to take into account the different number of units applied to various TSK columns.

identity of these growth factors can be ascertained only after sufficient material is available for amino acid analysis and sequencing.

SUMMARY AND CONCLUSION

Milk contains growth factors that stimulate DNA synthesis and cell division in 3T3 cells. There appear to be at least three growth factors for 3T3 cells in human milk. The predominant species, which represents about 75% of the total growth factor activity for 3T3 cells in human milk, is a polypeptide with a molecular weight of about 6,000 and a pI of between 4.4 and 4.7. This growth factor can be purified to homogeneity by a combination of acid precipitation, gel filtration, anion exchange chromatography, and HPLC on a size exclusion column. Biochemical and immunological studies indicate that this growth factor probably is a form of human EGF. The other two growth factors found in human milk differ from the EGF-like growth factor in that they have larger molecular weights and are inactivated irreversibly by treatment with dithiothreitol. It should be pointed out that the three growth factors are detected in the 3T3 cell bioassay. A different distribution of growth factor activity and even possibly other growth factors might be detected if other target cells such as epithelial cells were used to measure growth factor activity.

The presence of growth factors in human milk suggests that this biological fluid may be able to stimulate cellular proliferation. Recent studies indicate that milk can be used as a replacement for serum in medium used to support cell growth in culture. For example, Madin-Darby canine kidney epithelial (MDCK) cells, an established cell line that preserves the structure and function of kidney epithelium [Misfeldt et al., 1976], will grow readily in medium supplemented with human milk rather than with serum (Fig. 13). The ability of human milk to support cellular proliferation in vitro suggests that this might occur in vivo also. The stability of the three human milk growth factors to exposure to pH 1 and the presence of protease inhibitors in milk makes it plausible that these factors could survive transport through the gastrointestinal tract. Thus, milk-derived growth factors may play a role in the development of tissues in the newborn or perhaps in the development of the mammary gland of the mother. A major goal in this area of research would be to purify all of the human milk growth factors and study their possible involvement in the growth process.

REFERENCES

Carpenter G (1980): Epidermal growth factor is a major growth-promoting agent in human milk. Science 210:198-199.
Carpenter G, Cohen S (1979): Epidermal growth factor. Ann Rev Biochem 48:193-216.

Gregory H (1975): Isolation and structure of urogastrone and its relationship to epidermal growth factor. Nature 257:325-327.

Hollenberg MD (1979): Epidermal growth factor—urogastrone, a polypeptide acquiring hormonal status. Vitam Horm 37:69-110.

Klagsbrun M (1978): Human milk stimulates DNA synthesis and cellular proliferation in cultured fibroblasts. Proc Natl Acad Sci USA 75:5057-5061.

Klagsbrun M (1980): Bovine colostrum supports the serum-free proliferation of epithelial cells but not of fibroblasts in long-term culture. J Cell Biol 84:808-814.

Klagsbrun M, Neumann J (1979): The serum-free growth of Balb/c 3T3 cells in medium supplemented with bovine colostrum. J Supramol Struct 11:349-359.

Klagsbrun M, Shing YW (1984): Growth promoting factors in human and bovine milk. In Gordon G (ed): "Growth and Maturation Factors." Philadelphia: John Wiley and Sons (in press).

Klagsbrun M, Smith S (1980): Purification of a cartilage-derived growth factor. J Biol Chem 255:10859-10866.

Klagsbrun M, Langer R, Levenson R, Smith S, Lillehei C (1977): The stimulation of DNA synthesis and cell division in chondrocytes and 3T3 cells by a growth factor isolated from cartilage. Exp Cell Res 105:99-108.

Laemmli UK (1970): Cleavage of structural proteins during the assembly of the head of bacteriophage T4. Nature 227:680-685.

Misfeldt DS, Hamamoto ST, Pitelka DR (1976): Transepithelial transport in cell culture. Proc Natl Acad Sci 73:1212-1215.

Oakley BR, Kirsch DR, Morris NR (1980): A simplified ultrasensitive silver stain for detecting proteins in polyacrylamide gels. Anal Biochem 105:361-363.

Savage CR, Cohen S (1972): Epidermal growth factor and a new derivative: Rapid isolation procedures and biological and chemical characterization. J Biol Chem 247:7609-7611.

Savage CR, Harper R (1981): Human epidermal growth factor/urogastrone: Rapid purification procedure and partial characterization. Anal Biochem 111:195-202.

Sereni A, Baserga R (1981): Routine growth of cell lines in medium supplemented with milk instead of serum. Cell Biol Int 5:338-345.

Steimer KS, Klagsbrun M (1981): Serum-free growth of normal and transformed fibroblasts in milk: Differential requirements for fibronectin. J Cell Biol 88:294-300.

Steimer KS, Packard R, Holden D, Klagsbrun M (1981): The serum-free growth of cultured cells in bovine colostrum and in milk obtained later in the lactation period. J Cell Physiol 109:223-234.

Taylor JM, Mitchell WM, Cohen S (1972): Epidermal growth factor: Physical and chemical properties. J Biol Chem 247:5928-5934.

Methods for Preparation of Media, Supplements, and Substrata
for Serum-Free Animal Cell Culture, pages 181–194
© 1984 Alan R. Liss, Inc., 150 Fifth Avenue, New York, NY 10011

9
Purification of Type β Transforming Growth Factors From Nonneoplastic Tissues

Anita B. Roberts, Charles A. Frolik, Mario A. Anzano, Richard K. Assoian, and Michael B. Sporn

Transforming growth factors (TGFs) represent a relatively new addition to the set of hormone-like modulators of cell growth and cell phenotype (for a recent review see Roberts et al. [1983b]). Operationally, TGFs are defined by their ability to induce in nonneoplastic indicator cells the following phenotypic characteristics commonly associated with neoplastic behavior: 1) loss of density-dependent inhibition of cell growth in monolayer; 2) overgrowth of cells in monolayer; 3) change in morphology to resemble the neoplastic phenotype; and 4) acquisition of anchorage independence, with the resultant ability to grow in soft agar [Roberts et al., 1981]. All of these changes are reversible upon removal of the TGFs [De Larco and Todaro, 1978a] and thus represent phenotypic modulation. The property of anchorage-independent growth, which has been shown to have a particularly high correlation with neoplastic growth in vivo [Kahn and Shin, 1979; Cifone and Fidler, 1980], has been adopted for routine assay of this family of TGFs.

TGFs were first described by De Larco and Todaro [1978a], who isolated sarcoma growth factor from the conditioned media of Moloney murine sarcoma virus-transformed mouse 3T3 cells. Since that time, TGFs have been isolated from conditioned media and by direct extraction of many cells and tissues, both neoplastic and nonneoplastic, of several different genomes including mouse, rat, bovine, and human (reviewed in Roberts et al. [1983b]). The recent finding of TGFs in extracts of human and bovine blood platelets

Laboratory of Chemoprevention, National Cancer Institute, National Institutes of Health, Bethesda, Maryland 20205

[Childs et al., 1982; Assoian et al., 1983] is particularly relevant to the use of these factors for growth of cells under serum-free conditions.

It is now clear that TGF-dependent phenotypic transformation requires the concerted action of two distinct types of TGFs that can be distinguished on the basis of both biological and chemical properties [Anzano et al., 1982a; Roberts et al., 1983b]. Type α TGFs are mitogenic polypeptides that are characterized by their ability to bind to the epidermal growth factor (EGF) receptor and to induce the formation of large colonies (>62 μm diameter) in soft agar in the presence of TGF-β. This subset includes several well-characterized TGFs such as sarcoma growth factor [De Larco and Todaro, 1980], human melanoma TGF [Marquardt and Todaro, 1982], and rat sarcoma TGF [Twardzik et al., 1982a]. Since EGF, in the presence of TGF-β, can also fully potentiate the colony-forming response [Anzano et al., 1982a], it may also be classified as a TGF-α. Certain other TGFs derived from embryonic tissue [Twardzik et al., 1982b], possibly including "embryonic EGF" [Nexø et al., 1980], also appear to fit into this category. All of the characterized TGF-αs are single-chain polypeptides in the range of 6,000–7,000 MW, and all require intact disulfide bonds for activity. Members of the second subset, which are designated TGF-β, do not compete with EGF for receptor binding, are not potent mitogens, and require the presence of a type α TGF to induce expression of the transformed phenotype [Roberts et al., 1981, 1982; Anzano et al., 1982a]. Purified TGF-βs from human, mouse, and bovine genomes contain two polypeptide chains linked by disulfide bonds. Their behavior on SDS-polyacrylamide gel electrophoresis (SDS-PAGE) shows them to be approximately 23,000–25,000 MW under nonreducing conditions and 12,500 MW under reducing conditions [Assoian et al., 1983; Frolik et al., 1983; Roberts et al., 1983a]. Like TGF-α, TGF-β requires intact disulfide bonds for biological activity. Present evidence suggests that both TGF-α and TGF-β receptors must be occupied to induce expression of the neoplastic phenotype in the nonneoplastic cell and that the soft agar colony-forming activity can be regulated by control of the concentration of either TGF-α or TGF-β [Anzano et al., 1982a].

The finding of TGF-β in all tissues examined thus far, including blood platelets [Childs et al., 1982; Assoian et al., 1983], suggests that, in addition to its described role in the promotion of anchorage-independent growth in vitro, TGF-β might also play an essential role in normal cellular physiology. It is extremely potent on a molar basis: under optimal assay conditions where only TGF-β is of limiting concentration, it can induce colony formation in soft agar at concentrations of less than 0.1 ng/ml (10^{-11} to 10^{-12} M). In this chapter, we describe methods for purification to homogeneity of type β TGFs from a variety of sources.

ASSAYS FOR TGF ACTIVITY

Routine assay of the family of TGFs generally involves quantitation of colony growth in soft agar, although morphological and behavioral changes associated with the transformed phenotype can be observed in monolayer cultures as well. Type β TGF must be assayed in the presence of a type α TGF [Anzano et al., 1982a]; EGF (2-5 ng/ml) is usually used for this purpose. To date, no species selectivity has been observed within the TGF family; TGFs from all genomes studied thus far appear to be equally active in the assays.

Soft Agar Assay

Although several cell types including Rat-1 cells [Ozanne et al., 1980] and AKR-2B cells [Moses et al., 1981] have been used for assay of TGF activity, our laboratory routinely uses the NRK-2B, clone 49F cell line cloned by De Larco and Todaro [1978b]. These cells were obtained frozen at passage 3 after cloning and are typically carried no longer than passage 25 in culture. Cells are grown in a 5% CO_2 atmosphere in Dulbecco's modified Eagle's medium (DMEM, Gibco) supplemented with 10% calf serum (Gibco), penicillin (100 units/ml), and streptomycin (100 μg/ml). Subconfluent cultures are used for soft agar assay. Cells are trypsinized by washing the cell layer two times with trypsin (Gibco), removing all trypsin, and waiting 3-5 min for cells to round up; 10 ml supplemented media is immediately added and the cells are triturated 3-4 times before counting in a hemacytometer. Cells used for assay are diluted in media to 3×10^4 cells per milliliter. Samples to be assayed are aliquoted into sterile, capped 12 mm $\times$ 75 mm test tubes (Falcon), and lyophilized to dryness. Sample solutions in 1 M acetic acid, dilute hydrochloric acid, or organic solvents acidified to pH 2 (high-performance liquid chromatography fractions) do not support bacterial growth and require no filtration prior to testing.

For soft agar assay, 10-30 ml of a sterile stock solution of 5% agar (Difco, Noble Agar; prepared in 20- to 30-ml aliquots, autoclaved, and stored at room temperature) is melted in a boiling water bath and diluted 1:10 with supplemented medium maintained at 45°C. Base layers are made by pipetting 0.7 ml of the diluted agar (0.5%) onto 35-mm petri dishes. The lyophilized sample is dissolved in 0.2 ml 4 mM HCl containing 1 mg/ml bovine serum albumin as carrier (HB, sterilized by filtration); TGFs have been shown to be more soluble in acid solution than at neutral pH. When assaying TGF-β, EGF is added to the HB at a concentration of 20-50 ng/ml (the final concentration of both TGF-β and EGF in the assay will be 0.1 the initial concentration in this sample solution). Sequential additions of 0.6 ml 0.5% agar (at 45°C) and

0.2 ml cell suspension are combined with the sample (0.2 ml) and vortexed, and 0.7 ml of the resultant mixture (0.3% agar) is pipetted onto the hardened bottom layer. Plates are incubated in a humidified 5% CO_2 atmosphere at 37°C for 7 days without further feeding. Sterile staining solution, consisting of 2-(p-iodophenyl)-3-(p-nitrophenyl)-5-phenyltertrazolium chloride (0.5 mg/ml in water, autoclave to dissolve; stable to storage at room temperature for 1-2 months), is then layered over the agar and the incubation is continued for an additional 24 h [Schaeffer and Friend, 1976]. The plates may be scored directly [Roberts et al., 1980] or can be sealed in a plastic bag and stored at 4°C for 2-4 weeks before counting.

Colonies are counted and measured by a Bausch and Lomb Omnicon Image Analysis System with automatic stage. The field is displayed two-dimensionally, and all values are derived from the area of the colony cross section. The image analyzer is programmed to size colonies on a logarithmic scale from 43 to 107 μm diameter. A total of 5 cm^2 is analyzed for each plate. A unit of TGF-β colony-forming activity is defined as the ED_{50} of the dose-response curve of colonies greater than 62 μm in diameter when assayed in the presence of EGF.

Assay for Phenotypic Changes in Monolayer Growth

Cells, NRK clone 49F, are plated in 35-mm petri dishes at a density of 2 × 10^4 cells per 2 ml per dish, using DMEM supplemented as for the soft agar assay. Four h after plating, TGFs to be assayed are added in 25-50 μl HB. Growth and morphology are assessed at 3 days and at 5 days; replicate plates can be trypsinized and counted in a Coulter counter. The sensitivity of the assay in monolayer is similar to that observed in the soft agar assay. Again, only the combined action of TGF-α and TGF-β on the cell gives rise to overgrowth, loss of density-dependent inhibition of growth, and morphological changes associated with the transformed phenotype.

Assay for EGF-Binding Competition

This assay is employed to distinguish between TGF-α and TGF-β. TGF-α, by definition, competes for binding to the EGF receptor, but TGF-β does not. EGF is iodinated by a Chloramine T procedure [Carpenter and Cohen, 1976] to a specific activity of approximately 200 μCi/μg. Either NRK or mink lung epithelial cells (CCL 64) are seeded in 24-well 16-mm Costar plates at a density of 8 × 10^4 cells per millilter in DMEM, 10% calf serum. Approximately 24 h later, cells are used in a standard radio-receptor assay for binding competition with ^{125}I-EGF as described by De Larco and Todaro [1980]. As in other assays, TGFs to be tested are dissolved in a small volume of HB and then diluted appropriately in binding buffer (DMEM containing 100 mg bovine serum albumin and 1.1 g N,N-bis[2-hydroxyethyl]-2-aminoethane sulfonic acid per 100 ml, adjusted to pH 6.8). Assays are routinely carried out by simultaneous addition of the competing substance. The ^{125}I-EGF binding is measured after a 2 h incubation at room temperature.

PURIFICATION OF TYPE β TGFs

Procedures developed for purification of TGF-βs from cells and tissues are based on two important physical properties: 1) their stability under acidic conditions; and 2) their relatively low molecular weight (23,000-25,000). TGF treated with 1 M acetic acid for 30 h at room temperature retains full biological activity [De Larco and Todaro, 1978a; Roberts et al., 1980; Moses et al., 1981; Ozanne et al., 1980]. Therefore an acidic extraction procedure has been developed for isolation of these polypeptides from tissues and cells. Because of their low molecular weight, these proteins can be easily chromatographed on conventional high-performance liquid chromatography (HPLC) columns. Taking these and other properties into account, TGF-βs from human placenta and platelets and from bovine kidney have recently been purified to homogeneity [Frolik et al., 1983; Assoian et al., 1983; Roberts et al., 1983a].

Extraction

A modification of an acid-ethanol procedure, used for extraction of insulin from the pancreas [Davoren, 1962], has been employed for the extraction of TGF-βs from cells and tissues. The tissue (500-1,000 g human placenta or bovine kidney), which has been stored at $-70°$ or colder, is placed at $-20°$ overnight. It is then sliced into pieces and allowed to thaw in an acid-ethanol solution (4 ml/g tissue) containing 792 ml 95% ethanol, 192 ml water, and 16 ml concentrated HCl plus 53 mg of phenylmethylsulfonyl fluoride and 3 mg of pepstatin A as protease inhibitors. The tissue is homogenized in a commercial Waring blender and then stirred at room temperature for 2-3 h prior to centrifugation at 11,000g for 15 min. The resulting supernatant is adjusted to pH 3.0 with concentrated ammonium hydroxide. The protein is then precipitated with four volumes of cold anhydrous ether and two volumes of cold ethanol. The mixture is allowed to stand overnight at 4°C and the precipitate is collected by filtration through Whatman No. 1 paper and redissolved in 1 M acetic acid (1 ml/g tissue). Insoluble material is removed by centrifugation and the supernatant is lyophilized. The final residue is stored at $-20°$ until ready for subsequent column chromatography.

Human platelets (20-30 g) are extracted by a similar method. Platelet concentrates are initially centrifuged at 3,200g for 30 min at 0° to remove residual plasma proteins. The pellet is washed twice with 500-ml portions of Tris-citrate buffer, pH 7.5 as described by Antoniades et al. [1979] and centrifuged. The final platelet pellet is either stored in liquid nitrogen or directly extracted in the acid-ethanol solution (3 ml/g) in a Dounce homogenizer. The resulting homogenate is then carried through the extraction procedure detailed above.

The yield of crude residue from 1 g of tissue is 27 mg from human placenta, 13 mg from bovine kidney, and 4 mg from human platelets. For both the placenta and platelets, the extract contains 0.9–1.0 g of protein per gram of residue; for the kidney this figure is closer to 0.5 g per gram of extract. Measurement of the crude extract soft agar colony-forming activity in the presence of EGF showed an ED_{50} for TGF-β of 8–12 μg/ml for both placenta and kidney and a specific activity of 0.06–0.10 units/μg protein. The colony-forming activity of the platelet extract, on the other hand, is 100 times higher than the placenta and kidney in both the ED_{50} of the TGF (0.05–0.01 μg/ml) and the specific activity (10–20 units/μg). Most of the soft agar colony-forming activity of the crude extracts can be characterized as TGF-β; when the assay is performed in the absence of EGF, the ED_{50} increases by a factor of 150 to 1,000.

In experiments with crude extract from human placenta and bovine kidney, it was noted that if the concentration of protein in the soft agar assay is increased above 75–100 μg/ml, the number of large colonies decreases sharply. Extensive dialysis of the extract at 4° against 0.17 M acetic acid (Spectrapor tubing, 3,500 MW cutoff) removes the inhibitory substance. In the placenta and kidney extract both the ED_{50} and specific activity is unaffected by dialysis, only the activity at high concentration of protein showing a change. However in some cases, for example, in the crude extract of the human rhabdomyosarcoma A673 cell line, the soft agar colony-forming activity cannot be measured without prior dialysis. Dialysis of the extract of human platelets does not measurably affect the TGF-β activity.

Gel Filtration of TGF-β

Gel filtration has been used routinely as an early step in the purification of TGF-β from a wide variety of tissues. Acid-ethanol extracts of the tissues (see above) are dissolved in 1 M acetic acid and applied to columns of Bio-Gel (Bio-Rad Laboratories) that have been equilibrated in the sample solvent. The conditions for gel filtration of TGF-β from extracts of bovine kidney, human placenta, and human platelets are as follows:

1. *TGF-β from bovine kidney:* The solubilized extract (40–50 g protein in 1.5–2 liters of 1 M acetic acid) is applied to a sectional column (Pharmacia) of

Bio-Gel P-30 composed of five sections, each containing 16 liters (37 cm ×
15 cm). The sample is gel-filtered with upward flow at a rate of 1.2–1.5 liters/
h. Fractions containing 1 liter are collected and aliquots (50 μl) of alternate
fractions are assayed for TGF-β activity (Fig. 1A).

   2. *TGF-β from human placenta:* The solubilized extract (150 g protein in
3 liters of 1 M acetic acid) is gel-filtered at a downward flow rate of 1.6 liters/
h on a column (35.6 cm × 90 cm, Amicon) of Bio-Gel P-30. Fractions
containing 0.8 liter are collected and aliquots (50 μl) of alternate fractions are
assayed for TGF-β activity (Fig. 1B).

   3. *TGF-β from human platelets:* The solubilized extract (100 mg protein in
10 ml of 1 M acetic acid) is gel-filtered on a column (4.4 cm × 115 cm) of
Bio-Gel P-60 at a flow rate of 15–20 ml/h. Fractions containing 5 ml are
collected and 5-μl aliquots of every fifth fraction are subjected to TGF-β
bioassay (Fig. 1C).

   Figure 1 shows the elution profiles of acid-ethanol extracts from bovine
kidney (panel A), human placenta (panel B), and human platelets (panel C).
The profiles for the three different columns have been normalized by plotting
K_{av} (abscissa) vs. number of colonies per 5 cm^2 (ordinate). The extracts from
each tissue show a single peak of TGF activity. With a larger aliquot from the
fractions, the profile of placental TGF-β shows two peaks of biological activity
[Frolik et al., 1983]. Control studies demonstrate that the low level of activity
between the two peaks is due to an inhibitory substance that coelutes with
placental TGF-β. Dialysis of the extract prior to gel filtration will remove this
material, or it can be removed from placental TGF-β after gel filtration by
applying the Bio-Gel P-30 pool to a column of Bio-Gel P-6. For all three
elution profiles, little activity was detected when the assay was performed in
the absence of EGF, confirming that these TGFs are type β. Purified TGF-β
from each of these tissues has a molecular weight of 23,000–25,000 daltons
(see Fig. 2A), but, as is apparent in Figure 1, the protein is retarded during
gel filtration on Bio-Gel P-30 or P-60. Moreover, its apparent molecular
weight as determined by gel filtration can range from 6,000 daltons (panel B)
to greater than 15,000 daltons (panel C). The nature and amount of contaminating proteins as well as conditions of chromatography appear to modify the
elution characteristics of impure TGF-β.

   Pooled TGF-βs from the Bio-Gel columns have been analyzed by SDS-
PAGE and silver staining. At this stage of purification a distinct protein migrating at 25,000 daltons is not detectable in partially purified TGF-β preparations
from bovine kidney or human placenta. However, the Bio-Gel pool from
platelet extracts shows a minor component at 25,000 daltons as well as large
amounts of contaminating protein at 13,000 daltons (Fig. 1, panel C, inset).

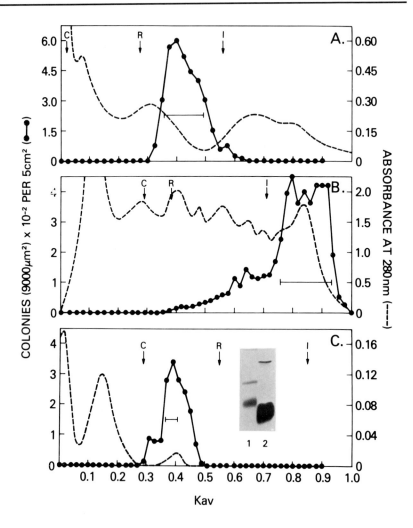

Fig. 1. Gel filtration of crude TGF-β. Panels A, B, and C show the profiles resulting from gel filtration of acid ethanol extracts of bovine kidney, human placenta, and human platelets, respectively. To normalize the column profiles, results are plotted as K_{av}. The dashed lines show the protein profiles as determined by absorbance at 280 nm, and the solid lines show the TGF-β activity. The vertical arrows indicate, from left to right for each panel, the elution positions of chymotrypsinogen A (C), RNase (R), and insulin (I), respectively. Pooled fractions are indicated by the horizontal bars. The inset to panel C shows the proteins present in the Bio-Gel P-60 pool of platelet-derived TGF-β as determined by SDS-PAGE and silver staining (lane 2). Lane 1 of the inset shows, from top to bottom, the positions of soybean trypsin inhibitor and cytochrome C, respectively.

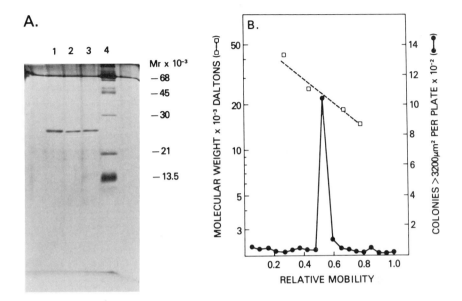

Fig. 2. SDS-polyacrylamide gel electrophoresis of purified type β TGFs. A. 50 ng each of purified TGF-β from bovine kidney (lane 1), human placenta (lane 2), and human platelets (lane 3) were run on a polyacrylamide gel (1.5 mm, 12.5% acrylamide) under nonreducing conditions and stained with silver (Upjohn Diagnostics). Molecular weight standards (lane 4) were bovine serum albumin (68,000), ovalbumin (45,000), carbonic anhydrase (30,000), soybean trypsin inhibitor (21,000), and cytochrome C (13,500). B. Human placental TGF-β (240 ng) was applied to a gel (1.5 mm, 12.5% polyacrylamide). After electrophoresis, gel slices of 2 nm were extracted and assayed for colony-forming activity as described in the text. ^{14}C-protein standards run in an adjacent well consisted of ovalbumin (43,000), α-chymotrypsinogen (25,700), β-lactoglobulin (18,400), and lysozyme (14,300).

The contaminants can be completely removed by gel filtration of the P-60 pool (about 5 mg protein per 20-30 units platelets; lyophilized from 1 M acetic acid and redissolved in 0.5 ml of 1 M acetic acid, containing 8 M ultrapure urea) on a column (1.6 cm × 85 cm) of Bio-Gel P-60 equilibrated in the sample solvent and run at 3 ml/h. (The presence of urea eliminates the aberrant elution of TGF-β during gel filtration.) As determined by SDS-PAGE of aliquots (25 μl) from selected column fractions (0.5 ml) the growth factor is separated from the lower-molecular-weight contaminants by 15-20 fractions. Pooled TGF-β (approximately 95% pure; 10 μg per 20-30 units of washed platelets) is dialyzed against 1 M acetic acid to remove urea and lyophilized in the presence of BSA prior to incubation with cells [Assoian et al., 1983].

Ion-Exchange Chromatography

Although platelet TGF-β is homogeneous after a second gel filtration step in the presence of urea, the human placental and bovine kidney TGF-β preparations require further chromatography to achieve homogeneity. The lyophilized fractions from the P-30 column (2–3 g protein) are redissolved in dilute (0.01–0.1 M) acetic acid (30–50 ml/g protein) by stirring overnight at 4°. The pH of the sample is then adjusted to 4.5 either with 1 N sodium hydroxide or by a 1:1 (vol./vol.) dilution with 0.1 M sodium acetate, pH 4.5. The conductivity of the sample is adjusted to 1.5 mS/cm, or lower, by dilution with water. The sample is then applied at a flow rate of 50 ml/h to a cation-exchange column (CM-Trisacryl M, LKB, 5 cm × 10 cm) equilibrated in 0.05 M sodium acetate, pH 4.5 (buffer A). The column is eluted with 100–300 ml of buffer A (100–150 ml/h) followed by a linear sodium chloride gradient to 0.7–0.85 M sodium chloride in buffer A at 0.8 mM/min (100–150 ml/h) collecting 20-ml to 30-ml fractions. The column is stripped with approximately 800 ml 1 M sodium chloride, pH 2–2.5 and finally reequilibrated with buffer A. Aliquots are removed from alternate fractions for determination of protein and TGF-β activity. The sample can be analyzed in the soft agar assay directly after lyophilization if the size of the aliquot is kept below 10 μl. Larger aliquots should be dialyzed in a microdialysis unit (BRL) against 1 M acetic acid at room temperature before lyophilization to avoid cytotoxicity in the assay due to a high salt concentration. The peak of activity from the column is pooled for further analysis by HPLC.

For both human placenta and bovine kidney preparations, a single peak of soft agar colony-forming activity is detected that elutes at a conductivity of 11–13.5 μS/cm. In general, a fourfold to 15-fold purification of TGF-β is achieved by this step. More importantly, however, ion-exchange chromatography reduces the amount of protein to levels that can be adequately handled by conventional HPLC. Although 85–95% of the applied protein is recovered routinely from the column, only 10–50% of the applied TGF-β activity is detected. Whether this loss of activity is due to denaturation of the TGF, to specific loss of the TGF protein, or to separation of the TGF from an activator is at this time still under investigation.

High-Performance Liquid Chromatography

HPLC was performed with a Varian 5060 or an Altex liquid chromatography system. Solvents used are HPLC grade whenever available and are thoroughly degassed by sonication. The flow rates range from 0.8 to 1.2 ml per minute and 1- or 2-min fractions are collected. The effluent is monitored at 280 nm or 210 nm by means of a variable wavelength Schoeffel spectroflow monitor SF 770 with the outputs coupled to a Hewlett-Packard F130A dual pen recorder.

The pooled fractions of human placenta or bovine kidney TGF-β from the ion-exchange column are made 10% in acetonitrile, 0.1% (vol./vol.) trifluoroacetic acid (TFA), and the pH is adjusted to 2. The sample (100–200 ml) is then pumped onto a Waters μBondapak C18 column (10 μm particle size, 0.78 cm × 30 cm) equilibrated with acetonitrile:water:TFA (10:90:0.1), pH 2. Lyophilization of pooled fractions should be avoided, as it decreases protein recovery and promotes association of the TGF-βs with contaminating proteins. The sample is washed onto the column with 50 ml of the initial solvent and is eluted at a flow rate of 1.2 ml/min with a 60-min linear gradient from 25:75:0.1 to 45:55:0.1 acetonitrile:water:TFA, pH 2. After 75 fractions (1.2 ml/fraction) the column is stripped with 80:20:0.1 acetonitrile:water:TFA, pH 2. Aliquots (0.5–5 μl) are removed for soft agar assay. TGF-β activity, assayed in the presence of 2 ng/ml EGF, is eluted at 35% acetonitrile.

The peak of TGF-β activity from the C18 column is pooled and diluted with an equal volume of 0.1% TFA in water (as in the previous step, lyophilization should be avoided). The sample (6–20 ml) is then applied to a Waters μBondapak CN column (10-μm particle size, 0.38 cm × 30 cm) equilibrated with n-propanol:water:TFA (30:70:0.1). The column is then eluted at a flow rate of 1.1 ml/min with a 110-min linear gradient from 30:70:0.1 to 50:50:0.1 n-propanol:water:TFA, pH 2 followed by a 10-min linear gradient strip to 70:30:0.1 n-propanol:water:TFA, pH 2. Aliquots (0.5–5 μl) are removed for soft agar assay.

TGF-β activity is eluted at 45% n-propanol and corresponds to a strong absorbance peak at 280 or 210 nm [Anzano et al., 1982b; Frolik et al., 1983]. Subsequent analysis of the TGF-β region by SDS-PAGE shows both the human placenta and bovine kidney TGF-βs to be homogeneous and, like the human blood platelet TGF-β, to have a molecular weight of approximately 23,000–25,000 daltons (Fig. 2A). Final yield of TGF-β from either human placenta or bovine kidney was approximately 3–10 μg/kg tissue extracted (see Table I).

Extraction of Colony-Forming Activity From SDS-Polyacrylamide Gels

Type β TGF colony-forming activity can be shown to be associated with the protein band at 23,000–25,000 daltons (Fig. 2A) by electrophoresis of the purified TGF-β on SDS-PAGE [Laemmli, 1970] and extraction of the protein from the gel. For the elution of TGF-β activity, polyacrylamide gels are fractionated into 2-mm slices and transferred into pipet tips that have been

TABLE I. Summary of Purification of Type β TGFs From Human and Bovine Sources

Source of TGF	Apparent molecular weight (kilodaltons)		Soft agar colony-forming activity ED_{50} (ng/ml)[a]		Yield (ng/g)
	Bio-Gel[b] (crude extract)	SDS-PAGE[c]	Acid/ethanol extract	Final	
Human platelets	15-18	23-25	50-100	$(3-10) \times 10^{-2}$	400
Human placenta	3-6	23-25	$(8-12) \times 10^3$	$(3-10) \times 10^{-2}$	10
Bovine kidney	8-12	23-25	$(8-12) \times 10^3$	$(3-10) \times 10^{-2}$	3-5

[a]The ED_{50} is determined from the dose-response curve of colonies >62 μm in diameter assayed under standard conditions in the presence of 2-5 ng/ml EGF.
[b]As determined from chromatography of the acid-ethanol extract on Bio-Gel P-30 or Bio-Gel P-60 (Fig. 1).
[c]As determined for the purified TGF-βs under nonreducing conditions (Fig. 2A).

sealed at the end and plugged with glass wool. The extraction solution (0.7 ml of 1 M acetic acid containing 200 μg/ml bovine serum albumin) is added and the gel slices are crushed into small pieces with a metal or glass rod. After a 24-h extraction at room temperature, the pipet tips are cut and the solubilized TGF is collected in the eluate whereas the acrylamide particles are retained in the glass wool. The particles are rinsed twice with 0.5 ml of extraction solution and the eluates are combined. Eluates are dialyzed in 1 M acetic acid overnight with a microdialyzer, are lyophilized, and are assayed for soft agar activity. The molecular weight of the TGF-β-associated colony-forming activity can be determined by comparing its migration to that of ^{14}C-protein standards run in an adjacent well (Fig. 2B). The molecular weight of human placental TGF-β determined in this manner (Fig. 2B) corresponded to that previously determined by silver staining of the purified TGF-βs after SDS-PAGE (Fig. 2A). Recovery of TGF-β activity ranged from 5% to 20%.

Summary of Purification of Type β TGFs

As shown in Table I, the properties of the purified type β TGFs from human blood platelets and placenta are nearly identical to those from bovine kidney. Each has a molecular weight of 23-25 kilodaltons as determined by SDS-PAGE under nonreducing conditions (Fig. 2A) and an apparent molecular weight of approximately 12.5 kilodaltons under reducing conditions (data not shown). In addition, each of these type β TGFs has a similar specific activity in the colony-forming assay (ED_{50} 0.03-0.1 ng/ml assayed in the presence of EGF). Although blood platelets clearly represent the most concentrated source of type β TGFs (400 ng/g), availability and low cost of the bulk tissue sources (human placenta, 10 ng TGF-β per g; bovine kidney, 3-5 ng TGF-β per g) make them viable alternative starting materials. No species specificity has been observed in the assay systems employed to date, suggesting that these polypeptides are highly conserved proteins.

CONCLUSIONS

Type β TGFs represent a distinct subset of a new family of growth factors whose function in normal physiology is not yet understood. Clearly the presence of type β TGFs in all tissues examined thus far suggests an essential physiological role for these factors that is unrelated to the development of neoplastic behavior. It is already known that one member of the TGF-α family, namely EGF, is present in serum [Byyny et al., 1974], and it has now been shown that TGF-β can be extracted from blood platelets [Childs et al., 1982; Assoian et al., 1983]. These facts, together with the intense biological activity of type β TGFs (10^{-11} to 10^{-12} M), suggest that TGFs are prime candidates for testing as components of serum-free media.

ACKNOWLEDGMENTS

We thank Linda Dart, Lois Lamb, Dorothea Miller, Diane Smith, and Joseph Smith for technical assistance and Ellen Friedman and Erin Crawford for assistance in the preparation of this manuscript.

REFERENCES

Antoniades HN, Sher CD, Stiles CD (1979): Purification of human platelet-derived growth factor. Proc Natl Acad Sci USA 76:1809-1813.
Anzano MA, Roberts AB, Meyers CA, Komoriya A, Lamb LC, Smith JM, Sporn MB (1982a): Synergistic interaction of two classes of transforming growth factors from murine sarcoma cells. Cancer Res 42:4776-4778.
Anzano MA, Roberts AB, Lamb LC, Smith JM, Sporn MB (1982b): Purification by reverse-phase high performance liquid chromatography of an epidermal growth factor-dependent transforming growth factor. Anal Biochem 125:217-224.
Assoian RK, Komoriya A, Meyers CA, Sporn MB (1983): Transforming growth factor-β in human platelets: Identification of a major storage site, purification and characterization. J Biol Chem 258:7155-7160.
Byyny RL, Orth D, Cohen S, Doyne ES (1974): Epidermal growth factor: Effects of androgens and adrenergic agents. Endocrinology 95:776-782.
Carpenter G, Cohen S (1976): ^{125}I-Labeled human epidermal growth factor. J Cell Biol 71:159-171.
Childs CB, Proper JA, Tucker RF, Moses HL (1982): Serum contains a platelet-derived transforming growth factor. Proc Natl Acad Sci USA 79:5312-5316.

Cifone MA, Fidler IJ (1980): Correlation of patterns of anchorage-independent growth with in vivo behavior of cells from a murine fibrosarcoma. Proc Natl Acad Sci USA 77:1039-1043.

Davoren PR (1962): The isolation of insulin from a single cat pancreas. Biochim Biophys Acta 63:150-153.

De Larco JE, Todaro GJ (1978a): Growth factors from murine sarcoma virus-transformed cells. Proc Natl Acad Sci USA 75:4001-4005.

De Larco JE, Todaro GJ (1978b): Epithelioid and fibroblastic rat kidney cell clones: Epidermal growth factor (EGF) receptors and the effect of mouse sarcoma virus transformation. J Cell Physiol 94:335-342.

De Larco JE, Todaro GJ (1980): Sarcoma growth factor (SGF): Specific binding to epidermal growth factor (EGF) membrane receptors. J Cell Physiol 102:267-277.

Frolik CA, Dart LL, Meyers CA, Smith DM, Sporn MB (1983): Purification and initial characterization of a type β transforming growth factor from human placenta. Proc Natl Acad Sci USA (in press).

Kahn P, Shin S (1979): Cellular tumorigenicity in nude mice. J Cell Biol 82:1-16.

Laemmli UK (1970): Cleavage of structural proteins during the assembly of the head of bacteriophage T4. Nature 227:680-685.

Marquardt H, Todaro GJ (1982): Human transforming growth factor: Production by a melanoma cell line, purification, and initial characterization. J Biol Chem 257:5220-5225.

Moses HL, Branum EL, Proper JA, Robinson RA (1981): Transforming growth factor production by chemically transformed cells. Cancer Res 41:2842-2848.

Nexø E, Hollenberg MD, Figueroa A, Pratt RM (1980): Detection of epidermal growth factor—Urogastrone and its receptor during fetal mouse development. Proc Natl Acad Sci USA 77:2782-2785.

Ozanne B, Fulton RJ, Kaplan PL (1980): Kirsten murine sarcoma virus transformed cell lines and a spontaneously transformed rat cell line produce transforming factors. J Cell Physiol 105:163-180.

Roberts AB, Lamb LC, Newton DL, Sporn MB, De Larco JE, Todaro GJ (1980): Transforming growth factors: Isolation of polypeptides from virally and chemically transformed cells by acid/ethanol extraction. Proc Natl Acad Sci USA 77:394-398.

Roberts AB, Anzano MA, Lamb LC, Smith JM, Sporn MB (1981): New class of transforming growth factors potentiated by epidermal growth factor: Isolation from non-neoplastic tissues. Proc Natl Acad Sci USA 78:5339-5343.

Roberts AB, Anzano MA, Lamb LC, Smith JM, Frolik CA, Marquardt H, Todaro GJ, Sporn MB (1982): Isolation from murine sarcoma cells of novel transforming growth factors potentiated by EGF. Nature (London) 295:417-419.

Roberts AB, Anzano MA, Meyers CA, Wideman J, Blacher R, Pan Y-CE, Stein S, Lehrman SR, Smith JM, Lamb LC, Sporn MB (1983a): Purification and properties of a type β transforming growth factor from bovine kidney. Biochemistry 22:5692-5698.

Roberts AB, Frolik CA, Anzano MA, Sporn MB (1983b): Transforming growth factors from neoplastic and non-neoplastic tissues. Fed Proc 42:2621-2626.

Schaeffer NI, Friend K (1976): Efficient detection of soft agar grown colonies using a tetrazolium salt. Cancer Lett 1:259-262.

Twardzik DR, Todaro GJ, Marquardt H, Reynolds FH, Stephenson JR (1982b): Transformation induced by abelson murine leukemia virus involves production of a polypeptide growth factor. Science 216:894-897.

Twardzik DR, Ranchalis JE, Todaro GJ (1982b): Mouse embryonic transforming growth factors related to those isolated from tumor cells. Cancer Res 42:590-593.

Methods for Preparation of Media, Supplements, and Substrata
for Serum-Free Animal Cell Culture, pages 195–205
© 1984 Alan R. Liss, Inc., 150 Fifth Avenue, New York, NY 10011

10
Preparation of Endothelial Cell Growth Factor

Thomas Maciag and Robert Weinstein

The cultivation and serial propagation of mammalian cells in culture is dependent upon a complex synergism between physical, chemical, and biological vectors. In general, most cell culture systems attempt to provide a constant physical environment upon which is superimposed an aqueous environment composed of nutrient medium and serum. The role of the media is to provide the cells with a complex array of nutritional components for cultivation, and the role of serum is to provide a complex library of hormones, growth factors, transport factors, and attachment factors without which cells cannot traverse the cell cycle. It has been a recent accomplishment of modern cell biology to define the biological nature of these components such that the serum-free cultivation of many mammalian cells is possible under biochemically defined conditions [Hayachi and Sato, 1976]. However, many mammalian cells do not proliferate in serum-supplemented media. The human umbilical vein endothelial cell (HUV-EC) is only one example of such a fastidious cell type that cannot be serially propagated in a serum-supplemented environment [Maciag et al., 1981]. Since serum contains a complex of factors requisite for cell growth, it can be argued that the failure of fastidious mammalian cells to proliferate in a serum-supplemented environment may be due to the limitations of serum or plasma in providing the families of growth factors that are required for the growth of the fastidious mammalian cell. The absence of a specific factor in serum for those cell types may be a result of many influences including 1) chemical lability of the growth factor, 2) biochemical modification of the growth factor by serum-derived enzymes, and 3)

Departments of Pathology and Medicine, Charles A. Dana Research Institute, Harvard-Thorndike Laboratory, Harvard Medical School, Beth Israel Hospital, Boston, Massachusetts 02215

low titer of the growth factor in plasma. The failure of serum to provide the requisite growth factor(s) for HUV-EC growth has been evaluated in detail and has resulted in the identification of a new family of growth factors present in bovine neural tissue. These factors are collectively termed endothelial cell growth factor (ECGF) in reference to their mitogenic activity on HUV-EC.

BIOLOGICAL PROPERTIES OF ECGF

Populations of viable HUV-EC can be established as primary cultures by the traditional methods developed by Jaffe et al. [1973] and Gimbrone et al. [1974]. HUV-EC cultures established in this manner can be cultivated at high-density cell seed ratios in Medium 199 and 20% fetal bovine serum (FBS). The serial propagation of these cells has proved to be frustrating for most laboratories [Gimbrone, 1976; Schwartz, 1978]. Although an occasional success has been reported for the serial propagation of HUV-EC [Haudenschild, 1975], this has certainly not been a routine accomplishment for most investigators interested in endothelial cell biology.

We have reported the successful long-term propagation of HUV-EC in vitro with Medium 199, 20% FBS, and 150 μg/ml of a neutral extract prepared from bovine brain, hypothalamus, or pituitary glands [Maciag et al., 1973]. The active component in these extracts has been identified, purified, and characterized as ECGF. Our success in stimulating low seed density HUV-EC growth can be directly attributed to the addition of ECGF to the FBS-supplemented medium. We have shown that ECGF is the only mitogen capable of stimulating the growth of quiescent populations of HUV-EC [Maciag et al., 1982a]. None of the classical growth factors (epidermal growth factor, fibroblast growth factor [cationic form], insulin-like growth factor I, insulin-like growth factor II, nerve growth factor, platelet-derived growth factor, thrombin, growth hormone, insulin, transferrin, and hydrocortisone) are able to substitute for the biological activity for ECGF [Maciag et al., 1982a]. Thus the stimulation of low seed density HUV-EC growth by ECGF is remarkably specific as a biological assay. Our success in achieving long-term serial propagation of HUV-EC to 50–60 cumulative population doublings can be ascribed to the use of a human fibronectin (HFN) matrix in conjunction with ECGF and FBS-supplemented medium [Maciag et al., 1981a].

HUV-EC can be cultivated in vitro on a plastic surface at cell seed densities greater than 10^4 cells per 1 cm^2 surface area in the presence of ECGF. However, optimum growth of HUV-EC at cell seed densities less than 10^4 cells per 1 cm^2 is difficult to achieve in the presence of ECGF. Furthermore, only 30–40 cumulative population doublings can be achieved on plastic surfaces even in the presence of ECGF. The introduction of an HFN matrix to the cell culture system corrects these deficiencies in vitro. The investigator can routinely achieve up to 50 cumulative population doublings in the presence of an HFN matrix. It is also possible to cultivate the HUV-EC at clonal cell seed densities [Maciag et al., 1981a]. We have determined that the optimal concentration of HFN for HUV-EC attachment is 10 $\mu g/cm^2$ [Maciag et al., 1981a]. The plating efficiency at cell densities greater than 10^4 cells per 1 cm^2 is approximately 40–50% after 10 min exposure to the HFN matrix, whereas the plating efficiency at 1.25 cells per 1 cm^2 is approximately 10% [Maciag et al., 1981a]. We have suggested that the role of HFN is to simply provide a surface for optimal HUV-EC attachment and does not influence HUV-EC growth.

Another attribute of ECGF is the ability of the growth factor to reduce the concentration of FBS required for HUV-EC proliferation. We have demonstrated that half-maximum HUV-EC growth can be achieved at concentrations of FBS as low as 2.5% [Maciag et al., 1981a]. We have attempted to utilize the growth response to FBS in the presence of ECGF to screen a complete library of hormones and growth factors for their ability to stimulate low-density HUV-EC growth in the presence of ECGF and 2.5% FBS. While insulin, transferrin, epidermal growth factor, and thrombin were able to promote HUV-EC growth under these conditions, platelet-derived growth factor, fibroblast growth factor (cationic-form), insulin-like growth factor I, insulin-like growth factor II, hydrocortisone, and nerve growth factor (7S form) did not stimulate HUV-EC growth [Maciag et al., 1982a]. These results are consistent with the observation that endothelial cell growth in vivo does occur in hypophysectomized rats [Tiell et al., 1978] and HUV-EC growth can occur in vitro in the presence of hypophysectomized serum supplemented with ECGF [Maciag et al., 1981a]. We have suggested that the use of platelet-poor plasma-derived hypophysectomized serum and ECGF may offer a selective endocrine advantage for endothelial cell growth in primary culture since the common cell culture contaminants, i.e., the smooth muscle cell [Weinstein et al., 1981] and fibroblast [Weinstein et al., 1982], require hydrocortisone, the somatomedins, and platelet-derived growth factor for optimum growth [Maciag et al., 1981a].

The final and probably most significant biological attribute of ECGF is the ability to support the serial propagation of HUV-EC populations in vitro [Maciag et al., 1981a]. We have studied the long-term cultivation of the HUV-EC in vitro and have routinely achieved between 50 and 60 cumulative population doublings with the use of ECGF. This response in vitro is similar to the in vitro aging response of normal human fibroblasts. We have observed that as the HUV-EC population ages in vitro, the HUV-EC become larger. We have expressed this response in quantitative terms by demonstrating a linear decrease in the number of HUV-EC at confluence as a function of cumulative population doublings [Maciag et al., 1981a]. The extracellular concentration of factor VIII:Ag does not vary significantly with in vitro age although immunofluorescence staining for factor VIII:Ag does reveal a qualitative decrease in the intracellular content of the endothelial cell marker after approximately 40 cumulative population doublings [Maciag et al., 1981a]. We suggest that populations of the HUV-EC can be used for cellular biochemical studies if the investigator employs HUV-EC populations with an in vitro age of less than 35 cumulative population doublings. Information obtained with these cells should prove to be valuable to the study of the developmental and pathophysiological biology of the vessel wall. A second biological activity can also be attributed to ECGF: the ability to stimulate DNA synthesis in Balb/c 3T3 cells [Maciag et al., 1979; Lemmon et al., 1982]. We have demonstrated that ECGF stimulates the incorporation of ^{3}H-thymidine into quiescent Balb/c 3T3 cells at nanomolar concentrations [Lemmon et al., 1982; Maciag et al., 1982]. This biological activity has proved valuable during the final stages of ECGF purification.

PHYSICAL AND CHEMICAL PROPERTIES OF ECGF

Considerable effort has been expended toward the preparation of homogeneous ECGF for the anticipated study of receptor:ligand interaction and mechanism of action. The purification of ECGF has proved to be difficult because of the lability of the biological activity of ECGF. Although the biological activity is quite stable as a lyophilized preparation in crude form, once 5×10^3-fold to 10^4-fold purity has been achieved, a significant decrease in biological activity is consistently observed upon storage. Although this singular situation has prevented the dissemination of homogenous ECGF to the endothelial cell and growth factor community, we have increased our efforts to understand and solve this problem and have generated physical and chemical evidence that unequivocally demonstrates the unique character of ECGF.

Recent efforts from this laboratory have demonstrated that ECGF exists in two forms: a high-molecular-weight (HMr) form and a low-molecular-weight (LMr) form [Maciag et al., 1982b]. The physical properties of both forms of ECGF are summarized in Table I. HMr-ECGF is defined as a fraction of biological activity possessing a Mr greater than 75,000 daltons and an isoelectric point (pI) between pH 4 and pH 6. LMr-ECGF is defined as a protein that possesses a Mr of 22,000 daltons and a pI between pH 5 and pH 6 [Maciag

TABLE I. Comparison of Mitogens From Bovine Neural Tissue

	FGF	ECGF
Apparent isoelective point	9.6 [Gospodarowicz et al., 1982]; 9.0–8.0 [Lemmon et al., 1982]	4.0–6.0 [Maciag et al., 1982b; Barritault et al., 1981] 4.5–4.7 [D'Amore, 1982]
Apparent molecular weight	13K [Gospodarowicz et al., 1978]	22K [Maciag et al., 1982b; D'Amore, 1982; Barritault et al., 1981]
Target (assay) cell	Balb/c 3T3 [Gospodarowicz et al., 1978]	HUV-EC [Lemmon et al., 1982; Maciag et al., 1982b]

Original references are in brackets.

et al., 1982b]. Thus HMr-ECGF and LMr-ECGF may be related since they appear to have the same anionic pI and biological activities as HUV-EC and Balb/c 3T3 cell mitogens. It is not surprising that ECGF should possess both HMr and LMr forms, since this has been a common feature of the classical growth factors, i.e., epidermal growth factor [Taylor et al., 1970; Frey et al., 1979], nerve growth factor [Berger and Shooter, 1977], and the insulin-like growth factors [Knauer and Smith, 1981]. Although these data suggested a relationship between the HMr and LMr forms of ECGF, the relationship was actually demonstrated when it was shown that LMr-ECGF (Mr = 22,000; pI 5–6) can be generated from HMr-ECGF by the acidification of HMr-ECGF with dilute (1-mM) concentrations of acetic acid, by increasing the ionic strength, or by ethanol precipitation [Maciag et al., 1982b]. The result of the HMr-to-LMr-ECGF transition is the generation of a 10^3-fold increase in specific activity in both the HUV-EC growth and the ^{3}H-thymidine Balb/c 3T3 cell assay [Maciag et al., 1982b]. An additional 10-fold increase in purity can be achieved by isoelectric focusing of the LMr-ECGF prepared from the HMr-ECGF form [Maciag et al., 1982b]. Although LMr-ECGF prepared by this method [Maciag et al., 1982b] is active in the HUV-EC growth assay at 25–50 ng/ml and in the Balb/c 3T3 cell assay at 10–20 ng/ml, these preparations are still heterogeneous. Since highly purified LMr-ECGF is not stable under optimal conditions, we suggest that crude preparations of ECGF should be used for the routine propagation of HUV-EC strains.

The unique biochemical and biological properties of ECGF suggest that it does not belong to any of the traditional classes of growth factors. Other laboratories have noted similar biological activities in extracts of pituitaries

[Peehl and Ham, 1980], retina [D'Amore, 1982], and the eye [Barritault et al., 1981]. Most notable is the similarity in physical and biological properties shared by these mitogens. Eye-derived growth factor (EDGF) has been characterized as an endothelial cell growth factor with an apparent Mr of approximately 22,000 daltons and a pI of 5.4 [Barritault et al., 1981]. D'Amore has purified growth factor for bovine cornea endothelial cells with similar physical properties and suggests that EDGF belongs to the ECGF family of growth factors [D'Amore, 1982]. Partially purified preparations of EDGF have also been demonstrated to be mitogenic for human keratinocytes [Barritault et al., 1981]. We have observed similar human keratinocyte growth-promoting activity in crude preparations of ECGF [Maciag et al., 1981b] but have been able to successfully separate these activities [Maciag et al., 1982].

The physical and biological properties of ECGF demonstrate that the growth factor is not identical to bovine brain fibroblast growth factor (FGF) [Lemmon et al., 1982]. FGF purified from bovine brain remains a cationic protein with a Mr of approximately 13,000 daltons [Lemmon et al., 1982; Gospodarowicz et al., 1978]. The endothelial cell growth factor activity that is present in nonhomogeneous preparations of FGF can be readily separated by isoelectric focusing. Although the relationship between bovine brain FGF and fragments of myelin basic protein still remain unclear [Gospodarowicz et al., 1982; Gospodarowicz and Mescher, 1981], recent physical, immunological, and biological evidence strongly suggests the absence of homology between these two proteins [Lemmon et al., 1982; Thomas et al., 1980].

PREPARATION OF ECGF FOR MAINTENANCE PROPAGATION OF HUV-EC IN CULTURE

Although the investigator can use bovine brain, pituitary, or hypothalamus from commercial supply houses for the preparation of ECGF, we recommend the use of bovine hypothalamus as the starting material, since there appears to be more biological activity present in the hypothalamus than in either the pituitary or brain [Maciag et al., 1979]. Approximately 1 kg of organ is weighed frozen and then thawed overnight at 4°C. The tissue is washed with at least 3 liters of ice-cold distilled water, and excessive blood clots are removed manually. The tissue is drained, blot-dried, and homogenized in a Waring blender in three to four batches. Homogenization should not be longer than 3 min in ice-cold 0.15 M NaCl. We recommend that a maximum of 1.25 liters of 0.15 M NaCl per kilogram of tissue be utilized to complete the homogenization.

The crude homogenates are combined and extracted with mechanical agitation for 2 h at 4°C. The homogenate is centrifuged at a minimum of

13,000g for 40 min at 4°C and the supernatants are pooled. Care should be exercised during the decanting procedure to minimize collection of the buoyant white lipid precipitate. The volume of the clear red supernatant is measured and subjected to streptomycin sulfate extraction. The pH of the extract before the addition of streptomycin sulfate should be between pH 6.5 and 7.4.

The streptomycin sulfate solution is prepared in advance. We recommend using 7.5 g of streptomycin sulfate per kilogram of tissue. The streptomycin sulfate is weighed and dissolved in 50 ml of distilled water. The pH of the solution is adjusted to pH 8.0 with 1 N NaOH and the solution is cooled. The streptomycin sulfate solution is added to the clear red supernatant with stirring drop by drop. The final pH of the extract after the addition of streptomycin sulfate should be approximately pH 6.8 $\pm$ 0.3. The mixture is allowed to stand at 4°C for 18 h (generally overnight) and the mixture is centrifuged at 27,000g for 30 min.

The total volume and $A_{280\,nm}$ of the neutral extract is accurately measured. The material should be lyophilized as quickly as possible to minimize loss of ECGF activity. We recommend that the crude ECGF be stored in quantities of 100 mg per vial using an extinction coefficient of 10 ($E_{1\%/280\,nm} = 10$). If the lyophilized crude ECGF proves difficult to reconstitute and sterilize by filtration through a 0.22-μm filter, we recommend centrifugation of the preparation at 27,000g for 20 min to remove lipid vesicles. However, it has been our experience that if the streptomycin sulfate precipitation is performed correctly, the concentration of lipid vesicles will be minimized.

The biological efficacy of the crude ECGF is monitored by the HUV-EC growth assay. The crude ECGF is reconstituted with Medium 199 and sterile-filtered. A 100-mm stock dish of confluent HUV-EC is harvested with 0.05% trypsin-0.02% EDTA and the number of viable cells quantitated. We recommend using populations of HUV-EC of 10-35 cumulative population doublings for this assay. The HUV-EC are seeded into Medium 199 containing 20% FBS, and various concentrations of crude ECGF (1.0-500 μg/ml) at a viable cell seed density of 3×10^3 cells per 1 cm^2. The use of an HFN matrix (10 μg/cm^2) is optional and if used, the viable cell seed density should be reduced to 1×10^3 cells per 1 cm^2. The assay should be performed in duplicate. Cell culture dishes containing Medium 199 and 20% FBS without crude ECGF serve as negative controls. The dishes are fed with all supplements every 2-3 days and the assay is terminated on day 10 of growth. The assay dishes are harvested with trypsin-EDTA, number of viable cells per dish is quantitated, and the data are expressed as viable HUV-EC cells versus micrograms ECGF per 1 ml. Representative assay data are shown in Figure

1. The minimal concentration of ECGF that stimulates maximal HUV-EC growth should be noted and utilized in the maintenance of HUV-EC stocks. Although variation from preparation to preparation is possible, this value usually is between 100 and 300 µg crude ECGF per 1 ml.

Lyophilized preparations of crude ECGF should be stored at temperatures below −20°C and are stable for up to 6 months. We recommend that the assay be repeated every 2-3 months to ensure the viability of the growth factor. Once the lyophilized material is reconstituted with Medium 199, the sample should be stored at 4°C. Reconstituted ECGF should not be stored frozen, since a significant loss in biological activity is observed with multiple

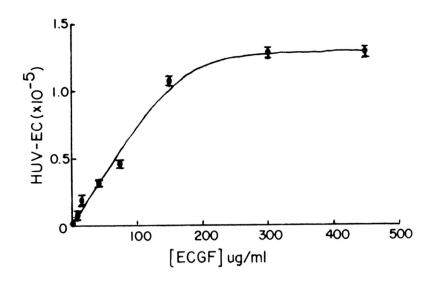

Fig. 1. Growth response of human umbilical vein endothelial cells to variable concentrations of crude bovine endothelial cell growth factor. Crude preparations of ECGF were prepared from bovine hypothalamus as presented in the text. HUV-EC (passage 6) were plated at a cell seed density of 1×10^3 cells per 1 cm^2 onto a purified human fibronectin matrix (10 µg/cm^2) in Medium 199 containing 10% FBS (vol./vol.). The dishes were also supplemented with varied concentrations of crude ECGF. The cells were fed every 2-3 days with a complete medium change. After 10 days in culture, the cells were harvested with trypsin/EDTA and the number of viable cells per dish was determined. The data are expressed as number of viable HUV-EC per 35-mm dish as a function of the ECGF concentration in micrograms per milliliter. Note that this preparation of crude ECGF is maximally active at concentrations greater than 200 µg/ml.

freeze-thawing. The reconstituted crude ECGF is stable at 4°C for 2 weeks and should not be used beyond this time.

Crude preparations of ECGF are being marketed by a number of companies under a variety of trade names. Endothelial cell growth supplement (ECGS) is sold by Collaborative Research, Inc. (Lexington, MA), and endothelial cell mitogen (ECM) can be purchased from Bethesda Research Laboratories, Inc. (Gaithersburg, MD) and KOR Biologicals, Inc. (Cambridge, MA). In all cases, these products represent crude preparations of the HMr form of ECGF. Since ECGF is such a labile mitogen, the biological efficacy of these commercial preparations has been variable. However, the recent introduction of ECGF by Seragen, Inc. (Boston) may obviate these concerns, since the biological activity of ECGF is closely monitored by the HUV-EC growth assay.

IDENTIFICATION OF OTHER BIOLOGICAL ACTIVITIES FROM EXTRACTS OF BOVINE NEURAL TISSUE

We have demonstrated that preparations of bovine brain and hypothalamus contain potent mitogens for two additional fastidious human cell types, the keratinocyte [Maciag et al., 1981b] and the melanocyte [Wilkins et al., 1982]. Similar human keratinocyte growth-promoting activities have been reported from the pituitary gland [Peehl and Ham, 1980] and the eye [Barritault et al., 1981]. Preliminary evidence suggests that the keratinocyte and melanocyte mitogens are unique and do not possess any biochemical homology to the LMr form of ECGF [Maciag et al., 1982b]. These additional biological activities have proved to be important in the formulation of serum-free media for the selective growth of the keratinocyte [Gilchrest et al., 1982] and melanocyte [Wilkins et al., 1982] from fetal and adult human epidermis. We therefore anticipate that additional mitogenic biological activities for other fastidious human cell types will be identified in crude extracts of bovine neural tissue and that such studies should be encouraged. Furthermore, we anticipate that the second edition of this book will contain a detailed protocol for the purification of biologically stable homogenous ECGF by preparative and analytical high-performance liquid chromatography from either bovine neural tissue or genetically manipulated microorganisms.

REFERENCES

Barritault D, Arruti C, Courtois V (1981): Is there a ubiquitous growth factor in the eye? Differentiation 18:29–42.

Berger EA, Shooter EM (1977): Evidence for pro-beta-nerve growth factor, a biosynthetic precursor to beta-nerve growth factor. Proc Natl Acad Sci USA 74:3647–3651.

D'Amore PA (1982): Purification of a retina-derived endothelial cell mitogen/angiogenic factor. J Cell Biol 95:192a.

Frey P, Forand R, Maciag T, Shooter EM (1979): The biosynthetic precursor of epidermal growth factor and the mechanism of its processing. Proc Natl Acad Sci USA 76:6294–6298.

Gilchrest BA, Calhoun JK, Maciag T (1982): Attachment and growth of human keratinocytes in a serum-free environment. J Cell Physiol 112:197–206.

Gimbrone MA (1976): Culture of vascular endothelium. Prog Hemostasis Thromb 3:1–28.

Gimbrone MA, Cotran RS, Folkman J (1974): Human vascular endothelial cells: Growth and DNA synthesis. J Cell Biol 60:673–684.

Gospodarowicz D, Mescher AL (1981): Fibroblast growth factor and vertebrate regeneration. Adv Neurol 29:149–171.

Gospodarowicz D, Bialecki H, Greenburg G (1978): Purification of fibroblast growth factor from bovine brain. J Biol Chem 253:3736–3743.

Gospodarowicz D, Lui G-M, Cheng J (1982): Purification in high yield of brain fibroblast growth factor by preparative isoelective focusing at a pH of 9.6. J Biol Chem 257:12266–12276.

Haudenschild CC, Cotran RS, Gimbrone MA, Folkman J (1975): Fine structure of vascular endothelium in culture. J Ultrastruct Res 50:22–32.

Hayashi I, Sato GH (1976): Replacement of serum by hormones permits growth of cells in defined medium. Nature 259:132–134.

Jaffe EA, Nachman RL, Becker CG, Minick CR (1973): Culture of human endothelial cells derived from umbilical veins. Identification by morphologic and immunologic criteria. J Clin Invest 52:2745–2756.

Knauer DJ, Smith GL (1981): Inhibition of biological activity of multiplication stimulating activity by binding to its carrier protein. Proc Natl Acad Sci USA 77:7252–7256.

Lemmon SK, Riley MC, Thomas KA, Hoover GA, Maciag T, Bradshaw RA (1982): Bovine fibroblast growth factor: Comparison of brain and pituitary preparations. J Cell Biol 95:162–169.

Maciag T, Kelley B, Cerundolo J, Ilsley S, Kelley PR, Forand R (1979): An endothelial cell growth factor from bovine hypothalamus: Identification and partial characterization. Proc Natl Acad Sci USA 76:5674–5678.

Maciag T, Hoover GA, Stemerman MB, Weinstein R (1981a): Serial propagation of human endothelial cells in vitro. J Cell Biol 91:420–426.

Maciag T, Nemore RE, Weinstein R, Gilchrest BA (1981b): An endocrine approach to control of epidermal growth: Serum-free cultivation of human keratinocytes. Science 211:1452–1454.

Maciag T, Hoover GA, van der Spek J, Stemerman MB, Weinstein R (1982a): Growth and differentiation of human umbilical vein endothelial cells in culture. In Sato GH, Pardee AB, Sirbasku DA (eds): "Growth of Cells in Hormonally Defined Media." Cold Spring Harbor Conf Cell Prolif 9:525–538.

Maciag T, Hoover GA, Weinstein R (1982b): High and low molecular weight forms of endothelial cell growth factor. J Biol Chem 257:5333–5336.

Peehl DM, Ham RG (1980): Growth and differentiation of human keratinocytes without a feeder layer or conditioned medium. In Vitro 16:516–525.

Schwartz SM (1978): Selection and characterization of bovine aortic endothelial cells. In Vitro 14:966–980.

Taylor JM, Cohen S, Mitchell WM (1970): Epidermal growth factor: High and low molecular weight forms. Proc Natl Acad Sci USA 67:164–171.

Thomas KA, Riley MC, Lemmon SK, Baglan NC, Bradshaw RA (1980): Brain fibroblast growth factor: Nonidentity with myelin basic protein. J Biol Chem 255:5517-5520.

Tiell ML, Stemerman MB, Spaet TH (1978): The influence of the pituitary on arterial intimal proliferation in the rat. Circ Res 42:644-649.

Weinstein R, Stemerman MB, Maciag T (1981): Hormonal requirements for growth of arterial smooth muscle cells in vitro: An endocrine approach to atherosclerosis. Science 212:818-820.

Weinstein R, Hoover GA, Majure J, van der Spek J, Stemerman MB, Maciag T (1982): Growth of human foreskin fibroblasts in a serum-free, defined medium without platelet-derived growth factor. J Cell Physiol 110:23-28.

Wilkins LM, Szabo G, Connell L, Gilchrest BA, Maciag T (1982): Growth of enriched human melanocyte cultures. In Sato GH, Pardee AB, Sirbasku DA (eds): "Growth of Cells in Hormonally Defined Media." Cold Spring Harbor Conf Cell Prolif 9:929-936.

Methods for Preparation of Substrata

Methods for Preparation of Media, Supplements, and Substrata
for Serum-Free Animal Cell Culture, pages 209–213
© 1984 Alan R. Liss, Inc., 150 Fifth Avenue, New York, NY 10011

11
Use of Basic Polymers as Synthetic Substrata for Cell Culture

Wallace L. McKeehan

Most normal somatic mammalian cells are anchorage-dependent and require an adequate substrate on which to attach, grow, and function [Grinnell, 1978]. Transformed, tumor-like cells are less dependent on the culture substrate. However, normal anchorage-independent cells [Bargellesi et al., 1976] and tumor cells are often also cultured on a fixed substrate. Cell attachment to various substrates that are most commonly used for culture of living cells is often inadequate. Even if attachment is adequate, the culture substrate is sometimes nonpermissive to adequate levels of cell proliferation and function. This is especially amplified at low cell density and in serum-free medium [McKeehan and Ham, 1976]. Negatively charged polystyrene is currently the most common cell culture substrate. This evolved empirically as a practical substitute for the negatively charged glass substrate that was adapted to the culture of mammalian cells from the culture of bacteria and lower organisms. Although widely applicable, presently employed negatively charged polystyrene is not necessarily the optimum substrate for all cultured cell types and experimental objectives. In recent years, chemical modifications to result in positively charged substrates for cell culture have been employed for specific applications [Levine et al., 1977]. Attempts to modify polystyrene chemically to yield a positively charged surface usually results in a loss of optical clarity. However, a simple coating of various tissue culture substrates with basic polymers results in improved attachment and cell proliferation rate and still preserves the optical clarity of polystyrene substrates [McKeehan and Ham, 1976].

W. Alton Jones Cell Science Center, Lake Placid, New York 12946

METHODS

The following procedure can be adapted to treat the surface of any size culture vessel and type of substrate:

1. Perform all steps aseptically.
2. Prepare the basic polymer solution at a concentration of 0.10 mg per milliliter of tissue culture grade water.
3. Sterilize the basic polymer solution by passage through 0.22 μM membrane filters.
4. Pipette 50 μl of basic polymer solution per 1 cm^2 of surface area to be coated. This can be increased or decreased to fit the application. However, be sure enough solution is used to completely wet the entire surface.
5. After 5 min at room temperature, remove the coating solution. For replicate vessels, collection and disposal of the used coating solution with an aspirator is convenient.
6. Add 100 μl of sterile H$_2$O per 1 cm^2 surface area and wash the surface thoroughly. Remove the H$_2$O wash solution by aspiration. Be sure to remove all the solution to avoid dilution of culture medium added in the next step.
7. Add the appropriate culture medium and cells and then incubate according to individual application.

RESULTS

Basic polymer-coated culture substrates have been demonstrated to improve both attachment and the proliferation rate of cells once they have attached to the substrate [Goodman, 1982; Ham, 1982; McClure et al., 1982; McKeehan and Ham, 1976; Michler-Stuke and Bottenstein, 1982; Puymirat et al., 1982; Salomon et al., 1982]. A variety of basic polymers work equally well (Table I) although polylysine is most commonly employed. The D or L isomers of polylysine are equally effective. The D isomers of amino acid polymers are preferred to avoid interference of any biologically active L-amino acids produced by degradation of the polymer. A minimum polymer length of about 10 amino acids appears necessary in the case of polylysine. Monomeric lysine or dilysine will not work (Table I). The coating of tissue culture polystyrene by the described method is rapid and complete within minutes. Although the procedure is normally performed at ambient room temperature, the coating effect appears to be independent of temperature. Although the above procedure is described for immediate use of the coated substrate in cell culture experiments, the procedure is adaptable to large-scale treatment and storage of culture vessels. All steps can be performed without using aseptic procedure and the coated vessels can then be sterilized by irradiation and

TABLE I. Effect of Different Culture Surface Coating Agents on Clonal Growth of WI-38 Cells

Coating agent	Percentage of clonal growth on polylysine-coated plates[a]
None	54
DEAE-dextran	72
Histone[b]	118
L-lysine	55
L-lysyl-L-lysine	55
Poly-L-arginine·HCl (Mr 15,000-50,000)	95
Poly-L-histine (Mr 5,000-15,000)	94
Poly-D-lysine·HBr (Mr 30,000-70,000)	100
Poly-DL-ornithine·HBr (Mr 3,000-15,000)	111
Protamine[c]	140
Sodium alginate	30
Dextran sulfate (Mr 500,000)	33
Heparin (Na salt)	44
Hyaluronic acid (Na salt)[d]	46
Poly-D-glutamic acid (Na salt) (Mr 50,000-100,000)	26
Dextran (Mr 500,000)	47
Methylcellulose (15 centipoise)	28
Poly-L-asparagine (Mr 5,000-10,000)	46
Poly-L-proline (Mr 10,000-30,000)	55
Polyvinylpyrrolidone (Mr 360,000)	36
Bovine serum albumin	50
Collagen	75
FBSP	55
Gelatin	62

[a]Cell culture plastic petri dishes were treated with the indicated polymer as described in Methods. WI-38 cells were grown into colonies in medium MCDB 104 containing 500 μg per 1 ml fetal bovine serum protein (FBSP) in the treated dishes. Colonies were fixed and stained, and colony size was measured photometrically. Untreated dishes and dishes treated with poly-D-lysine were included as controls in all experiments. For direct comparison of responses from different experiments, all colony size measurements are reported as percentages of the values obtained for the poly-D-lysine controls in each experiment. The value for no treatment is an average of controls from all experiments.
[b]Sigma type II-A from calf thymus.
[c]Sigma grade IV from salmon.
[d]Sigma grade IIIS from human umbilical cord.

From McKeehan and Ham [1976], with permission.

stored for future use for at least 3 months. A critical step in the procedure is to remove and wash all soluble polymer from the culture vessels after the coating procedure. Soluble basic polymers in the culture medium inhibit proliferation of human fibroblasts [McKeehan and Ham, 1976].

The efficacy of the basic polymer-coated surface for either attachment or an effector of cell proliferation and other cell functions is dependent on cell type and the nutritive and hormonal environment. Fibroblasts, some tumor-derived cells, and other mesenchymal-derived cells appear most responsive to the basic polymer-coated surface, whereas no reports of positive effects on normal epithelial cells has appeared. The basic polymer-coated surface has also been applied in combination with other extracellular matrix elements as collagen and fibronectin [McClure et al., 1982; Michler-Stuke and Bottenstein, 1982; Salomon et al., 1982]. The effect of basic polymer-coated surfaces on the proliferation of fibroblasts and tumor cell lines is clearly more apparent at low cell density and low serum protein concentrations [Goodman, 1982; Ham, 1982; McClure et al., 1982; Michler-Stuke and Bottenstein, 1982; Puymirat et al., 1982; Salomon et al., 1983]. This may indicate that the coated substrate is substituting for serum proteins or cell-derived factors that play a role similar to that of the basic polymer-coated surface.

The role of the basic polymer-coated surface in cell attachment and cell proliferation is largely unknown. Cell spreading and migration is overtly more extensive on the coated substrate [McKeehan and Ham, 1976; Letourneau, 1975a,b]. This may increase the cell surface area to affect transport and receptor sites for nutrients and hormones in the culture medium. Basic polymers have profound effects on other cell membrane functions, which include polymer uptake [Ryser and Hancock, 1965], cell aggregation [Duskin et al., 1970, 1975], and proteoglycan metabolism [Gillard et al., 1970, 1980; Poole et al., 1980].

REFERENCES

Bargellesi A, Damiani G, Kuehl WM, Scharf MD (1976): Synthesis of immunoglobulin by substrate attached mouse myeloma cells. J Cell Physiol 88:247-252.

Duskin D, Katchalski E, Sachs L (1970): Specific aggregation of SV40-transformed cells by ornithine, leucine copolymers. Proc Natl Acad Sci USA 67:185-192.

Duskin D, Katchalski-Katzir E, Sachs L (1975): Interaction of the basic copolymer-poly(ornithine, leucine) with normal and transformed cells. FEBS Lett 60:21-25.

Gillard GC, Birnbaum P, Reilly HC, Merrilees MJ, Fling MH (1979): The effect of charged synthetic polymers on proteoglycan synthesis and sequestration in chick embryo fibroblast cultures. Biochim Biophys Acta 584:520-528.

Gillard GC, Thompson BM, Flint MH (1980): The chemical synthesis of (^{3}H) poly (L-lysine) and its interaction with and degradation by fibroblasts in culture. Connect Tissue Res 7:203.

Goodman R (1982): Growth and differentiation of pheochromocytoma cells in chemically defined medium. In Sato GH, Pardee AB, Sirbasku DA (eds): "Growth of Cells in Hormonally Defined Media." Cold Spring Harbor, NY: Cold Spring Harbor Laboratory, Vol 9, pp 1053-1068.

Grinnell F (1978): Cellular adhesiveness and extracellular substrata. Int Rev Cytol 53:65-144.

Ham RG (1982): Importance of the basal nutrient medium in the design of hormonally defined medium. In Sato GH, Pardee AB, Sirbasku DA (eds): "Growth of Cells in Hormonally Defined Media." Cold Spring Harbor, NY: Cold Spring Harbor Laboratory, Vol 9, pp 39-60.

Letourneau PC (1975a): Possible roles for cell-to-substratum adhesion in neuronal morphogenesis. Dev Biol 44:77-91.

Letourneau PC (1975b): Cell-to-substratum adhesion and guidance to axonal elongation. Dev Biol 44:92-101.

Levine DW, Wong JS, Wang DIC, Tully WG (1977): Microcarrier cell culture: New methods for research-scale application. Somat Cell Genet 3:149-155.

McClure DB, Hightower MJ, Topp WC (1982): Effect of SV40 transformation on the growth-factor requirements of the rat embryo cell line REF52 in serum-free medium. In Sato GH, Pardee AB, Sirbasku DA (eds): "Growth of Cells in Hormonally Defined Media." Cold Spring Harbor, NY: Cold Spring Harbor Laboratory, Vol 9, pp 345-364.

McKeehan WL, Ham RG (1976): Stimulation of clonal growth of normal fibroblasts with substrata coated with basic polymers. J Cell Biol 71:727-734.

Michler-Stuke A, Bottenstein JE (1982): Defined media for the growth of human and rat glial-derived cells. In Sato GH, Pardee AB, Sirbasku DA (eds): "Growth of Cells in Hormonally Defined Media." Cold Spring Harbor, NY: Cold Spring Harbor Laboratory, Vol 9, pp 959-971.

Poole CA, Slack C, Flint MH (1980): An electron microscopy study of the effects of poly(L-lysine) on chick embryo fibroblasts and chondrocyte cultures. Connect Tissue Res 7:204.

Puymirat J, Loudes C, Faivre-Bauman A, Tixier-Vidal A, Bourre JM (1982): Expression of neuronal functions by mouse fetal hypothalamic cells cultured in hormonally defined medium. In Sato GH, Pardee AB, Sirbasku DA (eds): "Growth of Cells in Hormonally Defined Media." Cold Spring Harbor, NY: Cold Spring Harbor Laboratory, Vol 9, pp 1033-1051.

Ryser HJP, Hancock R (1965): Histones and basic polyamino acids stimulate the uptake of albumin by tumor cells in culture. Science 150:501-504.

Salomon DS, Liotta LA, Foidart JM, Yaar M (1982): Synthesis and turnover of basement-membrane components by mouse embryonal carcinoma cells in serum-free hormone-supplemented medium. In Sato GH, Pardee AB, Sirbasku, DA (eds): "Growth of Cells in Hormonally Defined Media." Cold Spring Harbor, NY: Cold Spring Harbor Laboratory, Vol 9, pp 203-207.

Methods for Preparation of Media, Supplements, and Substrata
for Serum-Free Animal Cell Culture, pages 215–230
© 1984 Alan R. Liss, Inc., 150 Fifth Avenue, New York, NY 10011

12
Preparation of Cellular Fibronectin

Kenneth M. Yamada and Steven K. Akiyama

The adhesive glycoprotein fibronectin exists in at least two major forms, termed cellular fibronectin and plasma fibronectin (recently reviewed by Mosher [1980], Saba and Jaffe [1980], Mosesson and Amrani [1980], Pearlstein et al. [1980], Wartiovaara and Vaheri [1980], Ruoslahti et al. [1981a], Hynes [1981], Hynes and Yamada [1982], and Yamada [1983]). Although plasma fibronectin appears to be produced by the liver [Owens and Cimino, 1982], cellular fibronectin can be synthesized by a wide variety of fibroblastic and epithelial cell types [Mosher, 1980; Saba and Jaffe, 1980; Mosesson and Amrani, 1980; Pearlstein et al., 1980; Wartiovaara and Vaheri, 1980; Ruoslahti et al., 1981a; Hynes, 1981; Hynes and Yamada, 1982; Yamada, 1983]. Other forms of fibronectin that differ substantially in carbohydrate content and/or in apparent molecular weight have been described from certain embryonic, amniotic, or tumor cells [Crouch et al., 1978; Ruoslahti et al., 1981b; Wilson et al., 1981; Hassell et al., 1979; Cossu and Warren, 1983]; it is not known whether the latter forms differ only in glycosylation or whether there are more basic differences.

Cellular and plasma fibronection are similar but not identical [Yamada and Kennedy, 1979]. They share nearly identical amino acid and carbohydrate compositions, secondary and tertiary structures, organization of structural domains, and certain biological activities as attachment proteins (Table I) (reviewed by Hynes and Yamada [1982] and Yamada [1983]). In certain other biological assays measuring cell-to-cell interactions, however, they show threefold to 200-fold differences in specific activity. For example, highly

Membrane Biochemistry Section, Laboratory of Molecular Biology, National Cancer Institute, National Institutes of Health, Bethesda, Maryland 20205

**TABLE I. Comparison of Cellular and Plasma
Fibronectins**

Biological activity tested	Ratio of activities (cellular/plasma)
Cell attachment to type I collagen	$1 \times$
Cell spreading on plastic	$1 \times$
Restoration of normal morphology to transformed cells	$3\text{-}50 \times$
Hemagglutination	$150\text{-}200 \times$
Phagocytosis of gelatinized particles	$1 \times$
Increased saturation density	$+$[a]

[a]Plasma fibronectin lacks activity in the presence of serum.

purified cellular fibronectin is 50 times more active in restoring a more normal shape and cell-cell interaction pattern to certain malignant cells [Yamada and Kennedy, 1979]. Similarly, although cellular fibronectin can stimulate the overgrowth of 3T3 cells in serum-containing medium, plasma fibronectin is without detectable effect [Yamada et al., 1982]. These findings indicate the importance of examining the activity of cellular fibronectin in cell culture systems, especially when plasma fibronectin fails to produce a specific biological effect desired.

Subtle but significant biochemical differences have also been detected between cellular and plasma fibronectins. Cellular fibronectin is a relatively insoluble protein at physiological pH [Yamada et al., 1977], although dimeric cellular fibronectin present in low concentrations in spent culture medium may be more soluble. In contrast, purified plasma fibronectin is a soluble protein that is present in plasma at 0.3 g/liter [Mosher, 1980; Mosesson and Amrani, 1980]. In addition, the two types of fibronectin generally differ in their extent of polymerization, since plasma fibronectin is normally a dimer and cellular fibronectin is primarily a multimer. It is not yet clear whether the degree of polymerization affects the solubility and activity of fibronectin on cultured cells. It should be noted that in cell culture, a minority of plasma fibronectin molecules can also gradually polymerize and form fibrils similar to those formed by cellular fibronectin [McKeown-Longo and Mosher, 1982]; whether this process requires the concomitant presence of cellular fibronectin is not known.

Monoclonal antibodies can distinguish between cellular and plasma fibronectins, and the site of difference does not involve carbohydrates [Atherton and Hynes, 1981]. Direct comparisons of proteolytic fragments of cellular and plasma fibronectins reveal at least three sites of difference between the two forms of fibronectin [Hayashi and Yamada, 1981]. All of these three difference regions are located at internal sites within the molecule, ruling out the possibility that the differences between fibronectins are simply the result of differences in proteolytic processing [Hayashi and Yamada, 1981]. In addition, subtle carbohydrate differences exist between these two major forms [Fukuda et al., 1982]. All of these findings suggest that there are significant structural differences between cellular and plasma fibronectins and that they must be viewed as similar but not interchangeable molecules.

The implications of these differences between cellular and plasma fibronectins in respect to cell culture are clear: Although it is usually simplest to use plasma fibronectin to obtain simple cell-to-substrate attachment activities, particularly since it is much simpler to isolate, there are situations in which it is necessary to test and to utilize cellular fibronectin. To date, no differences in biological activities of cellular fibronectins from different species have been identified. It therefore appears reasonable to use the cellular fibronectin that is the simplest for the individual investigator to obtain for such experiments.

This chapter will first compare published methods for the isolation of cellular fibronectin, then will describe two specific, detailed, recommended protocols. Detailed procedures for the isolation and further purification of plasma fibronectin are available elsewhere [Chen and Mosesson, 1977; Engvall and Ruoslahti, 1977; Yamada, 1982].

APPROACHES TO ISOLATING CELLULAR FIBRONECTIN

There are three different sources of cellular fibronectin: monolayers of cells, spent culture medium, and tissues. Each source has advantages and disadvantages, as discussed briefly below. In general, purified cellular fibronectins must be maintained at alkaline pH (pH 11) to remain soluble. They are neutralized only at the time of final use.

Most laboratories have chosen to use fibronectin isolated from cultured monolayers of cells because of the relative simplicity and high yields and purity with this approach, although other sources can be more useful for specific applications.

Extraction From Cell Cultures

Cellular fibronectin can be isolated after inducing its release from the cell surface by low concentrations of urea; alternatively, it can be isolated from intact monolayers. The most commonly used source of cellular fibronectin is a 1 M urea extract of cultured cells, and this protocol will be described in detail below. The initial extract appears to contain roughly half of fibronectin on the cell surface, and consists of a mixture of monomers, dimers, and multimers

[Yamada et al., 1977]. Although dimers and multimers predominate, the extract probably contains a lower percentage of multimers than total cell surface fibronectin, since the latter may form large, highly insoluble disulfide-linked aggregates. The mechanism of fibronectin release by urea is unknown, and variable amounts of contaminating proteins and proteoglycans can be present in the initial extract. Subsequent purification is usually necessary unless a partially purified preparation is satisfactory.

Fibronectin is a major component of extracellular matrices, and such matrices can be isolated by detergent extraction of cell monolayers [Chen et al., 1978; Hedman et al., 1979; Gospodarowicz et al., 1980]. One approach is to use strong detergents and to solubilize fibronectin from the extract with 8 M urea; this procedure permits the isolation of purified fibronectin from hamster fibroblasts [Carter and Hakomori, 1979]. Another approach, however, is to use gentle detergents that leave a three-dimensional matrix that appears to be structurally intact. The advantage of such a matrix is that it retains the morphology of normal matrices, and can possess biological activities distinct from purified fibronectin [Chen et al., 1978; Hedman et al., 1979; Gospodarowicz et al., 1980]. For example, neural crest cell migration is more rapid on such matrices than on simple fibronectin substrates, even though fibronectin appears to be the critical element [Rovasio et al., 1983]. Such matrices also contain varying amounts of proteoglycans and collagen, with the concomitant advantage of providing important modulating or structural molecules and the disadvantage of using a biochemically impure system.

Cell Culture Medium

Cells release fibronectin into culture medium in substantial quantities, and it can be recovered by various methods [Ali et al., 1977; Fukuda and Hakomori, 1979]. This fibronectin tends to be primarily in the form of dimers, which may account for its relative solubility before isolation. Fibronectin isolated from culture medium could be less active than fibronectin isolated from the cell surface in comparison to plasma fibronectin in certain assays for effects on transformed cells, although this possibility has not been fully tested. A major theoretical advantage to using this source of fibronectin is that cells often release much more fibronectin into culture medium than they retain on the cell surface. In practice, however, fibronectin isolated from culture medium may have significantly higher contamination with other macromolecules such as procollagen and plasma fibronectin from serum. These problems can be decreased by culturing cells in serum-free medium as described below.

FIBRONECTIN FROM TISSUES

A potentially major source of fibronectin might be tissues. Isolation of relatively pure tissue fibronectin has been reported from placenta by extraction with 1 M urea as described below, followed by gelatin affinity chromatography [Fisher and Laine, 1982]. Yields of fibronectin from tissues can be increased by adding heparin to extraction solutions [Bray et al., 1981]. The major concern with these preparations is their yield, purity, and freedom of contamination with plasma fibronectin, which should be evaluated carefully.

COMMERCIAL SOURCES OF FIBRONECTIN

Another "source" of cellular fibronectin is, of course, a commercial company. Although plasma fibronectin is available at reasonable cost from many suppliers, cellular fibronectin appears to be available from only one company as yet (Bethesda Research Laboratories). In tests by our laboratory, several preparations of chick and human cellular fibronectin were found to be moderately pure and to have biological activities similar to our own preparations. The disadvantages of this source of fibronectin are the high price and the absence of further purification beyond the initial urea-extraction step. Although the use of commercially obtained fibronectin appears to be reasonable for initial screening, preparation as described below may be more attractive from the points of view of cost, continual availability, and confidence in the final product.

PROTOCOLS FOR PURIFICATION OF CELLULAR FIBRONECTIN FROM MONOLAYER CULTURES

Two excellent sources of cellular fibronectin are chick and human fibroblasts. The highest yields and purity are usually obtained from chick embryo fibroblasts. Nevertheless, with suitable precautions, cellular fibronectin can be isolated in relatively pure form from human fibroblasts. The current methods of choice appear to be to extract cellular fibronectin from monolayers of chick embryo fibroblasts (this section), or to isolate it from serum-free culture media of human fibroblasts (later section). Although the preparation of chicken cell surface fibronectin is relatively simple if chick embryo fibroblasts are available, human cellular fibronectin may be preferred for studies requiring human material.

Cells

Chick embryo fibroblasts are cultured according to the methods of Vogt [1969]. Ten-day-old chicken embryos (obtained from a local breeder such as Truslow Farms, Chestertown, MD, or SPAFAS, Norwich, CT) are dissociated with 0.25% trypsin and established as primary cultures in 10% serum. Culture medium for primary culture consists of 10% heat-inactivated calf serum (56°C for 30 min), 10% tryptose phosphate broth, 0.5% beef embryo extract (can be omitted), 0.056% additional sodium bicarbonate, 50 units/ml penicillin, and 50 μg/ml streptomycin in Ham's F-10 medium containing glutamine. Cells are cultured for 3 to 4 days without change of culture medium in order to select for fibroblasts. Cells are passaged to secondary cultures with 0.25% trypsin and cultured for 2 more days. Cells are then passaged to tertiary cultures in 850-cm^2 plastic disposable roller bottles (Falcon) at concentrations of 2.5×10^7 to 4×10^7 cells in 100 ml of the above medium, except that the concentration of serum is reduced to 5% to decrease the tendency of the monolayers to detach. Large plastic tissue culture dishes can also be used, with appropriate modifications in amounts of media and extraction solutions.

The medium in roller bottle cultures is changed on days 3, 4, and 5 of culture, and fibronectin is usually extracted from the cell monolayers on day 6. Because it is crucial that the cell monolayers be very heavily confluent before extraction, it may be necessary to delay extraction until cells reach that density. On the other hand, cultures that become too dense will overgrow and detach, since chick fibroblasts show little density-dependent inhibition of growth under these culture conditions.

Human fibroblasts can be obtained from foreskins or other sources, including established lung fibroblast lines such as WI38. Cell lines can vary substantially in their yields of fibronectin, for reasons that are not yet clear. In our experience, the best yields are obtained from low-passage primary fibroblasts. To establish monolayers for urea extraction, these fibroblasts are cultured in 10% fetal calf serum in Dulbecco-Vogt modified Eagle's medium plus glutamine and antibiotics. Supplementation of this medium with 45% Ham's F-10 medium and the use of serum screened for supporting cell growth at low density (e.g., from Hyclone) can increase growth rates and possibly yields. Cells are plated into roller bottles at relatively high initial densities (1:2 or 1:3 split ratio; i.e., cells from about two 150-cm^2 flasks per roller bottle). Cultures are fed every 2-3 days; since these cells show strong density-dependent inhibition of growth, they can be allowed to become heavily confluent with little danger of detachment of monolayers.

Technical points. The purpose of passaging chick embryo cells several times is to select against nonfibroblastic cells, to attempt to reduce rates of procollagen synthesis, and to minimize total work. The numbers of cells plated into roller bottles may need adjustment depending on the rate of cell proliferation as there are changes in the flock of laying hens; more cells should be plated if bottles are not confluent within 6-7 days. If cells grow poorly, media can be supplemented with fetal calf serum (5%, although monolayers may tend to detach from the substrate more easily in such high serum concentrations).

Extraction of Cellular Fibronectin From Monolayers

Reagents needed: 1. The protease inhibitor phenylmethanesulfonyl fluoride (PMSF, available from Calbiochem, Sigma, and other sources— different lots vary in purity and ease of solubilization). A freshly made stock solution at 34.8 mg/ml in 95% ethanol is added to solutions at a ratio of 1:100 with initial vigorous stirring to avoid precipitation; the final concentration will be 2 mM.

2. Ultrapure urea and ammonium sulfate (Schwarz/Mann or BRL).

3. 0.2 M CAPS (cyclohexylaminopropane sulfonic acid; Sigma) buffer, pH 11.0 at 4°C.

4. Buffer A: 0.15 M NaCl, 1 mM $CaCl_2$, 10 mM CAPS, readjusted to pH 11.0. It is simplest to make a $10\times$ concentrate of this solution and to add it to distilled or deionized water at 4°C.

5. Extraction media: Hanks' balanced salt solution and Dulbecco-Vogt modified Eagle's medium.

6. Materials for further purification (if needed): a. Sepharose CL-4B (Pharmacia) and buffer B, 0.1 M NaCl, 1 mM $CaCl_2$, 50 mM CAPS, pH 11.0; b. gelatin-Sepharose, prepared as described elsewhere [Cuatrecasas and Anfinsen, 1971] or purchased from Pharmacia or Pierce.

Protocol. The media listed below should be warmed to 37°C immediately before extraction of cells.

1. Hanks' balanced salt solution (approximately 200 ml per roller bottle).

2. Serum-free Dulbecco-Vogt modified Eagle's medium containing 4 mM glutamine and freshly added 2 mM PMSF (addition after warming to 37°C reduces chances of precipitation), incubated for 15 min prior to use in order to irreversibly inactivate proteases (approximately 100 ml of medium per roller bottle, stored in a beaker with minimal airspace above the liquid, and sealed tightly with Parafilm).

3. 1 M ultrapure urea in the Dulbecco-Vogt medium with PMSF (approximately 25 ml per roller bottle, prepared about 15 min before actual use with the medium in step 2).

In general, considerable care should be taken to avoid exposing materials to proteases from any source. For example, only glassware known to be protease-free should be used, and items should be handled with disposable gloves to avoid the traces of proteases on hands.

Cell monolayers are gently rinsed four times with 25–50 ml of Hanks' salt solution per roller bottle. After they have drained well (care being taken not to allow drying of cells), 25 ml of serum-free Dulbecco-Vogt medium with PMSF is added to each roller bottle and incubated for 1 h at 37°C with rotation at 1 rpm. This step depletes monolayers of adsorbed serum proteins.

The serum-free medium is decanted, and the roller bottles are then rinsed once with the PMSF-containing Dulbecco-Vogt medium left over from the original preparation. After draining well, the bottles then receive 20 ml of freshly prepared 1 M urea in Dulbecco-Vogt medium with PMSF. Bottles are incubated for 2 h at 37°C rotating at 1 rpm. Human cellular fibronectin is often substantially more efficiently extracted with 2 M urea, although there appears to be increased cell damage.

After the extraction with urea, the medium is decanted into protease-free polypropylene centrifuge tubes (50 ml capacity with caps) and centrifuged for 15 min at 25,000g to remove cells and particulates. The supernatant solutions are decanted and either frozen in a polypropylene bottle in dry ice–ethanol for storage below −70°C or processed further. Chick embryo fibroblasts can be saved for a second extraction on the following day, although the fibronectin is often less pure. The medium on cells to be saved is changed to 0.5 M urea in serum-free medium for approximately 30 min; the bottles are then rinsed well with serum-free medium and the medium is replaced with normal culture medium.

The fibronectin-containing extract is precipitated with 70% saturated ammonium sulfate (0.472 g of ultrapure ammonium sulfate crystals per 1 ml of extract, dissolved at 4°C with a stirring bar). After solubilization of the ammonium sulfate, the solution is allowed to stand without mixing for at least 1 h at 0-4°C. The purified fibronectin is collected by centrifugation at 25,000g for 15 min in capped polypropylene tubes.

The supernatant solution is decanted and discarded, and the tubes are allowed to drain well after inversion for at least 15 sec. The walls of the tubes are wiped free of residual ammonium sulfate with a piece of tissue wound tightly around a pair of long forceps. Pellets are then resuspended in the equivalent of 1 ml of 0.2 M CAPS buffer at pH 11.0 per original roller bottle extract (20 ml). It is very important to resuspend the pellet rapidly by quickly squirting in the CAPS and mixing vigorously with a plastic pipet tip, then immediately adjusting the pH to 11 with 5 N NaOH with a protease-free pH meter electrode. (The electrode can be washed ahead of time by soaking in 0.1 HCl for at least several minutes, followed by rapid alternation between 0.1 N NaOH and 0.1 N HCl solutions to release contaminating proteins and proteases.) If the pellet is not resolubilized and adjusted sufficiently rapidly in pH 11, the fibronectin sometimes forms an insoluble, clear gel that adamantly resists solubilization.

The resuspended fibronectin will be pink to purple from the residual phenol red in the serum-free medium. It is placed into a flattened dialysis bag (to increase the surface area of the bag) and dialyzed against buffer A with vigorous stirring at 4°C. The usual container is a 4-liter beaker covered with plastic wrap pressed down onto the surface of the liquid and held in place with a sheet of aluminum foil molded so as to exclude all air above the liquid. This precaution helps to reduce the entry of CO_2 from room air into the alkaline solution, which would otherwise lower the pH and form a calcium carbonate precipitate.

There are two possibilities at this point: a) If the purity of fibronectin is not critical, it can be dialyzed further and then used. The usual minor contaminants are procollagens and proteoglycans (although amounts of the latter are decreased by the ammonium sulfate precipitation step). b) If additional purity is needed, samples are purified by molecular sieve chromatography or (for human fibronectin) gelatin affinity chromatography.

If purity is not critical, dialysis is continued, the solution being changed a total of 3-4 times over a period of 24-48 h until the fibronectin solution is colorless. An aliquot is taken for protein determination and the sample is frozen as described below under Storage.

For routine further purification, the fibronectin is reprecipitated after dialysis for only 16-18 h by the addition of ammonium sulfate to 50% saturation (0.313 g/ml of original solution, neutralized to pH 7). After collecting by centrifugation, it is resuspended to approximately 5 mg/ml (often a fourfold to sixfold increase in concentration from the original dialyzed material). After dialysis overnight against buffer A, the fibronectin is centrifuged at 25,000g for 15 min, then applied to a 2.5-cm × 100-cm Sepharose CL-4B column preequilibrated with buffer B. This cross-linked agarose column is stable to alkaline pH, and the increased buffer concentration compared to buffer A helps to decrease pH fluctuation during fraction collection. (Occasionally, columns can nonspecifically bind fibronectin in large amounts even at pH 11. This problem can be alleviated by pretreating the column with 1 mg/ml crystalline bovine serum albumin, followed by extensive washing before use.)

In a routine preparation, 40-50 mg of fibronectin in 8 ml is chromatographed over a period of 18-24 h; 6-ml fractions are collected in polystyrene or polypropylene tubes. Aliquots are taken from each tube to determine protein by the method of Lowry et al. [1951]; although protein can be measured by absorbance, there is considerable danger of misleading readings as a result of fibronectin adsorption to the quartz walls of cuvettes. A sample protein elution profile is shown in Figure 1 (see Yamada et al. [1977] for details). The peak in the void volume fractions consists primarily of fibronectin

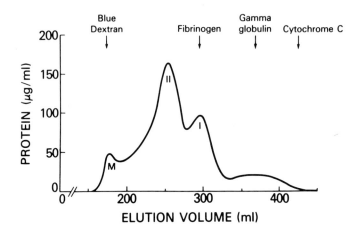

Fig. 1. Purification of chicken cellular fibronectin by gel filtration. Fibronectin was resolved by molecular sieve chromatography on a 2.5-cm × 100-cm Sepharose CL-4B column eluted by buffer B. Elution of reference standard peaks is indicated by the vertical arrows. M, multimer; II, dimer; I, monomer. A detailed SDS gel analysis of this fractionation procedure is provided by Yamada et al. [1977].

multimers (M), plus residual procollagen and proteoglycan contaminants. The major peak contains fibronectin dimers (II). The trailing peak or shoulder contains fibronectin monomers (I). We usually save the fibronectin dimer peak, and occasionally also save the multimer peak for comparative purposes.

The fractions to be saved are neutralized with 1 N HCl, and the fibronectin is precipitated as described above with 70% ammonium sulfate. The pellet is rapidly resuspended in CAPS buffer as above to a concentration of 1-2 mg/ml. The solution is then dialyzed against two changes of buffer A for 18-30 h.

Further purification of extracted human cellular fibronectin. Preparations of human cellular fibronectin after extraction from monolayers with 2 M urea are often contaminated with various cellular proteins that must be removed prior to use. The most effective procedure is affinity chromatography on gelatin-Sepharose affinity columns. Detailed instructions for washing the gelatin and preparing such columns are presented elsewhere [Yamada, 1982], or gelatin-agarose can be purchased (Pharmacia or Pierce). Columns are prewashed with 8 M urea, then equilibrated with phosphate-buffered saline. Extracts containing fibronectin are diluted to 0.5 M urea and applied to an affinity columns as described for plasma fibronectin [Engvall and Ruoslahti, 1977]. Cellular fibronectin is unusual in that recoveries from affinity columns tend to be substantially greater if the ratio of fibronectin to gelatin is high. The fibronectin bound to the affinity column is washed extensively with column buffer, then eluted in 8 M (not 4 M) urea in 10 mM Tris-Cl, pH 7.0. After dialysis for several hours against 0.15 M NaCl, 10 mM Tris-Cl, pH 7.0, the fibronectin is precipitated with 70% ammonium sulfate and further processed as described above for chick cellular fibronectin.

PROTOCOL FOR ISOLATION OF HUMAN CELLULAR FIBRONECTIN FROM SERUM-FREE MEDIUM

Large amounts of human cellular fibronectin can be isolated from roller bottle cultures of human foreskin fibroblasts maintained in serum-free medium (S.K. Akiyama and K.M. Yamada, manuscript in preparation). It is safest to use serum-free media, since the concentration of plasma fibronectin in serum is approximately 100 μg/ml, whereas the concentrations of cellular fibronectin secreted into media are only 1-3 μg/ml. Affinity purification of fibronectin from serum-containing media would produce preparations containing primarily plasma rather than cellular fibronectin, as well as varying amounts of other contaminating serum proteins.

Cells in Serum-Free Medium

The strategy is to culture cells to confluence in serum-containing medium, then to transfer them to serum-free medium for long-term maintenance and harvesting of fibronectin. Although the cells do not appear to proliferate extensively in the latter media, they remain viable and produce large amounts of fibronectin for weeks to months of culture, even with daily harvests of medium for fibronectin; it is therefore possible to obtain many milligrams of fibronectin from each roller bottle culture.

Human foreskin fibroblasts are cultured in plastic tissue culture flasks in 45% Dulbecco-Vogt modified Eagle's medium, 45% Ham's F-12 medium, and 10% fetal calf serum (Sterile Systems, Logan, UT) supplemented with 2 mM additional glutamine and antibiotics (50 units/ml penicillin and 50 μg/ml streptomycin). After the cells reach confluence, cultures are passaged approximately 1:3 into 850-cm^2 plastic roller bottles (Falcon); i.e., cells from two 150-cm^2 flasks per roller bottle. The cells are cultured in this medium with 10% serum until densely confluent. The cells are then rinsed three times in serum-free medium, and cultured in CRCM-30 serum-free medium (Gibco) containing an additional 2 mM glutamine and penicillin-streptomycin as above. After 24 h, this medium is discarded and replaced with fresh serum-free medium. Conditioned medium containing fibronectin is then harvested at intervals of 24-48 h.

Technical points. In our experience, as human fibroblasts are repeatedly passaged with trypsin, their rate of fibronectin production declines; cells at passages 15 and older tend to secrete much less fibronectin. An additional concern is that fibronectin from senescent cells is probably abnormal in structure [Chandrasekhar and Millis, 1980]. Although other serum-free media can be used, CRCM-30 appeared the most effective of those tested for long-term production of fibronectin. After many weeks or months, the cells often begin to detach from the substrate. Such cultures are either discarded or grown back to confluence in serum-containing medium and the harvesting process is repeated.

Purification of Human Fibronectin From Serum-Free Medium

Reagents needed: 1. Materials listed in Reagents Needed in the section on purification from monolayers, except the extraction media, Sepharose CL-4B, and buffer B.

2. Buffer C: 4 M urea, 0.5 M NaCl, 50 mM CAPS (cyclohexylaminopropane sulfonic acid), pH 11.0.

3. BSA-treated gelatin-Sepharose: Gelatin-Sepharose is washed with 3 bed volumes of 5 mg/ml crystalline bovine serum albumin in 0.15 M NaCl, 10 mM Tris-HCl, pH 7.0, followed by 5 volumes of 8 M urea, 50 mM Tris-HCl, pH 7.0, and finally by 0.15 M NaCl, 10 mM Tris-HCl, pH 7.0.

Serum-free medium (100 ml per roller bottle) collected after incubation with human fibroblasts for 24–48 h is decanted and adjusted immediately to 2 mM PMSF (using a 1:100 dilution from ethanol as described above). After centrifugation at 25,000g for 15 min at 4°C, the supernatant solution is decanted and either processed further or stored frozen below −70°C.

Cellular fibronectin is isolated from the medium by incubation with 1 ml (packed volume) of BSA-treated gelatin-Sepharose per liter of conditioned medium. The beads are maintained in suspension for 24 h at 4°C, preferably by rotation in polypropylene bottles (e.g., in four nearly filled 2-liter polypropylene bottles rotating at 3 rpm on a roller bottle apparatus). After the 24 h rotation period for binding of fibronectin to the beads, the beads are separated from the medium a) by allowing them to settle by gravity and decanting the medium, b) by sedimenting the beads by centrifugation, or c) by filtering out the beads. A sequential combination of steps (a) and (b) is simplest. The fibronectin-depleted medium is then discarded, and the beads are poured into a column, e.g., a 1.5-cm-diameter column for 10 ml of beads.

The column is washed at room temperature with 4 bed volumes of 0.15 M NaCl, 10 mM Tris-HCl, pH 7.0. The fibronectin is eluted with buffer C. The protein peak is determined as described above, and the fractions containing protein are combined, dialyzed against buffer A, and stored as described above.

Technical points. Serum-free CRCM-30 conditioned for 48 h by early-passage human foreskin fibroblasts contains approximately 1-2 μg/ml cellular fibronectin, which can be readily recovered by binding to gelatin-Sepharose. This quantity of secreted fibronectin is approximately 10% of the total fibronectin present in the cell monolayer as determined by gel electrophoresis of SDS (sodium dodecyl sulfate) extracts of cells cultured in serum-free or serum-containing media. The amount of fibronectin present after incubation for only 24 h is usually slightly greater than half of the amount after 48 h; harvesting the medium at intervals greater than 2 days does not appreciably increase the concentration of fibronectin present in the medium under the culture conditions described.

The cellular fibronectin isolated as described above from serum-free medium is approximately 90% pure, as determined by Coomassie blue staining of SDS-polyacrylamide gels. The impurities appear to be fragments of intact fibronectin. In addition, small amounts of the fibronectin that bind to gelatin-Sepharose remain tightly bound, and this residual material is only partially removed even by 8 M urea.

The gelatin-Sepharose can be reused repeatedly if the urea is washed out after the elution of fibronectin and is stored in neutral buffer (0.15 M NaCl, 10 mM Tris-HCl, pH 7.0) at 4°C. Before each use, the gelatin-sepharose is washed with 8 M urea and reequilibrated in the NaCl/Tris buffer before use.

STORAGE OF CELLULAR FIBRONECTIN

After dialysis is completed, aliquots of the diffusate solution outside of the dialysis bag are saved as controls for biological experiments and as blanks for protein determinations. The fibronectin preparation is centrifuged at 25,000g for 10-15 min in a capped tube. Samples are taken for protein determination and aliquots in polypropylene tubes are frozen rapidly in liquid nitrogen; if none is available, powdered dry-ice can be substituted (though there is a risk of altering pH with the CO_2). Samples are stored at temperatures below $-70°$C or in liquid nitrogen. Samples cannot be stored for more than a few days at $-20°$C, and even at $-70°$C there may be some decrease in biological activity after several months. Samples in liquid nitrogen appear to be stable for at least several years.

Samples are defrosted in room temperature water with occasional gentle agitation. If insoluble material is noted, the pH is probably incorrect and all samples from that batch should be readjusted to pH 11 with 1 N NaOH immediately after defrosting. Cellular fibronectin tends to lose biological activity after a decrease in pH (e.g., if exposed to air in an uncapped tube) or after repeated freezing and thawing. It is safest to store the fibronectin in small aliquots that are defrosted only once for experiments.

ASSAY AND USE OF CELLULAR FIBRONECTIN

A variety of biological activities of purified cellular fibronectin have been described in the literature (see, for example, Hynes and Yamada [1982]), and it can be assayed by several possible biological assays. Two simple assays involve cell spreading and hemagglutination by published protocols [Grinnell and Feld, 1980; Yamada et al., 1975]. In our experience, different preparations of chick cellular fibronectin prepared as described above show little variation in biological activity as long as the precautions described under Storage are observed. As a quick test of activity, concentrations of 1-5 μg/ml can be added in serum-free medium to culture dishes, then the spreading of any conveniently available fibroblastic cell line with low amounts of cell surface fibronectin can be examined for fibronectin-induced spreading of cells after 30-60 min at 37°C (cells such as BHK, L929, and rounded transformed fibroblasts are good candidates, whereas HeLa and primary fibroblasts producing endogenous fibronectin are not).

Cellular fibronectin can be added to tissue cultures either as a coating to substrates or simply as an additive to routine culture medium. It appears to differ from plasma fibronectin in that the presence of serum proteins such as albumin does not inhibit its effects on cell behavior as assayed by alterations in morphology of transformed fibroblasts. If used to coat culture dishes, aliquots of fibronectin are added with sterile pipets to phosphate-buffered saline and allowed to adsorb to dishes for 5-60 min. After extensive rinsing, the substrates are used immediately without drying. If added directly to cultures, cellular fibronectin is added at final concentrations of 25-50 μg/ml with thorough mixing. If the culture medium is well buffered and the fibronectin stock was > 1 mg/ml, neutralization is not necessary. Medium containing fibronectin should not be Millipore-filtered, since fibronectin can bind to nitrocellulose. In our experience, culture media containing antibiotics have never shown contamination with microorganisms, perhaps because they are inactivated by the alkaline pH in which stocks of fibronectin are stored. Control cultures routinely receive equal amounts of the buffer A diffusate solution from the last dialysis of the same batch of fibronectin.

ACKNOWLEDGMENTS

S.K. Akiyama was supported by grant No. CA 06782 of the National Cancer Institute, National Institutes of Health.

REFERENCES

Ali IU, Mautner V, Lanza R, Hynes RO (1977): Restoration of normal morphology, adhesion and cytoskeleton in transformed cells by addition of a transformation-sensitive surface protein. Cell 11:115-126.

Atherton BT, Hynes RO (1981): Difference between plasma and cellular fibronectins located with monoclonal antibodies. Cell 25:133-141.

Bray BA, Mandel I, Turino GM (1981): Heparin facilitates the extraction of tissue fibronectin. Science 214:793-795.

Carter WG, Hakomori S (1979): Isolation of galactoprotein a from hamster embryo fibroblasts and characterization of the carbohydrate unit. Biochemistry 18:730-738.

Chandrasekhar S, Millis AJ (1980): Fibronectin from aged fibroblasts is defective in promoting cellular adhesion. J Cell Physiol 103:47-54.

Chen AB, Mosesson MW (1977): An improved method for purification of the cold-insoluble globulin of human plasma (CIg). Anal Biochem 79:144-151.

Chen LB, Murray A, Segal RA, Bushnell A, Walsh ML (1978): Studies on intercellular LETS glycoprotein matrices. Cell 14:377-391.

Cossu G, Warren L (1983): Lactosaminoglycans and heparan sulfate are covalently bound to fibronectins synthesized by mouse stem teratocarcinoma cells. J Biol Chem 258:5603-5607.

Crouch E, Balian G, Holbrook K, Duksin D, Bornstein P (1978): Amniotic fluid fibronectin: Characterization and synthesis by cells in culture. J Cell Biol 78:701-715.

Cuatrecasas P, Anfinsen CB (1971): Affinity chromatography. Methods Enzymol 22:345-378.

Engvall E, Ruoslahti E (1977): Binding of soluble form of fibroblast surface protein, fibronectin, to collagen. Int J Cancer 20:1-5.

Fisher SJ, Laine RA (1982): Studies on the major carbohydrate from human placental cellular fibronectin. In Hawkes S, Wang JL (eds): "Extracellular Matrix." New York: Academic, pp 247-251.

Fukuda M, Hakomori S (1979): Proteolytic and chemical fragmentation of galactoprotein a, a major transformation-sensitive glycoprotein released from hamster embryo fibroblasts. J Biol Chem 254:5442-5450.

Fukuda M, Levery SB, Hakamori S (1982): Carbohydrate structure of hamster plasma fibronectin: Evidence for chemical diversity between cellular and plasma fibronectins. J Biol Chem 257:6856-6860.

Gospodarowicz D, Vlodavsky I, Savion N (1980): The extracellular matrix and the control of proliferation of vascular endothelial and vascular smooth muscle cells. J Supramol Struct 13:339-372.

Grinnell F, Feld MK (1980): Spreading of human fibroblasts in serum-free medium: Inhibition by dithiothreitol and the effect of cold-insoluble globulin (plasma fibronectin). J Cell Physiol 104:321-334.

Hassell JR, Pennypacker JP, Kleinman HK, Pratt RM, Yamada KM (1979): Enhanced cellular fibronectin accumulation in chondrocytes treated with vitamin A. Cell 17:821-826.

Hayashi M, Yamada KM (1981): Difference in domain structures between plasma and cellular fibronectins. J Biol Chem 256:11292-11300.

Hedman K, Kurkinen M, Alitalo K, Vaheri A, Johansson S, Höök M (1979): Isolation of the pericellular matrix of human fibroblast cultures. J Cell Biol 81:83-91.

Hynes RO (1981): Fibronectin and its relation to cellular structure and behavior. In Hay ED (ed): "Cell Biology of the Extracellular Matrix." New York: Plenum, pp 295-334.

Hynes RO, Yamada KM (1982): Fibronectins: Multifunctional modular glycoproteins. J Cell Biol 95:369-377.

Lowry OH, Rosebrough NJ, Farr AL, Randall RJ (1951): Protein measurement with the Folin phenol reagent. J Biol Chem 193:265-275.

McKeown-Longo PJ, Mosher DF (1983): Binding of plasma fibronectin to cell layers of human skin fibroblasts. J Cell Biol 97:466-472.

Mosesson MW, Amrani DL (1980): The structure and biologic activities of plasma fibronectin. Blood 56:145-148.

Mosher DF (1980): Fibronectin. Prog Hemostas Thrombos 5:111-151.

Owens MR, Cimino CD (1982): Synthesis of fibronectin by the isolated perfused rat liver. Blood 59:1305-1309.

Pearlstein E, Gold L, Garcia-Pardo A (1980): Fibronectin: A review of its structure and biological activity. Mol Cell Biochem 29:103-127.

Rovasio RA, DeLouvee A, Yamada KM, Timpl R, Thiery JP (1983): Neural crest cell migration: Requirements for exogenous fibronectin and high cell density. J Cell Biol 96:462-473.

Ruoslahti E, Engvall E, Hayman EG (1981a): Fibronectin: Current concepts of its structure and functions. Collagen Res 1:95-128.

Ruoslahti E, Engvall E, Hayman EG, Spiro RG (1981b): Comparative studies on amniotic fluid and plasma fibronectin. Biochem J 193:295-299.

Saba TM, Jaffe E (1980): Plasma fibronectin (opsonic glycoprotein): Its synthesis by vascular endothelial cells and role in cardiopulmonary integrity after trauma as related to reticuloendothelial function. Am J Med 68:577-594.

Vogt PK (1969): Focus assay of Rous sarcoma virus. In Habel K, Salzman NP (eds): "Fundamental Techniques in Virology." New York: Academic, pp 198-211.

Wartiovaara J, Vaheri A (1980): Fibronectin and early mammalian embryogenesis. Dev Mammals 4:233-266.

Wilson BS, Ruberto G, Ferrone S (1981): Sulfation and molecular weight of fibronectin shed by human melanoma cells. Biochem Biophys Res Commun 101:1047-1051.

Yamada KM (1982): Isolation of fibronectin from plasma and cells. In Furthmayr H (ed): "Immunochemistry of the Extracellular Matrix." Boca Raton, FL: CRC Press, Vol 1, pp 111-123.

Yamada KM (1983): Cell surface interactions with extracellular materials. Annu Rev Biochem 52:761-799.

Yamada KM, Kennedy DW (1979): Fibroblast cellular and plasma fibronectins are similar but not identical. J Cell Biol 80:492-498.

Yamada KM, Yamada SS, Pastan I (1975): The major cell surface glycoprotein of chick embryo fibroblasts is an agglutinin. Proc Natl Acad Sci USA 72:3158-3162.

Yamada KM, Schlesinger DH, Kennedy DW, Pastan I (1977): Characterization of a major fibroblast cell surface glycoprotein. Biochemistry 16:5552-5559.

Yamada KM, Kennedy DW, Hayashi M (1982): Fibronectin in cell adhesion, differentiation, and growth. In "Growth of Cells in Hormonally Defined Media." Cold Spring Harbor, NY: Cold Spring Harbor Laboratory.

Methods for Preparation of Media, Supplements, and Substrata
for Serum-Free Animal Cell Culture, pages 231–238
© 1984 Alan R. Liss, Inc., 150 Fifth Avenue, New York, NY 10011

13
Isolation of Laminin

Steven R. Ledbetter, Hynda K. Kleinman, John R. Hassell,
and George R. Martin

Laminin is a large glycoprotein ($M_r = 10^6$) that has been demonstrated, by immunohistology, to be a constituent of all basement membranes [Timpl et al., 1979; Chung et al., 1979; Rohde et al., 1979; Foidart et al., 1980]. Laminin is of particular interest because it is thought to function as a specific cell attachment protein binding epithelial cells to basement membrane (type IV) collagen [Terranova et al., 1980]. Laminin may also function to mediate early stages in development, since it appears in the embryo at the morula stage [Leivo et al., 1980] and is the first basement membrane component observed in the embryogenesis of the nephrogenic mesenchyme [Ekblom et al., 1980].

Electron microscopy of laminin, after rotary shadowing, shows an extended cross-shaped structure, composed of four arms with globular domains connected by rod-like elements [Engel et al., 1981] (Fig. 1). Laminin is composed of three B-chains* (~200,000 daltons each) and one A-chain (~400,000 daltons) held together by disulfide bonds [Timpl et al., 1979] (Fig. 2). Although the A- and B-chains have similar amino acid composition [Sakashita et al., 1980], they are not identical and probably represent different gene products [Chung et al., 1979; Cooper et al., 1981, Liotta et al., 1981]. Physical and electron microscope studies have shown that the A- and B-chains are present in the same molecule [Engel et al., 1981]. The long arm of

Laboratory of Developmental Biology and Anomalies, National Institute of Dental Research, National Institutes of Health, Bethesda, Maryland 20205

*Rao et al. [1982] have used a different nomenclature, referring to the 200,000 component as the α-chain and the longer chain (Mr = 400,000) as the β-component.

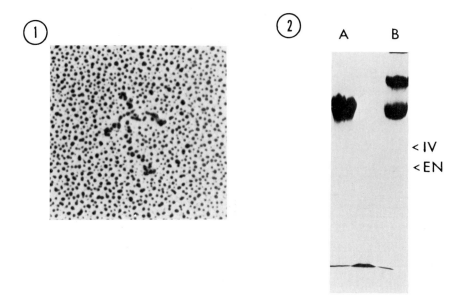

Fig. 1. Rotary shadowing image of a single laminin molecule (photograph courtesy of Dr. Lance Liotta).

Fig. 2. Fibronectin (A) and laminin (B) were electrophoresed on a 5% polyacrylamide gel in the presence of 2% SDS and DTT. The B-chain of laminin (~ 200,000 daltons) migrates slightly slower than fibronectin but is clearly separated from the A-chain of laminin (~ 400,000 daltons). Arrows indicate where the two chains of basement membrane (type IV) collagen and entactin (EN) would migrate.

laminin (75 nm) is apparently formed from one A-chain and the three apparently identical short arms (35 nm) are formed from three B-chains [Ott et al., 1982]. Circular dichroism measurements of laminin indicate a secondary structure with 30% α-helix, some β-structure, and the remainder (50%) as random coil [Ott et al., 1982]. The molecule shows a reversible thermal transition at 58°C.

The extended conformation of laminin may serve the biologic functions of this molecule by allowing multiple interactions with other basement membrane components as well as with the cell surface. For example, laminin binds to type IV collagen [Woodley et al., 1983] and to heparan sulfate proteoglycan [Sakashita et al., 1980; Del Rosso et al., 1981; Woodley et al., 1983], both of which are components of basement membranes. The binding site for heparin and presumably for heparan sulfate proteoglycan has been localized to the A-chain [Ott et al., 1982], whereas the collagen binding domain was localized to the B-chains [Rao et al., 1982; Terranova et al., 1983]. The cell binding site has been localized to the three intersecting short arms of laminin [Terranova et al., 1983].

Laminin has diverse effects on cells in culture. Laminin promotes the attachment and spreading of the epithelial cell line PAM 212 to type IV (basement membrane) collagen and to tissue culture plastic but not to other collagen types [Terranova et al., 1980]. It also promotes the attachment of hepatocytes [Carlsson et al., 1981; Johannson et al., 1981], intestinal epithelial cells [Burrill et al., 1981], and Schwann cells [Palm and Furcht, 1982] to plastic petri dishes. Under some conditions certain types of fibroblasts may also bind to laminin [Couchman et al., 1983]. Laminin can stimulate the growth of lens epithelial cells [Reddan et al., 1982] and an embryonal cell line [Rizzino et al., 1980]. In nervous tissue, laminin promotes the outgrowth of neurites and influences the adhesion, growth, morphology, and migration of Schwann cells [Baron-Von Evercorren et al., 1982; Palm and Furcht, 1982; McCarthy et al., 1982] and probably of other cells as well. Such activities could be important in the formation and repair of various tissues.

Laminin may promote the growth and differentiation of cells in culture by initiating the deposition of basement membrane, as recently demonstrated in cultured thyroid epithelial cells [Garbi and Wellman, 1982]. Since laminin appears early in embryonic development, prior to the appearance of visible basement membranes, it may serve as a template for the deposition of other basement membrane components. Recent studies have in fact shown that laminin precipitates type IV collagen in vitro [Kleinman et al., 1982].

Laminin is also of interest to clinicians because it has recently been shown that individuals infected with trypanosomes produce high levels of antibodies against laminin [Szarfman et al., 1982]. Such antibodies are observed in patients with Chaga disease and African trypanosomal diseases and could contribute to the fibrotic lesions in these diseases by eliciting an autoimmune response. Antibodies to laminin and type IV collagen, when injected into animals, localize in basement membranes in the lungs and kidneys and initiate immunologic reactions that cause pathologic changes similar to those seen in Goodpasture syndrome [Yaar et al., 1982].

Laminin can be isolated in native form from the Engelbreth-Holm-Swarm (EHS) tumor [Foidart et al., 1982] and from other such tumors [Wewer et al., 1981; Martinez-Hernandez et al., 1982]. It has been isolated from the medium of a variety of cell cultures [Chung et al., 1979; Cooper et al., 1982; Leivo et al., 1982; Salomon et al., 1982], and fragments of laminin can be isolated from placenta digested with pepsin [Risteli and Timpl, 1981]. There is little or no detectable laminin in serum. This chapter will deal with some of the procedures used to isolate laminin.

PURIFICATION
Purification of Laminin From the EHS Tumor

The tumor is grown subcutaneously or in the hind limb muscles of C57BL mice and harvested when approximately 1–2 cm in diameter [Orkin et al., 1977]. Tumors are excised, dissected free of capsular material, and homogenized (two volumes of solvent/volume of tumor) at 4°C in 3.4 M NaCl–0.05 M Tris-HCl, pH 7.4 containing 0.004 M EDTA and 0.002 M N-ethylmaleimide (NEM). This and all subsequent procedures (Fig. 3) are carried out at 4°C

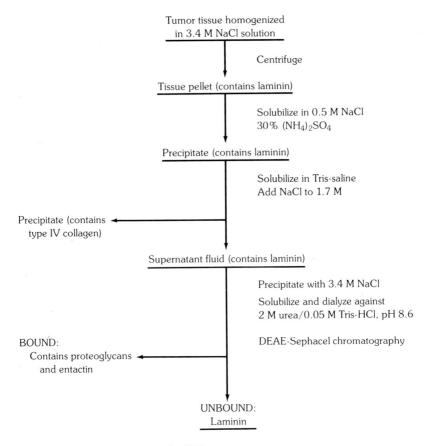

Fig. 3. Isolation of laminin from the EHS tumor.

to minimize proteolysis. Insoluble material, including laminin, is collected by centrifugation (47,500g for 20 min), and the precipitate is resuspended and recentrifuged twice more to remove serum and other soluble proteins. To solubilize laminin the residue is extracted overnight in two volumes of 0.5 M NaCl-0.05 M Tris-HCl, pH 7.4, containing EDTA and NEM, as above. The insoluble residue is again removed by centrifugation and reextracted in the same buffer. The supernatant fluids from the two extractions are pooled. Ammonium sulfate is added to the combined supernatant fluid to a final concentration of 30% of saturation (16.4 g/100 ml). After 1 h, the precipitate that forms is collected by centrifugation, resuspended by stirring for several hours in 0.15 M NaCl-0.05 M Tris, pH 7.4 (Tris-saline), and dialyzed against Tris-saline. Solid NaCl is added to the extract to give a final concentration of 1.7 M, stirred for 1 h, and centrifuged, and the supernatant fraction is collected. The small precipitate forming under these conditions contains any type IV collagen in the extract. Solid NaCl is dissolved in the supernatant fraction to a concentration of 3.4 M, and the precipitate that forms is collected by centrifugation. The pellet is dissolved in Tris-saline by stirring for several hours and is then dialyzed against 2 M urea containing 0.05 M Tris-HCl, pH 8.6 and 0.5 M NaCl and is diluted with an equal volume of 2 M urea-0.05 M Tris-Hcl, pH 8.6 to give a final NaCl concentration of 0.25 M. Sulfated proteoglycans, anionic glycoproteins, and nucleic acids are removed by passing the sample over a column of DEAE-Sephacel equilibrated in 2 M urea-0.05 M Tris-HCl, pH 8.6 containing 0.25 M NaCl, since they bind and laminin does not. The unbound fraction, containing laminin, is dialyzed against 2 M urea-0.05 M Tris-HCl, pH 8.6 and rechromatographed over a column of DEAE-Sephacel equilibrated in 2 M urea-0.05 M Tris-HCl, pH 8.6. Laminin is present in the unbound fraction and should be dialyzed against tris-saline or culture medium.

The EHS tumor yields 5-10 mg of laminin/g wet tissue weight as determined by the ELISA assay [Kleinman et al, 1982]. Purified laminin is active in promoting cell attachment at protein concentrations of 1-5 μg/ml/35 mm petri dish [Terranova et al., 1980]. Laminin can be used in cell culture either by drying it onto the surface of petri dishes or by adding soluble laminin to the culture medium. Laminin is not very stable in culture medium at 37°C, and addition of fresh laminin may be required if its effects are to be studied over several days.

Storage

Laminin in saline or culture medium may be stored in small (1-5 ml) aliquots in solution at −70°C without loss of biologic activity for up to 12 months, whereas storage at 4° or −20°C results in some loss of biologic activity. Alternatively, laminin can be dialyzed against 0.01 M CAPS (cyclo-hexylaminopropane sulfonic acid) buffer at pH 11.0 and lyophilized. Laminin prepared in this manner is available from Bethesda Research Laboratories, Gaithersburg, MD. Laminin is also available from EY Laboratories, San Mateo, CA.

Isolation of Laminin From Medium of Cultured Cells

There are a number of cell lines that are reported to synthesize and secrete laminin including the teratocarcinoma-derived mouse cell lines where laminin is the major secretory product [Chung et al., 1979; Sakashita and Ruoslahti, 1980; Howe and Solter, 1980; Leivo et al., 1982]. Purification of laminin from cell culture medium can be accomplished by taking advantage of the noncovalent binding of laminin to heparin [Sakashita et al., 1980]. The culture medium is dialyzed against saline, passed through a column of Sepharose 4B to eliminate nonspecific binding, and applied to a collagen-Sepharose affinity column to remove any fibronectin. The unbound sample is dialyzed against 0.05 M Tris-HCl, pH 7.4 containing 0.1 M NaCl and applied to a heparin-Sepharose column (Pharmacia) equilibrated in the same buffer. Laminin, in essentially pure form, can be eluted from the heparin column with 0.5 M NaCl in 0.05 M Tris-HCl, pH 7.4. The laminin can then be dialyzed as described above and stored.

Laminin can also be isolated by affinity chromatography on immobilized *Griffonia simplicifolia* I lectin [Shibata et al., 1982]. This lectin recognizes and binds terminal α-D-galactopyranosyl groups on laminin. It is eluted with 0.05 methyl α-D-glactopyranoside in saline, dialyzed, and treated as above. In general, as is the case with most materials prepared from cultured cells, little is obtained.

Isolation of Laminin From Tissues

Laminin has also been prepared from placenta and kidney following pepsin digestion [Risteli and Timpl, 1981]. The enzymatic digestion releases a large fragment (about 250,000 daltons) termed P1 that contains the cell-binding site and has antigenic similarity to laminin isolated from the EHS tumor. The final yield from human placenta is 10–15 mg/kg wet tissue weight.

REFERENCES

Baron-Van Evercooren A, Kleinman HK, Shinichi O, Marangos P, Schwartz J, Dubois-Dalcq M (1982): Nerve growth factor, laminin, and fibronectin promote neurite growth in human fetal sensory ganglia cultures. J Neurosci 8:179-193.

Burrill P, Bernardini I, Kleinman H, Kretchmer N (1981): Effect of serum, fibronectin, and laminin on adhesion of rabbit intestinal epithelial cells in culture. J Supramol Struct 16:385-392.

Carlsson R, Engvall E, Freeman A, Ruoslahti E (1981): Laminin and fibronectin in cell adhesion: Enhanced adhesion of cells from regenerating liver to laminin. Proc Natl Acad Sci USA 78:2403-2406.

~Chung A, Jaffe R, Freeman I, Vergnes JP, Braginski J, Carlin B (1979): Properties of a basement membrane-related glycoprotein synthesized in culture by a mouse embryonal carcinoma-derived cell line. Cell 16:277-287.   ✔ ℓ C℮ \2

~Cooper A, Kurkinen M, Taylor A, Hogan JB (1981): Studies on the biosynthesis of laminin by murine parietal endoderm cells. Eur J Biochem 119:189-197.   ✔ ℓ ℮ɯ G̶

Couchman J, Höök M, Rees D, Timpl R (1983): Adhesion, growth, and matrix production by fibroblasts on laminin substrates. J Cell Biol 96:177-183.

Del Rosso M, Cappelletti R, Viti M, Vannucchi S, Chiarugi V (1981): Binding of the basement membrane glycoprotein laminin to glycosaminoglycans. Biochem J 199:699-704.

Ekblom P, Alitalo K, Vaheri A, Timpl R, Saxén L (1980): Induction of a basement membrane glycoprotein in embryonic kidney: Possible role of laminin in morphogenesis. Proc Natl Acad Sci USA 77:485-489.

Engel J, Odermatt E, Engel A, Madri J, Furthmayr H, Rohde H, Timpl R (1981): Shapes, domain organizations and flexibility of laminin and fibronectin, two multifunctional proteins of the extracellular matrix. J Mol Biol 150:97-120.

Foidart JM, Bere E, Yaar M, Rennard S, Gullino M, Martin GR, Katz S (1980): Distribution and immunoelectron microscopic localization of laminin, a noncollagenous basement membrane glycoprotein. Lab Invest 42:336-342.

Foidart JM, Timpl R, Furthmayr H, Martin GR (1982): Laminin, a glycoprotein from basement membranes. In Furthmayr H (ed): "Immunochemistry of the Extracellular Matrix." Cleveland: CRC Press, pp 135-141.

Garbi C, Wollman SH (1982): Basal lamina formation on thyroid epithelia in separated follicles in suspension culture. J Cell Biol 94:489-492.

Howe C, Solter D (1980): Identification of noncollagenous basement membrane glycopeptides synthesized by a mouse parietal entoderm and an entodermal cell line. Dev Biol 77:480-487.

Johansson S, Kjellen L, Höök M, Timpl R (1981): Substrate adhesion of rat hepatocytes: A comparison of laminin and fibronectin as attachment proteins. J Cell Biol 90:260-264.

Kleinman HK, Woodley DT, McGarey ML, Gehron Robey P, Hassell J, Martin GR (1982): Interactions and assembly of basement membrane components. In Hawkes S, Wang JL (eds): "Extracellular Matrix." New York: Academic, pp 45-54.

Leivo I, Vaheri A, Timpl R, Wartiovaara J (1980): Appearance and distribution of collagens and laminin in the early mouse embryo. Dev Biol 76:100-114.

~Leivo I, Alitalo K, Risteli L, Vaheri A, Timpl R, Wartiovaara J (1982): Basal lamina glycoproteins laminin and type IV collagen are assembled into a fine-fibered matrix in cultures of a teratocarcinoma-derived endodermal cell line. Exp Cell Res 137:15-23.

Liotta L, Goldfarb R, Terranova V (1981): Cleavage of laminin by thrombin and plasmin: Alpha thrombin selectively cleaves the beta chain of laminin. Thromb Res 21:663-673.

McCarthy JB, Palm S, Furcht L (1982): Migration of the Schwannoma cell line RN22F to laminin. J Cell Biol 95:133a.

Martinez-Hernandez A, Miller EJ, Damjanov I, Gay S (1982): Laminin-secreting yolk sac carcinoma of the rat: Biochemical and electron immunohistochemical studies. Lab Invest 47:247-257.

FX S𝒟|

Orkin R, Gehron P, McGoodwin E, Martin GR, Valentine T, Swarm R (1977): A murine tumor producing a matrix of basement membrane. J Exp Med 145:204-220.

Ott U, Odermatt E, Engel J, Furthmayr H, Timpl R (1982): Protease resistance and conformation of laminin. Eur J Biochem 123:63-72.

Palm S, Furcht L (1982): The interaction of a Schwannoma cell line with laminin and fibronectin in vitro. J Cell Biol 95:133a.

Rao C, Margulies I, Tralka T, Terranova V, Madri J, Liotta L (1982): Isolation of a subunit of laminin and its role in molecular structure and tumor cell attachment. J Biol Chem 257:9740-9744.

Reddan J, Dziedzic D, McGee S, Dehart D, Sackman J (1982): Influence of extracellular matrix, laminin and fibronectin on the growth of cultured mammalian lens epithelial cells. In Hawkes S, Wang JL (eds): "Extracellular Matrix." New York: Academic, pp 121-176.

Risteli L, Timpl R (1981): Isolation and characterization of pepsin fragments of laminin from human placental and renal basement membranes. Biochem J 193:749-755.

Rizzino A, Terranova V, Rohrbach D, Crowley C, Rizzino H (1980): The effects of laminin on the growth and differentiation of embryonal carcinoma cells in defined media. J Supramol Struct 13:243-250.

Rohde H, Wick G, Timpl R (1979): Immunochemical characterization of the basement membrane glycoprotein laminin. Eur J Biochem 102:195-201.

Sakashita S, Ruoslahti E (1980): Laminin-like glycoproteins in extracellular matrix of endodermal cells. Arch Biochem Biophys 205:283-290.

Sakashita S, Engall E, Ruoslahti E (1980): Basement membrane glycoprotein laminin binds to heparin. FEBS Lett 116:243-246.

Salomon DS, Liotta LA, Rennard SI, Foidart JM, Terranova V, Yaar M (1982): Stimulation by retinoic acid of synthesis and turnover of basement membrane in mouse embryonal caranoma-derived endoderm cells. Collagen Rel Res 2:93-110.

Shibata S, Peters B, Roberts D, Goldstein I, Liotta L (1982): Isolation of laminin by affinity chromatography on immobilized Griffonia simplicifolia I lectin. FEBS Lett 142:194-198.

Szarfman A, Terranova VP, Rennard SI, Foidart JM, De Fatima Lima M, Scheinman J, Martin GR (1982): Antibodies to laminin in Chaga's disease. J Exp Med 155:1161:1171.

Terranova V, Rohrbach D, Martin GR (1980): Role of laminin in the attachment of PAM 212 (epithelial) cells to basement membrane collagen. Cell 22:719-726.

Terranova VP, Rao CN, Kalebic T, Marguilies IM, Liotta LA (1983): Laminin receptor on human breast carcinoma cells. Proc Natl Acad Sci USA 80:444-448.

Timpl R, Rohde H, Gehron Robey P, Rennard S, Foidart JM, Martin GR (1979): Laminin—a glycoprotein from basement membranes. J Biol Chem 254:9933-9937.

Wewer V, Albrechten R, Ruoslahti E (1981): Laminin a noncollagenous component of epithelial basement membranes synthesized by a rat yolk sac tumor. Cancer Res. 41:1518-1524.

Woodley D, Rao C, Hassell J, Liottta LA, Martin GR, Kleinman HK (1983): Interactions of basement membrane components. Biochim Biophys Acta (in press).

Yaar M, Foidart JM, Brown KS, Rennard SI, Martin GR, Liotta LA (1982): The Goodpasture-like syndrome in mice induced by intravenous injections of anti-type IV collagen and anti-laminin antibody. Am J Pathol 107:79-91.

Methods for Preparation of Media, Supplements, and Substrata
for Serum-Free Animal Cell Culture, pages 239–244
© 1984 Alan R. Liss, Inc., 150 Fifth Avenue, New York, NY 10011

14
Isolation of Chondronectin

Hugh H. Varner, A. Tyl Hewitt, and George R. Martin

Chondronectin is a large (Mr = 180,000) glycoprotein that mediates the attachment of chondrocytes to type II collagen [Hewitt et al., 1980]. Chondrocyte attachment activity was initially found in serum and shown to be distinct from fibronectin, the fibroblast attachment factor. Chondronectin was subsequently purified to homogeneity from chicken serum and shown to be active at concentrations as low as 5 ng/ml with maximal chondrocyte attachment observed in culture at about 50 ng/ml [Hewitt et al., 1982]. The concentration of chondronectin in serum is estimated to be between 1 and 20 μg/ml. Antibody against chondronectin does not cross-react with either fibronectin, the fibroblast attachment factor [Klebe, 1974; Pearlstein, 1976; Kleinman et al., 1981], or laminin, the epithelial cell attachment factor [Terranova et al., 1980]. Furthermore, this antibody inhibits the spontaneous attachment of chondrocytes to type II collagen, indicating that chondrocytes produce and utilize a similar factor [Hewitt et al., 1982]. Immunofluorescence localization of chondronectin in cartilage showed that the protein is present in a pericellular localization [Hewitt et al., 1982], a finding that is consistent with its role as an attachment factor.

It now appears that much of the binding of chondronectin to collagen and its ability to promote cell attachment depends on proteoglycans. Xylosides, inhibitors of proteoglycan synthesis, inhibit chondronectin-mediated cell attachment, and this inhibition can be overcome by the addition of cartilage proteoglycan to the substrate. Proteoglycan also promotes the binding of chondronectin to collagen, although some binding does occur in the absence

Laboratory of Developmental Biology and Anomalies, National Institute of Dental Research, National Institutes of Health, Bethesda, Maryland 20205

of proteoglycan, but slowly [Hewitt et al., 1983]. Chondronectin probably interacts with the glycosaminoglycan portion of the proteoglycan, since inhibition of attachment was also seen in the presence of the chondroitin sulfate glycosaminoglycan moiety of cartilage proteoglycan.

While limited studies have been carried out with chondronectin in culture systems, the available evidence suggests that chondronectin promotes the chondrocyte pheonotype. In the absence of fibronectin, chondrocytes remain differentiated, but in whole serum or upon addition of exogenous fibronectin they lose the chondrocyte pheonotype [Pennypacker et al., 1979; West et al., 1979]. Chondronectin, on the other hand, stimulates chondrogenesis in cultures of limb and mesenchyme and then helps to maintain their normal phenotype in culture [Hassell et al., unpublished observations; Grotendorst et al., 1982].

The application of affinity chromatography on collagen-Sepharose has provided a rapid, efficient method for the purification of plasma fibronectin [Hopper et al., 1976; Engvall and Ruoslahti, 1977; Vuento and Vaheri, 1979]. However, chondronectin does not readily bind to collagen, possibly because the initial interaction of the protein is with the chondrocyte itself or with proteoglycan [Hewitt et al., 1980]. Therefore, a more involved sequence of procedures was used in the purification of chondronectin [Hewitt et al., 1982].

CHONDROCYTE ATTACHMENT ASSAY

Chondronectin levels are measured at each step in the isolation procedure by assaying the material for chondrocyte attachment activity by an assay adapted from the procedure of Klebe [1974]. In these assays bacteriological petri dishes (35 mm) are coated with type II collagen by evaporating 1 ml of a 10-μg/ml solution of collagen in 0.1% acetic acid. Chondrocytes from 13- to 17-day old chick embryos are prepared by dissecting out sterna and carefully removing the perichondrium, followed by digestion for 3 h at 37° in 0.4% collagenase (Worthington, CLS II) in Hanks' balanced salt solution.

Samples to be tested for attachment activity are added to the collagen-coated dishes in 1 ml of Ham's F12 medium containing 200 μg/ml of bovine serum albumin for 60 min at 37°. Subsequently, cells (1.5 × 10^5) are added in 0.1 ml of medium and allowed to attach for 90 min in a humidified atmosphere of 5% CO_2/95% air at 37°. Unattached cells are removed by rinsing the plates with three 1-ml aliquots of phosphate-buffered saline. Attached cells are released from the dish with 0.1% trypsin–0.1% EDTA in phosphate-buffered saline and are counted with an electronic particle counter (Coulter Electronics).

PURIFICATION OF CHONDRONECTIN

Chicken serum (2 liters) is fractionated by ammonium sulfate precipitation, and the proteins that precipitate between 25% and 35% saturation contain more than 70% of the chondrocyte attachment activity. This precipitate is resolubilized in 200 ml of 50 mM Tris-HCl, 0.15 M NaCL, pH 7.4 (Tris-buffered saline), containing the protease inhibitors 6-aminohexanoic acid (25 mM) and phenylmethylsulfonyl fluoride (0.2 mM). Protease inhibitors are included in all solutions throughout the purification procedures, which are carried out at 4°. The sample is dialyzed against 50 mM Tris-HCl, pH 7.4 and clarified by centrifugation. The supernatant fluid is then chromatographed on a 5-cm × 20-cm column of DEAE-cellulose (Whatman) with bound material eluted by a 2-liter linear gradient of 0–0.2 M NaCl in 50 mM Tris-HcCl, pH 7.4. Fractions of 20 ml are collected and aliquots assayed for chondrocyte attachment activity. Chondronectin and fibronectin are readily separated by this procedure [Hewitt et al., 1980]. Fractions promoting chondrocyte attachment are pooled and dialyzed against 50 mM Tris-HCl, pH 7.4.

The pooled fractions are applied to a 2.6-cm × 20-cm column of Cibacron blue F3GA-agarose (Bethesda Research Laboratories) in line with a 1.6-cm × 10-cm column of DEAE-cellulose. The columns are then washed with 100 ml of 50 mM Tris-HCl, pH 7.4 to remove unbound proteins. Most serum proteins, including serum albumin, bind to the Cibacron blue matrix under these conditions. Chondronectin does not bind to the matrix, thus allowing for the removal of more than 90% of the contaminants in a single step. The small DEAE-cellulose column placed in series with the Cibacron blue F3GA-agarose column binds the chondronectin in the eluate, and concentrated chondronectin activity is subsequently eluted from the DEAE column with Tris-buffered saline.

Further purification is achieved by chromatography on a 1.6-cm × 10-cm column of wheat germ agglutinin-Sepharose (Pharmacia) equilibrated with Tris-buffered saline. The column is washed with buffer, and bound material is eluted with 100 mg/ml N-acetylglucosamine in Tris-buffered saline. The bound fraction contains electrophoretically homogeneous chondronectin (Fig. 1). The chondronectin is dialyzed against Tris-buffered saline and stored at −20°. The purification is summarized in Figure 2. Purified chondronectin at concentrations of 50 ng/ml gives maximal attachment and is approximately equivalent to that seen when 5% serum is used.

STABILITY AND STORAGE

Chondronectin has been stored frozen in Tris-buffered saline for up to 6 months with no significant loss of attachment activity. Occasionally a small

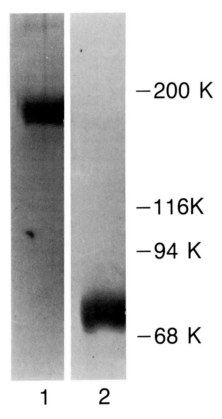

Fig. 1. SDS-PAGE of purified chondronectin from chicken serum. 1) Unreduced, 2) reduced.

precipitate forms that can be removed by centrifugation. Chondronectin can also be dialyzed against water and lyophilized. The protein is readily resolubilized with no apparent loss of activity. Chondronectin is quite stable to proteolysis. Treatment for 18 h at 37° in the presence of trypsin is required to totally destroy attachment activity (Varner et al., manuscript in preparation).

FURTHER CONSIDERATIONS

While chondrocyte attachment activity can be extracted from cartilage and rat chondrosarcoma, the amounts present are small relative to the amount in serum. This is not surprising, considering that cartilage contains very few cells

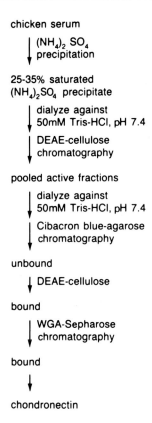

chicken serum

$(NH_4)_2 SO_4$ precipitation

25-35% saturated $(NH_4)_2SO_4$ precipitate

dialyze against 50mM Tris-HCl, pH 7.4

DEAE-cellulose chromatography

pooled active fractions

dialyze against 50mM Tris-HCl, pH 7.4

Cibacron blue-agarose chromatography

unbound

DEAE-cellulose

bound

WGA-Sepharose chromatography

bound

chondronectin

Fig. 2. Purification of chondronectin.

relative to other tissues and that the chondronectin in cartilage is localized pericellularly and is not present throughout the matrix [Hewitt et al., 1982].

REFERENCES

Engvall E, Ruoslahti E (1977): Binding of a soluble form of fibroblast surface protein, fibronectin to collagen. Int J Cancer 20:1-5.

Grotendorst GR, Kleinman HK, Rohrbach DH, Hewitt AT, Varner HH, Horigan EA, Hassell JR, Terranova VP, Martin GR (1982): In Sato GH, Pardee AB, Sirbasku DA (eds): "Cold Spring Harbor Conferences on Cell Proliferation," Vol 9: "Growth of Cells in Hormonally Defined Media." Cold Spring Harbor, NY: Cold Spring Harbor Laboratory, pp 403-413.

Hewitt AT, Kleinman HK, Pennypacker JP, Martin GR (1980): Identification of an adhesion factor for chondrocytes. Proc Natl Acad Sci USA 77:385-388.

Hewitt AT, Varner HH, Silver MH, Dessau W, Wilkes CM, Martin GR (1982): The isolation and partial characterization of chondronectin, an attachment factor for chondrocytes. J Biol Chem 257:2330-2334.

Hewitt AT, Varner HH, Silver MH, Martin GR (1983): The role of chondronectin and cartilage proteoglycan in the attachment of chondrocytes to collagen. In Kelley RO, Goetinck PF, MacCabe JA (eds): "Limb Development and Regeneration, Part B." New York: Alan R. Liss, pp 25-33.

Hopper KE, Adelmann BC, Gentner G, Gay S (1976): Recognition by guinea-pig peritoneal exudate cells of conformationally different states of the collagen molecule. Immunology 30:249-259.

Klebe RJ (1974): Isolation of a collagen-dependent cell attachment factor. Nature (London) 250:248-251.

Kleinman HK, Klebe RJ, Martin GR (1981): Role of collagenous matrices in the adhesion and growth of cells. J Cell Biol 88:473-485.

Pearlstein E (1976): Plasma membrane glycoprotein which mediates adhesion of fibroblasts to collagen. Nature (London) 262:497-500.

Pennypacker JP, Hassel JR, Yamada KM, Pratt RM (1979): The influence of an adhesive cell surface protein on chondrogenic expression in vitro. Exp Cell Res 121:411-415.

Terranova VP, Rohrbach DH, Martin GR (1980): Role of laminin in the attachment of PAM 212 (epithelial) cells to basement membrane collagen. Cell 22:719-725.

Vuento M, Vaheri A (1979): Purification of fibronectin from human plasma by affinity chromatography under non-denaturing conditions. Biochem J 183:331-337.

West CM, Lanza R, Rosenbloom J, Lowe M, Holtzer H, Avdalovic N (1979): Fibronectin alters the phenotypic properties of cultured chick embryo chondroblasts. Cell 17:491-501.

Methods for Preparation of Media, Supplements, and Substrata
for Serum-Free Animal Cell Culture, pages 245–268
© 1984 Alan R. Liss, Inc., 150 Fifth Avenue, New York, NY 10011

15
Human Serum Spreading Factor (SF): Assay, Preparation, and Use in Serum-Free Cell Culture

Janet Silnutzer and David W. Barnes

Serum spreading factor (SF) is a glycoprotein component of human plasma or serum that is capable of mediating attachment, spreading, migration, proliferation, and differentiative potential of a number of animal cell types in serum-free culture (Table I) [Barnes and Sato, 1979; Barnes et al., 1980, 1981, 1982, 1983, 1984; Barnes and Silnutzer, 1983; Silnutzer and Barnes, 1984; Barnes, 1984]. Circulating SF in the blood exists primarily in two independently active forms that migrate in sodium dodecylsulfate (SDS)-polyacrylamide gel electrophoresis in a manner consistent with molecular weights of 75,000–78,000 (SF75) and 65,000–70,000 (SF65) [Barnes et al., 1983; Barnes and Silnutzer, 1983; Hayman et al., 1982]. Thrombin cleavage of preparations of SF65 and SF75 generates a somewhat smaller molecule of molecular weight approximately 57,000 (SF57) that also is capable of promoting cell attachment and spreading in culture [Silnutzer and Barnes, 1984]. Although other cell spreading-promoting proteins, such as fibronectin, also are present in plasma or serum, SF is responsible for a significant portion of the total spreading-promoting activity of human serum [Barnes et al., 1983]. Purified SF is active in spreading-promoting assays at concentrations of 1 $\mu g/$ml or less [Barnes and Silnutzer, 1983; Silnutzer and Barnes, 1984].

Immunoassay with monoclonal antibody to human SF indicates that concentrations of SF in plasma or serum of normal adults range from about 100 $\mu g/ml$ to 400 $\mu g/ml$ with an average of approximately 250 $\mu g/ml$ [Barnes et al., 1983; Shaffer et al., 1984]. Concentrations of SF in plasma or serum

Department of Biological Sciences, University of Pittsburgh, Pittsburgh, Pennsylvania 15260

TABLE I. Cell Types Responding to Human Serum Spreading Factor In Vitro

Origin	Species	References
Cervical carcinoma (Hela)	Human	Holmes [1967]; Barnes and Silnutzer [1983]; Barnes et al. [1984]
Mammary carcinoma (MCF7)	Human	Barnes and Sato [1979, 1980c]; Barnes et al. [1981, 1983]; Basara et al. [1984]
Glioma (C6)	Rat	Barnes et al. [1980]; Wolfe et al. [1981]
Embryo/SV40-transformed (3T3/SV40-3T3)	Mouse	Barnes et al. [1980]; McClure [1983]
Embryonal carcinoma (F9, 1246, 1003)	Mouse	Barnes et al. [1980, 1982]; Darmon et al. [1981]; Serrero and Sato [1982]
Epidermoid carcinoma (A431)	Human	Barnes [1982]
Ovary (RF1)	Rat	Barnes et al. [1982]
Neuroblastoma (N18TG2)	Mouse	Wolfe and Sato [1982]
Embryonic lung fibroblast (WI38)	Human	Barnes et al. [1983]
Testes, Sertoli cell (primary)	Mouse	Mather and Phillips [this volume]
Fetal liver (primary)	Human	Salas-Prato et al. [1984]
Melanoma (1735)	Mouse	Basara et al. [1984]
Fibrosarcoma (HT1080)	Human	Basara et al. [1984]

from children, infants, and cord blood also fall within this range [Shaffer et al., 1984]. Little difference is observed in levels of immunoassayable SF in platelet-rich or platelet-poor plasma and serum derived from that plasma [Barnes et al., 1983; Shaffer et al., 1984]. In addition to its presence in significant concentrations in blood, SF is present in human urinary protein and amniotic fluid [Shaffer et al., 1984]. Tissue-associated (TASF) and platelet-associated (PASF) forms of the protein also have been reported [Barnes et al., 1983; Hayman et al., 1983].

Human serum SF is immunologically and biochemically distinct from fibronectin, chondronectin, and laminin, but it is immunologically related to human epibolin [Barnes et al., 1980, 1981, 1982, 1983; Stenn, 1981; Stenn et al., 1982; Stenn, this volume]. Although biochemical differences exist between human SF and a cell-spreading-promoting factor of similar molecular weight isolated from bovine sera [Whateley and Knox, 1980; Barnes and Silnutzer, 1983], the relationship of human SF to this protein and the precise relationship of human SF to human epibolin [Stenn, 1981; Stenn et al., 1982; Stenn, this volume] is not yet clear.

In this chapter, we describe methods for biologic assay and isolation of human serum SF (SF65 + SF75) and production of SF57 from preparations of SF65 + SF75 by incubation with human or bovine thrombin. We also describe methods for use of purified human serum SF in serum-free cell culture and point out some of the possible problems and areas of confusion that may arise in studies using this protein. Several of the approaches we have taken in characterization and quantitation of human SF involve the use of monoclonal antibodies to SF secreted by hybridomas developed in our laboratory [Barnes et al., 1983; Barnes and Silnutzer, 1983; Silnutzer and Barnes, 1984; Shaffer et al., 1984] (Figs. 1, 2). Procedures related to derivation, characterization, isolation, and use in immunoassay of monoclonal antibody to human serum SF have been presented elsewhere [Barnes et al., 1983; Barnes and Silnutzer, 1983; Shaffer et al., 1984] and will not be reviewed in detail in this chapter.

ASSAY OF CELL-SPREADING-PROMOTING ACTIVITY

The following assay has proved useful in assay of cell-spreading-promoting activity in serum or crude preparations as well as in assay of more purified material. It is particularly useful in some situations because the assay takes advantage of the strong plastic-binding property of some spreading-promoting factors, allowing assay of activities from preparations that may contain substances introduced during purification or experimental manipulation that may be potentially toxic to cells (detergents, denaturants, high salt concentrations, antimicrobial agents). Samples to be assayed may be added to tissue culture plates, the active material allowed to bind to the plastic, the soluble and potentially toxic material remaining removed, and the plate washed extensively. The assay is adaptable to a number of cell types; we have used the assay to quantitate effects on rat glioma [Barnes et al., 1980], mouse and human fibroblasts [Barnes et al., 1980] (Fig. 1), and human carcinoma of cervix and breast [Barnes and Sato, 1979] (Figs. 2, 3), as well as other types of cells. Occasionally, however, cell types are found that do not serve well in these assays because the rate of control spreading in the absence of spreading-promoting factors is too fast to allow sufficient quantitation of effects of the samples to be assayed. Cell types that are capable of attaching and spreading quickly in serum-free medium in the absence of exogenous spreading factors are probably synthesizing spreading-promoting proteins (particularly fibronectin) or mobilizing these proteins to the cell surface from internal sources at a rate that is sufficient to produce the biologic effect [Grinnell and Feld, 1980].

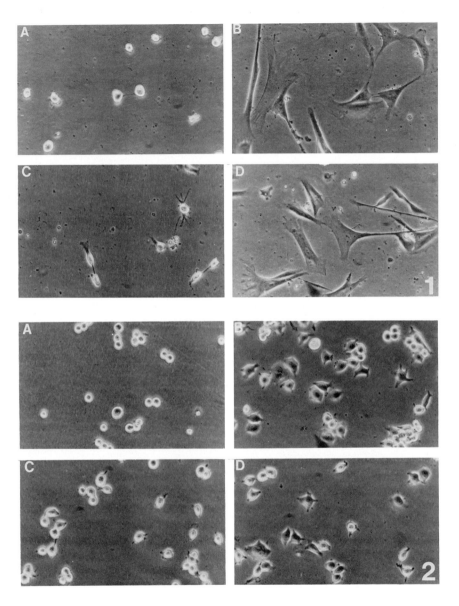

Fig. 1. Serum spreading factor-promoted spreading of WI38 human fibroblasts in culture. Cells were seeded onto plates previously incubated with PBS (A); 4 µg/ml serum spreading factor in PBS (B); serum spreading factor followed by 15 µg/ml monoclonal antibody to serum spreading factor (C); or serum spreading factor followed by 15 µg/ml nonspecific mouse IgG (D). All plates were incubated with PBS containing 1 mg/ml BSA between the incubations with serum spreading factor and the incubations with antibody. Cells were photographed 90 min after plating. From Barnes et al. [1983], with permission.

The assay described below is useful for quantitation of biologic effects of spreading-promoting factors other than SF, such as fibronectin or laminin, and substrata other than plastic, such as glass or collagen; it was developed in principle by others and has been used in various forms by a number of laboratories prior to our adaptations [Fisher et al., 1958; Grinnell, 1978; Klebe, 1974]. Procedures are as follows:

1) Pretreat dishes with samples to be assayed for spreading-promoting activity by incubation of 35-mm-diameter culture dishes with 1 ml of phosphate-buffered saline (PBS) containing the material to be tested for 1-6 h at room temperature. Remove the solution and wash plates three times with PBS immediately before plating the cells. We have found that use of 35-mm-diameter plates rather than larger vessels or smaller-diameter multiwell dishes provides the best compromise between economy of material and acceptable optical quality and field number for subsequent observation and quantitation.

2) Trypsinize cells from stock flasks with 0.1% trypsin-0.1 mM ethylene-diaminetetraacetate in phosphate-buffered saline at 37°.

3) Add an equal volume of 0.1% soybean trypsin inhibitor in culture medium (1-to-1 mixture of Ham's F12 and Dulbecco modified Eagle's medium supplemented with 1.2 gm/liter bicarbonate, 15 mM HEPES, pH 7.3, and antibiotics), after cells have detached from culture vessels.

4) Centrifuge cells from suspension and discard supernatant.

5) Resuspend cells in culture medium containing 1 mg/ml bovine serum albumin (BSA) at a density of about 10^5 cells/ml.

6) Plate 1 ml of cell suspension per 35-mm-diameter dish, incubate in 5% CO_2-95% air at 37°.

7) Observe and quantitate cell spreading over the next 0.5-24 h. If observation beyond this time is required, sterility of the preparations used to pretreat dishes may become a necessity. However, most cell types we have used in this assay will be well spread 45-90 min after plating. This procedure can also be adapted for quantitative studies of antibody effects on spreading-promoting proteins by inclusion of an additional incubation with an antibody solution in PBS containing 1 mg/ml BSA and subsequent washes before plating of cells. We have encountered some problems in these types of studies with nonspe-

Fig. 2. Serum spreading factor-promoted spreading of MCF7 human mammary carcinoma in culture. Procedures were as described in the legend to Figure 1. No substratum treatment (A); serum spreading factor pretreatment (B); serum spreading factor followed by monoclonal antibody to serum spreading factor (C); serum spreading factor followed by nonspecific mouse IgG (D). From Barnes et al. [1983], with permission.

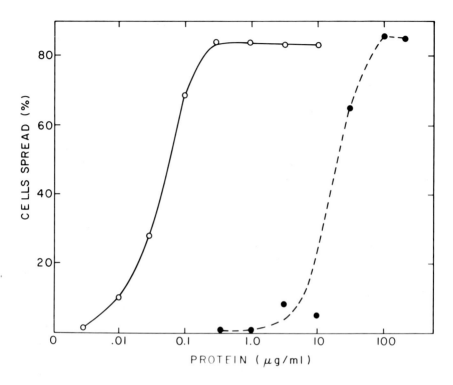

Fig. 3. Assay of cell-spreading-promoting activity of isolated human SF (○) and human serum (●). Methods are described in the text. HeLa human carcinoma cells were used in the assay. The percentage of spread cells was determined 45 min after plating. From Barnes and Silnutzer [1983], with permission.

cific spreading-inhibiting effects of immunoglobulin. These effects are reduced by including BSA in the antibody solution so that nonspecific binding of immunoglobulin to the culture dish is minimized, but appropriate controls to determine the extent of such a contribution to the overall inhibitory effect of the antibody preparation should be included in assays designed for these purposes [Barnes et al., 1983].

In general, we have found the method of quantitating the amount of spreading-promoting activity in a particular plate that is the least confusing and gives the most consistent results among different observers is to choose an arbitrary set of criteria to define a "spread cell" and score the percentage of cells that meet these criteria, rather than attempting to define a number of "stages" of spreading and quantitating the percentage of cells per plate at each stage. In many cases, the observer must consider the amount of control cell spreading on untreated plates. Cell spreading is a time-dependent phenomenon and often control cells plated on untreated plastic surfaces will eventually spread, although at a slower rate than those plated on dishes pretreated with spreading-promoting proteins. For this reason, it is best to minimize the time necessary to complete the quantitation step of the assay. Reducing the observation to a simple "spread" or "not spread" score helps prevent the time factor from becoming unwieldy. Reproducible results can be obtained if 200–500 cells, representing five or more microscopic fields, are counted on each dish.

If unacceptably high rates of control spreading are observed, making quantitation of spreading-promoting effects of the samples difficult, these problems sometimes can be reduced by lowering the plating density of the cells and by carrying out an additional wash of the cells with medium after trypsinization and before plating. These manipulations are designed to reduce the contribution of spreading-promoting factors synthesized by the cells themselves. Another occasional source of difficulties of this type is contaminating spreading-promoting activity in some preparations of albumin. Under these circumstances, it may be helpful to try other albumin preparations or to change the concentration of BSA used, within the range of about 250 μg/ml to 5 mg/ml. Depending on the cell type and assay conditions, it may be possible to observe and quantitate effects of spreading-promoting proteins on cell attachment, rather than cell spreading. We have not found quantitation of this effect very satisfactory, however, because of considerable variability from one experiment to another in the degree of cell attachment on untreated plates. Examples of the results of spreading-promoting assays employing several cell types and illustrating the principles described above are given in Figures 1–3.

OTHER BIOLOGICAL ASSAYS

Other assays have also been developed that allow quantitation of effects in vitro of collagens, serum spreading factor, fibronectin, and laminin. One such assay involves measurement of cell numbers under conditions of serum-free cell culture in which cell proliferation is dependent in a concentration-related manner on the presence of one or more of these factors [Orly and Sato, 1979; Barnes et al., 1980, 1982; Barnes and Sato, 1980a,b]. In general, these assays are carried out with serum-free media to which are added hormones, binding proteins like transferrin, and supplementary nutrients, and the details of the assay conditions depend on the particular requirements for these factors of the cell type to be used in the assay [Barnes and Sato, 1980a,b]. Although applicable to a large number of cell types, these assays usually require sterile samples of the factor to be assayed, and sterilization of these proteins sometimes presents problems. Also used in assay of biologic effects of spreading-promoting proteins are quantitation of the effects of these factors as mediators of cytokinesis, migration, and phagocytosis [Orly and Sato, 1979; McCarthy et al., 1983; van de Water et al., 1981; Barnes et al., 1982; Basara et al., 1984]. These assays are somewhat limited in that they can be carried out easily with only a limited number of cell types in culture.

PURIFICATION OF HUMAN SERUM SPREADING FACTOR

Described below are procedures for the purification of serum spreading factor from human serum, plasma, or plasma fractions by sequential column chromatography on glass beads, concanavalin A-Sepharose, DEAE-agarose, and heparin-agarose. These procedures represent some adaptations for large-scale preparations from our originally published procedure [Barnes and Silnutzer, 1983]. SF is eluted at each step in this procedure in a solution that allows direct loading of the material onto the next column in the series. The isolated SF obtained at the end of the procedure is in a solution of approximately physiologic pH and salt concentration and may be used directly for biochemical or cell culture experiments. In this way, we have avoided major losses in yields that result from absorption of SF to glass, quartz, or plastic cuvettes, collection tubes, tubing and dialysis bags. All chromatographic procedures are carried out at 4°.

Preparation of Starting Material for Isolation of Human SF by Column Chromatography

Either human plasma or serum derived from plasma can be used as starting material for chromatography [Barnes et al., 1983; Barnes and Silnutzer, 1983; Shaffer et al., 1984]. If plasma is used, 10 mM sodium citrate is added to the buffer solutions preceding and following the plasma on the first column in the purification series. Serum is prepared by dialysis of plasma (200 ml) overnight at 4° against 6 liters of 0.15 M NaCl followed by addition of 200 mg CaCl$_2$ to the dialyzed plasma. Clotting is allowed to occur at room temperature and the clot is removed by low-speed centrifugation. The resulting serum can be stored frozen at −20°. α-Toluene sulfonyl fluoride (PMSF; 2mM) is added before freezing. Although our initial work was carried out with serum prepared from freshly frozen plasma, we found that identical results can be obtained with outdated plasma as starting material.

SF is enriched in Cohn fraction IV (α-globulins) about threefold by immunoassay compared to human plasma or serum (Fig. 4) [Barnes and Silnutzer, 1983; Shaffer et al., 1984], and this blood fraction also can be used as starting

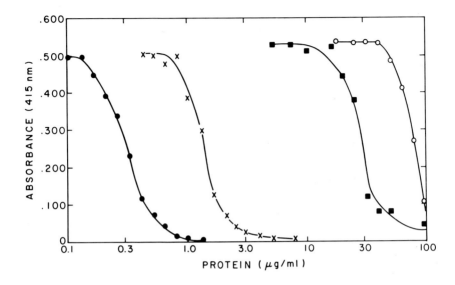

Fig. 4. Quantitative immunoassay of SF. A 100-μl volume of monoclonal antibody to serum SF (2 μg/ml in PBS with BSA) was incubated overnight in microtiter wells with 100 μl of samples to be assayed for SF, diluted in PBS with BSA to give the indicated amount of sample protein per well. At the end of this incubation, the amount of antibody remaining unbound to antigen in the incubation mixtures was determined by the following manipulations. The solutions were transferred to microtiter wells previously coated with a saturating amount of SF and the transferred solutions were incubated in the wells 1 h. The contents of the wells were then removed and the wells were washed three times. Wells were then incubated with peroxidase-conjugated goat anti-mouse IgG (1:1,000 in PBS) for 1 h, contents of the wells were removed, and the wells were washed three times. The wells were then incubated 15 min with phosphate-citrate buffer containing hydrogen peroxide and peroxidase-dependent chromogen 2,2'-azino-di(3-ethylbenzthiazelinesulfonic acid). Extent of the reaction was determined by spectrophotometric determination of absorbance at 415 nm after inhibition of the reaction with sodium fluoride and removal of the reaction mixture from the enzyme-containing wells. ●, Isolated human SF (SF65 and SF75); ×, partially purified human SF preparations from glass bead affinity column chromatography; ■, Cohn fraction IV from human plasma; ○, human serum. From Barnes and Silnutzer [1983], with permission.

material for glass bead chromatography of SF. Cohn fraction IV is prepared for chromatography by adding 1.5 gm to 100 ml of 0.15 M NaCl. The large amount of insoluble material is removed by centrifugation and 2 mM PMSF is added to the supernatant. This solution can be stored frozen until used.

Glass Bead Affinity Column Chromatography of Human SF

Glass microbeads (Number 1014, Class IV-A, Ferro, Cataphote Division, PO Box 2369, Jackson MS 39205) are washed with distilled water and then soaked overnight in an equal volume of a mixture of concentrated hydrochloric and nitric acid (3:1, vol./vol.) and washed five times with 10 volumes of distilled water followed by five washes with 10 volumes of 0.6 M NaHCO$_3$, pH 8.0. Beads in 0.6 M NaHCO$_3$ are poured into a column (Pharmacia K50/60) of dimensions 5 cm × 50 cm and equilibrated with 0.6 M NaHCO$_3$ overnight at 100 ml/h. The volume of a column of this size is approximately 1,000 ml and the void volume is approximately 400 ml.

Prior to application to the column, the sample is adjusted to pH 8.0 with 0.1 N NaOH and filtered. The sample (100ml) is applied to the column at 100 ml/h and the column is subsequently washed (100 ml/h) with the following buffers, all containing 2 mM PMSF:

1) 200 ml 0.6 M NaHCO$_3$, pH 8.0
2) 100 ml 0.6 M NaHCO$_3$-0.2 M Na$_2$CO$_3$, pH 9.3-9.5
3) 100 ml H$_2$O
4) 500 ml 0.2 M K$_2$CO$_3$, pH 11.0

Fractions containing material eluted with 0.6 M NaHCO$_3$-0.2 M Na$_2$CO$_3$ are pooled for further purification. The column is regenerated by the 0.2 M K$_2$CO$_3$ wash and may be reused a number of times if reequilibrated with 0.6 M NaHCO$_3$ before sample application. After a column has been regenerated and reused 5-10 times, sharpness of peaks becomes significantly reduced and the beads must be removed from the column and washed in water and acid and 0.6 M NaHCO$_3$ as before, and the column must be repoured.

Both purified fibronectin and SF are capable of binding to glass beads [Barnes et al., 1984] and the relative amount of each bound to glass beads from serum and eluted with 0.6 M NaHCO$_3$-0.2 M Na$_2$CO$_3$ is determined by the amount of serum put through the column (Figs. 5-8). At high ratios of serum volume to total glass bead surface area (glass bead column volume), very little fibronectin is found in the eluted peak (Fig. 7). As this ratio is decreased, fibronectin becomes a major component of the eluted material. The amount of SF as a percentage of total protein in these fractions is relatively constant regardless of the ratio of serum volume to bead surface area (glass bead column volume) (Fig. 8). Although we have used methods in the past that resulted in partially purified SF preparations isolated by glass bead column chromatography that contained negligible amounts of fibronectin [Barnes et al., 1980, 1981, 1982], the large-scale preparative techniques reported here may lead to significant amounts of fibronectin in the material eluted from the glass bead column with 0.6 M NaHCO$_3$-0.2 M Na$_2$CO$_3$. This contaminating fibronectin is removed in the later purification steps.

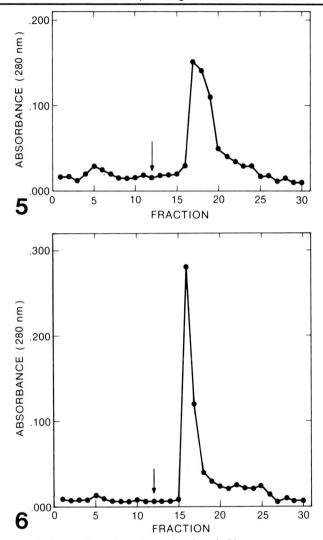

Fig. 5. Glass bead affinity column chromatography of purified human serum spreading factor. SF (1 ml at OD_{280} = 0.500 in PBS) isolated as described in the text was chromatographed on a glass bead column (0.7 cm × 25 cm, 20 ml/h) previously equilibrated with 0.6 M $NaHCO_3$, pH 8.0. One-milliter fractions were collected in polypropylene tubes. The sample was followed on the column by 10 ml 0.6 M $NaHCO_3$ (pH 8.0) and 10 ml 0.6 M $NaHCO_3$-0.2 M Na_2CO_3 (pH 9.3-9.5). Arrow denotes change of elution conditions from the former solution to the latter.
Fig. 6. Glass bead affinity column chromatography of purified human plasma fibronectin. Procedures were as described in the legend to Figure 5.

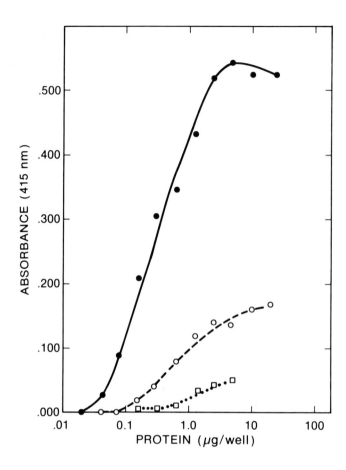

Fig. 7. Immunoassay of fibronectin in preparations isolated from human serum by glass bead affinity column chromatography at different ratios of serum volume to total glass bead surface area of the column. Procedures were as described in the legend to Figure 8 except that monoclonal anti-SF was replaced with rabbit anti-human fibronectin antiserum (1:200), and peroxidase-conjugated anti-mouse IgG was replaced with peroxidase-conjugated anti-rabbit IgG. □, Preparations obtained upon chromatography of 100 ml serum; ○, preparations obtained upon chromatography of 50 ml serum; ●, preparations obtained upon chromatography of 25 ml serum.

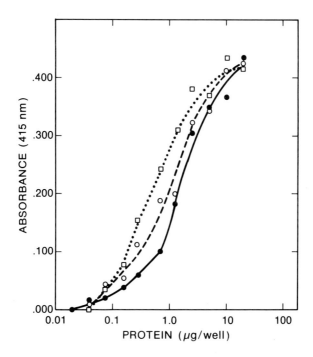

Fig. 8. Immunoassay of SF in preparations isolated from human serum by glass bead affinity column chromatography at different ratios of serum volume to glass bead surface area of the column. Human serum (100, 50, or 25 ml) was adjusted to pH 8.0 and chromatographed on a glass bead column (2.5 cm × 50 cm, 20 ml/h) previously equilibrated with 0.6 M NaHCO$_3$. The column was washed with 100 ml of 0.6 M NaHCO$_3$, and partially purified preparations of SF were eluted with 0.6 M NaHCO$_3$–0.2 M Na$_2$CO$_3$. Peak fractions were assayed for SF with monoclonal anti-SF and peroxidase-conjugated second antibody. Microtiter wells were incubated overnight with the partially purified preparations diluted in 100 μl of PBS to concentrations necessary to give the indicated amount of sample protein per well. At the end of the incubations, the contents of the wells were removed and the wells washed three times with PBS. Wells were then incubated 1 h with anti-SF (10 μg/ml in PBS with BSA), washed three times with PBS, and incubated 1 h with peroxidase-conjugated goat anti-mouse IgG (1:1,000 in PBS). After incubation with this antibody, the wells were washed three times and peroxidase-dependent chromogen was added, and the extent of reaction was determined as described in the legend to Figure 3. □, Preparations obtained upon chromatography of 100 ml serum; ○, preparations obtained upon chromatography of 50 ml serum; ●, preparations obtained upon chromatography of 25 ml serum.

The temperature at which the glass bead column is operated also may affect the nature of the proteins eluted with 0.6 M $NaHCO_3$-0.2 Na_2CO_3. Pilot experiments indicate that when the column is run at room temperature, considerably more SF65 than SF75 is eluted in these fractions (Fig. 9). Nearly equal amounts of both forms of SF are eluted when the column is run at 4°.

The nature and relative amounts of the major non-SF components of material eluted from the glass bead column are determined to some extent by the sample material applied (e.g., serum, plasma, or Cohn fraction IV), as well as by conditions under which the column is run. Major contaminants other than fibronectin tentatively identified in these preparations include α-2-macroglobulin, high-density lipoprotein (HDL) components, and fibrinogen. Use of plasma as the starting material results in fibrinogen as a major contaminant. Use of serum as the starting material results in HDL as a major

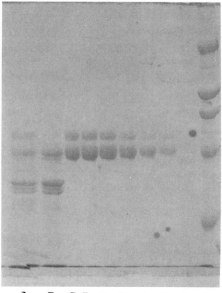

A B C D E F G H I J

Fig. 9. SDS-polyacrylamide gel electrophoresis of human SF preparations. Samples were reduced with mercaptoethanol, denatured at 100°C, and subjected to electrophoresis in a 7.5% polyacrylamide slab gel with a 2.5% stacking gel. A: Partially purified SF preparation eluted with 0.6 M $NaHCO_3$-0.2 M Na_2HCO_3 from glass bead column run at 4°C. B: Partially purified SF preparation eluted with 0.6 M $NaHCO_3$-0.2 M Na_2HCO_3 from glass bead column run at 22°. C: Purified SF, 20 μg. D: A second purified SF preparation, 20 μg. E: Purified SF, 17 μg. F: Purified SF, 12 μg. G: Purified SF, 8 μg. H: Purified SF, 6 μg. I: Blank. J: Molecular weight standards (top to bottom): Myosin (200,000), β-galactosidase (116,000), phosphorylase b (92,500), BSA (66,700), and ovalbumin (43,000).

contaminant. Use of Cohn fraction IV as the starting material results in greatly reduced levels of fibrinogen or HDL, but these preparations have α-2-macroglobulin as a major component.

Concanavalin A-Sepharose/DEAE-Agarose Column Chromatography

We have adapted our previous procedure involving sequential concanavalin A-Sepharose and DEAE-agarose chromatography [Barnes and Silnutzer, 1983] by using a single column in which a large volume of concanavalin A-Sepharose is poured above a smaller volume of DEAE-agarose. Loading of the fractions containing SF from the glass bead bead column and subsequent elutions results in the following sequence:

1) SF binds to concanavalin A.

2) SF is eluted from concanavalin A with mannose in a low-salt buffer at pH 6.0. Under these conditions, SF binds to the DEAE at the bottom of the column.

3) SF is eluted from DEAE by increasing the salt concentration. This material may be loaded directly onto heparin-Sepharose in the next step.

4) Remaining proteins more tightly bound to concanavalin A or DEAE are eluted with a buffer containing 0.5 M α-methylmannoside and 1 M NaCl. This step and a subsequent step in which the column is washed with divalent cation-containing buffer at physiologic pH and salt concentrations regenerates both the concanavalin A-Sepharose and DEAE-agarose portions of the column for reequilibriation and reuse.

The column (2.5 cm $\times$ 20 cm Biorad Econocolumn) is created by first pouring a bottom layer (2.5 cm $\times$ 3 cm) of DEAE-agarose (Biorad). Care must be taken to avoid adhesion of DEAE-agarose beads to the sides of the column above the level of the DEAE-agarose bed, and the DEAE-agarose layer should be as level as possible. Over this layer is poured a 2.5-cm $\times$ 16-cm concanavalin A-Sepharose layer (Pharmacia, or concanavalin A-agarose, Biorad). Care must be taken to prevent disturbance of the DEAE layer while pouring the second layer, and the final column should show a sharp demarcation between the two components. The column is then equilibrated with phosphate-buffered saline at 30 ml/h.

SF-containing fractions eluting from the glass bead column with 0.6 M $NaHCO_3$-0.2 M Na_2CO_3 are applied to the concanavalin A/DEAE column at 30 ml/h. About 200 ml of the pooled fractions can be loaded onto the column and the column is then eluted with the following:

1) 30 ml 0.25 M sodium phosphate (Na/Pi), pH 6.0, 30 ml/h.

2) 200 ml 25 mM Na/Pi, pH 6.0, 15 ml/h. It is important that the column be washed with this buffer until the salt concentration is equilibrated throughout in order to allow SF eluted in the next step to become trapped by the DEAE-agarose layer at the bottom of the column.

3) 250 ml 50 mM mannose in 25 mM Na/Pi, pH 6.0, 30 ml/h.

4) 30 ml 25 mM Na/Pi, pH 6.0, 30 ml/h.

5) 60 ml 100 mM NaCl in 25 mM Na/Pi, pH 6.0, 30 ml/h. Alternatively, this wash can be carried out at lower flow rate and larger volume and the column can be eluted overnight.

6) 100 ml 0.5 M α-methylmannoside with 1 M NaCl in 25 mM Na/Pi, pH 6.0, 30 ml/h.

7) 100 ml 0.15 M NaCl containing 1 mM $MgCl_2$, 1 mM $CaCl_2$, and 1 mM $MnCl_2$ in 20 mM Tris, pH 7.3.

SF-containing material elutes in a sharp peak with the 100 mM NaCl in 25 mM Na/Pi.

Heparin-Agarose Column Chromatography

SF preparations eluted from the concanavalin A/DEAE column are loaded onto a heparin-agarose column for final purification. A heparin-agarose (Biorad, or heparin-Sepharose, Pharmacia) column (0.7 cm × 10 cm) (Biorad Econocolumn) is poured and equilibrated (15 ml/h) with 100 mM NaCl in 25 mM Na/Pi, pH 6.0. The sample is loaded as individual fractions from the concanavalin A/DEAE column, without pooling, at 7.5 ml/h. This method of sample loading is tedious, but it improves yields. Fractions from several concancavalin A/DEAE runs can be loaded onto a single heparin-agarose column of the size indicated. The column is then eluted with the following:

1) 20 ml 100 mM NaCl in 25 mM Na/Pi, pH 6.0 (15 ml/h).

2) 20 ml 100 mM NaCl in 50 mM Na/Pi, pH 7.0 (4 ml/h). Alternatively, this step can be carried out with a larger volume and the column can be eluted overnight.

3) 20 ml 1 M NaCl in 50 mM Na/Pi, pH 7.4 (15 ml/h).

SF binds to heparin-agarose at pH 6.0 in 100 mM NaCl, 25 mM Na/Pi, and elutes when the pH is changed to 7.0 (Fig. 10). SF actually can be eluted from the column at lower pH (6.5-7.0), but we have chosen pH 7.0 to approximate physiologic conditions in the buffer in which the final material is obtained. Some gel filtration of the material eluted by the pH change to 7.0

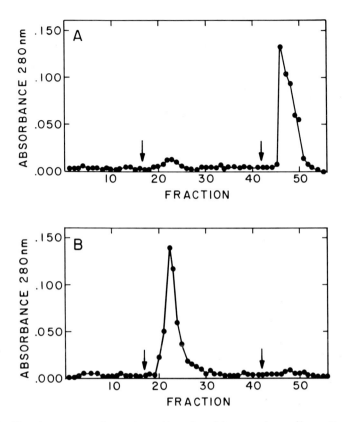

Fig. 10. Heparin-agarose column chromatography of human plasma fibronectin (A) and serum spreading factor (B). Purified human serum SF and human fibronectin were loaded onto identical heparin-agarose columns (0.7 cm × 2.5 cm) in 50 mM Na/Pi (pH 6.0) containing 100 mM NaCl. The columns were eluted with 15 ml of 100 ml NaCl in 50 mM Na/Pi (pH 6.0) containing 100 mM NaCl, 20 ml of 50 mM Na/Pi (pH 7.2) containing 100 mM NaCl, and 15 ml of 50 mM Na/Pi (pH 7.2) containing 1.5 M NaCl. Arrows indicate change in elution conditions: Fraction 17, change from pH 6.0 to 7.2 (100 mM NaCl, 50 mM Na/Pi); fraction 42, change from 100 mM NaCl to 1.5 M NaCl (50 mM Na/Pi, pH 7.2). From Barnes and Silnutzer [1983], with permission.

also occurs during the elution and occasionally high-molecular-weight trace contaminants appear in the earliest peak fractions. These fractions can be discarded and only the later fractions used if the presence of these contaminants is of concern. The column is regenerated by the 1 M NaCl wash. Periodically the column is washed with 50 mM Tris containing 8 M urea, pH 7.4 to remove any tightly binding material that may collect over a number of runs and affect column capacity, although we have not encountered this type of problem.

Both SF and fibronectin bind to heparin-agarose at pH 6.0 in 100 mM NaCl, although fibronectin does not elute at pH 7.0 unless the salt concentration is raised (Fig. 10) [Hynes and Yamada, 1982; Barnes and Silnutzer, 1983]. We have found that the capability of SF to bind to heparin-agarose at pH 6.0 and elute at pH 7.0 is not unique; many proteins will act similarly. The binding capability of the column at pH 6.0 may to some degree reflect heparin-agarose functioning in the same manner as an ionic resin. The use of heparin is of particular advantage in the purification of SF because 1) a major contaminant of samples prior to heparin-agarose chromatography is a protein of approximately 50,000 daltons that remains tightly bound to the column under conditions that elute SF and is removed from the column with the 1 M NaCl wash, and 2) any trace fibronectin that may be present at this purification stage will be removed.

If fibronectin contamination is of concern, the plasma used as the starting material also may be run through a gelatin-Sepharose column to remove this protein. SF does not bind to gelatin-Sepharose [Barnes et al., 1980] and it is possible to isolate SF from material passed over gelatin-Sepharose by the procedures described above with no reduction in yield. Including the gelatin-Sepharose step in the beginning allows the isolation of both SF and fibronectin independently from the same plasma sample. Because fibronectin and SF exhibit different binding properties on both gelatin-Sepharose and heparin-agarose and the gelatin-binding and heparin-binding sites of fibronectin are separable [Hynes and Yamada, 1982], the use of both affinity column procedures makes it unlikely that SF preparations made in this way will contain contaminating fibronectin or fibronectin fragments. In practice, we have not encountered fibronectin contamination as a problem at any purification stage except that of the material eluted from the glass bead affinity column under conditions of low ratio of serum volume to glass bead surface area.

SF prepared by the procedure employing chromatography on glass beads, concanavalin A, DEAE, and heparin-agarose as described above represents a purification of 250-fold to 300-fold from plasma (Figs. 3, 4) [Barnes and Silnutzer, 1983]. SF65 and SF75 are present in these preparations in similar proportions, although we estimate, on the basis of the relative intensity of Coomassie blue staining of the two bands after SDS-polyacrylamide gel electrophoresis of the preparations, that generally there is somewhat more SF65 than SF75 (Fig. 9) [Barnes and Silnutzer, 1983]. Also present in some preparations are small amounts of anti-SF antibody-binding material that behaves upon electrophoresis as if it consisted of multimeric forms (dimers or tetramers) of SF65 and SF75. The precise biochemical differences and the relationship between SF65 and SF75 are not yet clear. Both are biologically active and a single NH_2-terminal amino acid (aspartic acid or asparagine) is identified in preparations containing both SF65 and SF75 [Barnes and Silnutzer, 1983]. Occasionally, preparations of SF are obtained by the methods described that contain detectable amounts of a form of SF of molecular weight approximately 57,000 (SF57) that is biologically active and capable of binding monoclonal antibody to SF that inhibits the biologic activity of the factor (Fig. 11). SF57 in preparations isolated from serum is likely to be the result of thrombin cleavage of SF during the clotting process [Silnutzer and Barnes, 1984]. SF57 exhibits reduced plastic and glass-binding capability, compared to SF65 and SF75. Procedures for conversion of preparations of SF65 + SF75 to SF57 are described below.

PREPARATION OF A BIOLOGICALLY ACTIVE THROMBIN CLEAVAGE PRODUCT OF HUMAN SERUM SPREADING FACTOR

Although for many purposes, particularly those of cell culture, the SF65 + SF75 preparations made as described above are satisfactory, biochemical analysis is considerably easier on a single species, and we have found that human or bovine thrombin is capable of converting preparations that are a mixture of SF65 and SF75 to SF57, a single 57,000-dalton form of the molecule (Fig. 11) (Table II) [Silnutzer and Barnes, 1984].

SF57 is prepared by incubating SF preparations (100–500 µg/ml), obtained from the final heparin-agarose step described above, with highly purified human or bovine thrombin (Sigma, approximately 3,000 units/mg stored as a concentrated stock in 0.15 M NaCl with 50 mM sodium citrate, pH 6.5) diluted into the SF preparation to give a final thrombin concentration of 50 µg/ml. It may be possible to use much less thrombin, on the order of 5 µg/ml or less, and still obtain complete conversion to SF57. Incubate the mixture 6–8 h at 37° and determine degree of conversion by SDS-polyacrylamide gel electrophoresis.

We have not attempted to separate the added thrombin from the SF57 in these preparations at the end of the conversion, but several procedures are likely to be effective. These include use of glass bead affinity chromatography [Barnes et al., 1984], heparin-agarose chromatography [Barnes and Silnutzer, 1983], and filtration techniques for separation on the basis of size. Another approach that may be taken is to use insolubilized thrombin to carry out the conversion. We have observed formation of SF57 from SF65 + SF75 preparations by incubation with thrombin covalently attached to activated Sephar-

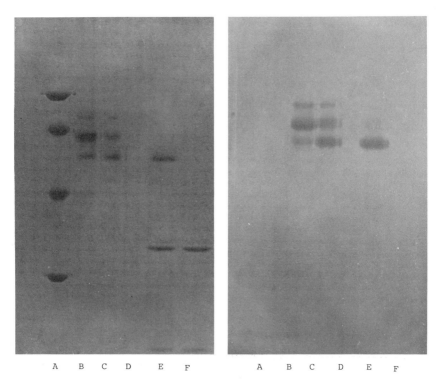

A   B   C   D   E F   A   B   C   D   E F

Fig. 11. Cleavage of serum SF by human thrombin and bovine thrombin-Sepharose. Incubations carried out for 6 h at 37° were stopped by addition of 2mM PMSF, and Sepharose beads were removed by centrifugation. Samples at the end of the incubations were reduced and analyzed by SDS-polyacrylamide gel electrophoresis. A duplicate gel was processed for immunoblotting as described elsewhere [Barnes et al., 1983]. Left Panel: Gel stained for protein. Right Panel: Immunoblot of duplicate gel. A: Molecular weight standards. B: A preparation of serum SF containing some endogenous SF57. C: The same preparation of serum SF further converted to SF57 by incubation with bovine thrombin covalently bound to Sepharose (equivalent to 6 μg thrombin per 1 ml). D: Thrombin-Sepharose incubated in the absence of serum SF. E: A preparation of serum SF converted to SF57 by incubation with human thrombin (20 μg/ml). F: Human thrombin incubated in the absence of serum SF. Molecular weight standards (top to bottom): phosphorylase b (92,000), BSA (66,700), ovalbumin (43,000), carbonic anhydrase (31,000). Electrophoresis was carried out in 10% polyacrylamide gels.

TABLE II. Cell-Spreading-Promoting Activity of SF57

Amount of sample μg/cm^2	Incubated with thrombin (SF57)	Incubated without thrombin (SF65 + SF75)	Unincubated (SF65 + SF75)
		Percentage spread cells	
0	0	0	0
0.03	23	38	55
0.06	55	62	80
0.18	70	Not determined	82

SF57 preparations were produced from SF65 + SF75 preparations by incubation with 20 μg/ml human thrombin for 6 h at 37°. Control preparations were incubated 6 h at 37° without thrombin. Hela cell-spreading-promoting assay was carried out as described in the text.

ose (Fig. 11), but we have been unable in this way as yet to achieve complete conversion of preparations to the SF57 form. Reasons for limited reaction are not clear; it is possible that trace amounts of potent thrombin inhibitors are present in some preparations of SF made by the procedures described above.

USE OF SERUM SPREADING FACTOR IN CELL CULTURE

Serum spreading factor in cell culture medium functions by adsorbing to the plastic or glass culture vessel and mediating the proper attachment of the cells to the substratum [Barnes et al., 1980, 1982]. Purified material is active at concentrations in the range of 0.1 μg/cm^2 [Barnes and Silnutzer, 1983; Barnes et al., 1984; Silnutzer and Barnes, 1984], although optimal concentrations vary somewhat with the cell type used. SF is similar in its properties and in procedures for manipulation in cell culture to those of fibronectin or laminin [Hynes and Yamada, 1982; Ledbetter et al., this volume; Yamada and Akiyama, this volume; Barnes, 1984]. Some affinity of purified SF for type IV collagen can be detected, but unlike fibronectin or laminin [Hynes and Yamada, 1982; Timpl et al., 1979; Ledbetter et al., this volume; Yamada and Akiyama, this volume] binding of SF to collagen is not as strong as binding of SF to plastic, and the physiologic significance of SF interaction with both collagen and heparin is unclear [Barnes and Silnutzer, 1983]. Some experimental evidence suggests that SF may form inactive multimers at high concentrations, and we tend generally to work with purified SF at concentrations in the range of 100–500 μg/ml as stocks. Lower concentrations present problems because of nonspecific adsorption of the highly sticky molecule to containers. SF is always stored in polypropylene tubes, and never in glass or polystyrene. Pipetting, solution transfer, and dilution should be kept to a minimum.

SF as isolated in the final step of the procedure may be diluted directly into culture medium or PBS for precoating of plates or exposure of cells to the factor. Serious problems exist, however, in sterilization of the purified material. Major losses are encountered upon membrane filtration, and we have found that the best way to expose cells to SF under sterile conditions is to precoat tissue culture vessels as described in the section Assay of Cell Spreading-Promoting Activity and, after washing the plate, briefly expose the treated culture surface to 70% ethanol under sterile conditions, remove the ethanol, and wash five times with sterile PBS. SF remains active and bound to the plastic under these conditions and cells in culture medium may be seeded directly onto these sterile, SF-treated plates. Although SF is effective if cells are seeded onto SF-treated plates, a pretreatment of the cells in suspension with SF in solution and removal of soluble SF before plating onto untreated dishes will not promote cell attachment or spreading [Barnes et al., 1980].

As a final note, it is important to remember that whenever any biologic effect is attributed to a particular molecular entity in a purified or partially purified preparation, some possibility exists that the activity actually may reside in another factor in the preparation. One might remember, then, when working with preparations of SF, that some biologic activities in cell culture have been attributed to molecules that may be contaminants of some SF preparations at various stages of isolation. These include α-2-macroglobulin [Salomon et al., 1982, this volume], high-density lipoprotein [Gospodarowicz, this volume], fibronectin, and possibly protease in inhibitors other than α-2-macroglobulin, such as heparin-binding antithrombin. We also have detected mitogenic activity that is separable from the spreading-promoting activity in some partially purified SF preparations [Barnes and Sato, 1980c; Barnes et al., 1980, 1981]. It is likewise necessary to consider situations in which SF may exert a biologic effect in culture as a minor contaminant of a preparation of some other factor. Among serum- or plasma-derived preparations used in cell culture medium, we have observed SF contamination in some preparations of albumin, α-2-macroglobulin, fibrinogen, and lipoprotein preparations (HDL, LDL, and VLDL). It is likely that the amount of SF contamination in such preparations may be quite variable, and it is unclear at this point if the contribution of contaminating SF to any of the reported biologic activities of these or other blood-derived materials is significant.

ACKNOWLEDGMENTS

This work was supported by American Cancer Society grant BC-368 and National Institutes of Health grant CA-35214. The authors thank B. Amos, E. Avner, L. Mousetis, C. See, M. Shaffer, and J. Zaslow for help and advice.

REFERENCES

Barnes D (1982): Epidermal growth factor inhibits growth of A431 human epidermoid carcinoma in serum-free cell culture. J Cell Biol 93:1-4.

Barnes D (1984): Attachment factors in cell culture. In Mather J (ed): "The Use of Serum-Free and Hormone-Supplemented Media." New York: Plenum, pp 195-237.

Barnes DW, Sato GH (1979): Growth of a human breast cancer cell line in serum-free medium. Nature 281:388-389.

Barnes D, Sato GH (1980a): Methods for growth of cultured cells in serum-free medium. Anal Biochem 102:255-270.

Barnes DW, Sato GH (1980b): Serum-free cell culture: A unifying approach. Cell 22:649-655.

Barnes D, Sato G (1980c): Factors that stimulate proliferation of breast cancer cells in vitro in serum-free medium. In McGrath C, Brennan M, Rich M (eds): "Cell Biology of Breast Cancer." New York: Academic, pp 277-287.

Barnes D, Silnutzer J (1983): Isolation of human serum spreading factor. J Biol Chem 258:12548-12552.

Barnes D, Wolfe R, Serrero G, McClure D, Sato G (1980): Effects of serum spreading factor on growth and morphology of cells in serum-free media. J Supramol Struct 14:47-64.

Barnes D, Darmon M, Orly J (1982): Serum-spreading factor: Effects on RF1 rat ovary cells and 1003 mouse embryonal carcinoma cells in serum-free media. Cold Spring Harbor Conf Cell Prolif 9:155-167.

Barnes DW, van der Bosch J, Masui H, Miyazaki K, Sato G (1981): The culture of human tumor cells. Methods Enzymol 79:368-391.

Barnes DW, Silnutzer J, See C, Shaffer M (1983): Characterization of human serum spreading factor with monoclonal antibody. Proc Natl Acad Sci USA 80:1362-1366.

Barnes DW, Mousetis L, Amos L, Silnutzer J (1984): Glass bead affinity chromatography of cell attachment and spreading-promoting factors of human serum. Anal Biochem 137:196-204.

Basara ML, McCarthy JB, Barnes DW, Furcht LT (1984): Tumor cell migration to serum spreading factor. Fed Proc 43 (in press).

Darmon M, Serrero G, Rizzino A, Sato G (1981): Isolation of myoblastic, fibroadipogenic, and fibroblastic clonal lines from a common precursor and study of their requirements for growth and differentiation. Exp Cell Res 132:313-327.

Fisher HW, Puck TT, Sato G (1958): Molecular growth requirements of single mammalian cells: The action of fetuin in promoting cell attachment to glass. Proc Natl Acad Sci USA 44:4-10.

Grinnell F (1978): Cellular adhesiveness and extracellular substrata. Int Rev Cytol 53:65-144.

Grinnell F, Feld MK (1980): Spreading of human fibroblasts in serum-free medium: Possible role of secreted fibronectin. Cell 17:117-129.

Hayman EG, Engvall E, A'Hearn E, Barnes D, Pierschbacher M, Ruoslahti E (1982): Cell attachment on replicas of SDS/polyacrylamide gels reveals two adhesive plasma proteins. J Cell Biol 95:20-23.

Hayman EG, Pierschbacher MD, Ohgren Y, Ruoslahti E (1983): Serum spreading factor (vitronectin) is present at the cell surface and in tissues. Proc Natl Acad Sci USA 80:4003-4007.

Holmes R (1967): Preparation from human serum of an alpha-one protein which induces the immediate growth of unadapted cells in vitro. J Cell Biol 32:297-308.

Hynes RO, Yamada KM (1982): Fibronectins: Multifunctional modular glycoproteins. J Cell Biol 95:369-378.

Klebe RJ (1974): Isolation of a collagen-dependent cell attachment factor. Nature 250:248-251.

McCarthy JB, Palm SL, Furcht LT (1983): Migration by haptotaxis of a Schwann cell tumor line to the basement membrane glycoprotein laminin. J Cell Biol 97:772-777.

McClure DB (1983): Anchorage-dependent colony formation of SV40 transformed BALB/c-3T3 cells in serum-free medium: Role of cell- and serum-derived factors. Cell 32:999-1004.

Mosher DF (1980): Fibronectin. In Spoet TH (ed): "Progress in Hemostasis and Thrombosis." New York: Grune and Stratton, pp 111-151.

Orly J, Sato G (1979): Fibronectin mediates cytokinesis and growth of rat follicular cells in serum-free medium. Cell 17:295-305.

Salas-Prato M, Tanguay JF, Lefebvre Y, Barnes DW, Grotenderst GR (1984): Fetal human liver cells in culture in a hormone-supplemented serum-free medium. In Vitro 20 (in press).

Salomon DS, Bano M, Smith KB, Kidwell WR (1982): Isolation and characterization of a growth factor (embryonin) from bovine fetuin and its resemblance to alpha-2-macroglobulin. J Biol Chem 257:14093-14101.

Serrero G, Sato G (1982): Growth and differentiation of teratoma-derived fibroadipogenic cell line in serum-free medium. Cold Spring Harbor Conf Cell Prolif 9:243-958.

Shaffer MC, Foley TP, Barnes DW (1984): Quantitation of spreading factor in human biological fluids. J Lab Clin Med (in press).

Silnutzer J, Barnes DW (1984): A biologically active cleavage product of human serum spreading factor. Biochem Biophys Res Commun 118:339-343.

Stenn KS (1981): Epibolin: A protein of human plasma that supports epithelial cell movement. Proc Natl Acad Sci USA 78:6907-6911.

Stenn KS, Madri JA, Tinghitella JA, Terranova VP (1982): Multiple mechanisms of dissociated epidermal cell spreading. J Cell Biol 96:63-67.

Timpl R, Heilwig R, Robey PG, Rennard SI, Foidart J, Martin GR (1979): Laminin—A glycoprotein from basement membranes. J Biol Chem 254:9933-9937.

van de Water L III, Schroeder S, Crenshaw EB, Hynes RO (1981): Phagocytosis of gelatin-latex particles by a murine macrophage line is dependent on fibronectin and heparin. J Cell Biol 90:32-39.

Whateley JG, Knox P (1980): Isolation of a serum component that stimulates the spreading of cells in culture. Biochem J 185:349-354.

Wolfe RA, Sato GH (1982): Continuous serum-free culture of the NG18TG-2 neuroblastoma, the C6BU-1 glioma, and the NG108-15 neuroblastoma × glioma hybrid cell lines. Cold Spring Harbor Conf Cell Prolif 9:1075-1088.

Wolfe RA, Sato GH, McClure DB (1981): Continuous culture of rat C6 glioma in serum-free medium. J Cell Biol 87:434-441.

Methods for Preparation of Media, Supplements, and Substrata
for Serum-Free Animal Cell Culture, pages 269–274
© 1984 Alan R. Liss, Inc., 150 Fifth Avenue, New York, NY 10011

16
Purification of Epibolin From Human Plasma

K.S. Stenn

The number of studies devoted to the cell-spreading function of serum is very small compared to the number devoted to the cell growth properties of serum. Nevertheless, attempts have been made to define and isolate those elements of serum that are responsible for its cell spreading properties. Proteins of serum that support the spreading of macrophages [Leonard and Skeel, 1976], of epidermal cells [Levine, 1972; Marks et al., 1972; Coombs et al., 1974; Ooka et al., 1975; Mitrani and Marks, 1978; Tammi and Jansen, 1980], and of cultured epithelial and fibroblastic cells [Holmes, 1967; Whateley and Knox, 1980; Wolf and Lipton, 1973; Lipton et al., 1971; Barnes et al., 1980, 1982] have been described. Currently there is some evidence that there is more than one protein in serum supporting cell spreading [Knox and Griffiths, 1980; Hayman et al., 1982]. One of these is generally accepted to be fibronectin [Ali and Hynes, 1978; Knox and Griffiths, 1980; Hayman et al., 1982; Grinnell and Hays, 1978].

In our isolation attempts we focused only on the properties of serum supporting the spreading of epidermal cells. That property of serum was assayed with primary cultures of epidermal cells or skin explants. Our initial studies showed that that property of human serum was not dialyzable, was resistant to serum-protease inhibitors, and was sensitive to heat and trypsin [Stenn, 1978; Stenn and Dvoretsky, 1979]. The epidermal cell-spreading effect of serum was rapid, independent of cell division, and separate from fibronectin [Federgreen and Stenn, 1980].

The isolation procedure developed is described in detail below. The isolated spreading function is a single-chain, glycoprotein of approximately 65 kd

Department of Dermatology, Yale University School of Medicine, New Haven, Connecticut 06510

molecular weight. Antibody to this protein blocks the epidermal cell-spreading activity of whole human serum or plasma. While the purified fractions of this protein are necessary for the spreading of primary cultured epidermal cells, this protein does not appear necessary for the spreading of cultured calf aortic endothelial cells or cultured human foreskin dermal fibroblasts (Stenn, Madri, Cohen, unpublished observation).

PURIFICATION PROCEDURE
Method of Preparation

The following is a modification of a purification method described earlier [Stenn, 1981]. For this purification only reagent grade chemicals and distilled water are used. Dialysis bags are prepared as previously described [Stenn and Blout, 1972]. All centrifugation is executed at 5,000g for 20 min. All buffers contain 0.02% NaN_3 and 0.1mM phenylmethylsulfonyl fluoride (PMSF). In this description "phosphate buffer" refers to a solution of 50 mM Na_2HPO_4/NaH_2PO_4, pH 6.0. The active fraction is quantitated by a described assay [Stenn, 1981] consisting of dissociated guinea pig epidermal cells. The number of spread cells per 100 attached cells per milligram protein is recorded and compared. It is necessary to emphasize at the outset that in developing this procedure 1) the spreading of these specific cells defined the fraction studied; 2) the cells were always of a primary culture and were not grown in culture before use in the assay; and 3) the cells were epithelial—specifically, epidermal cells.

1. To 1,000 ml of blood-blank plasma cleared by centrifugation is added dropwise, at 4°C and with slow stirring, 500 ml of saturated ammonium sulfate solution. After 2 h of mixing the solution is cleared by centrifugation and the precipitate is discarded. To the supernatant solution is added dropwise 1,500 ml of saturated ammonium sulfate (making the solution 66 2/3% saturated in ammonium sulfate). After 2 h of mixing, the solution is cleared by centrifugation (4°C). The supernatant is discarded and the precipitate is dissolved in 200 ml of phosphate buffer. All subsequent steps are performed at room temperature.

2. The above solution is dialyzed against three changes of 4 liters of phosphate buffer until the conductivity of the dialysand equals that of the initial phosphate buffer. The dialysand is poured over a DEAE Biogel A (100–200 mesh; Biorad) column of 8 cm × 23 cm and then washed with the phosphate buffer until the eluate has an absorbance at 280 nm of less than 0.05. The active fraction is eluted stepwise by the phosphate buffer containing 100 mM NaCl.

3. This eluate is dialyzed against two changes of 4 liters of 0.5 mM Na_2HPO_4 (pH 6.0). The dialysand is cleared by centrifugation and then titrated to pH 4.6 with 100 mM HCl. The precipitate formed is collected by centrifugation and then dissolved in 80 ml of phosphate buffer plus 100 mM NaCl. This solution is again dialyzed against 4 liters of 0.5 mM $N_aH_2PO_4$/ Na_2HPO_4. The dialysand is titrated to pH 5.3 and then cleared by centrifugation. The supernatant is brought to pH 4.6 and centrifuged. The pellet is dissolved in 25 ml of phosphate buffer plus 100 mM NaCl.

4. Four milliliters of the above solution is layered over a gel filtration column (Sephacryl S200, Pharmacia, 5 cm × 90 cm) and eluted with the phosphate buffer plus 100 mM NaCl. The active fraction elutes after approximately 540 ml have passed through the column (Ve/Vo = 1.64) in the trailing edge of the second large protein peak (see Fig. 1).

5. The active fraction is concentrated (Amicon filter PM-30) and fractionated further by polyacrylamide gel electrophoresis by the gel system of Davis [1964]. In this system the active fraction runs with an R_f value of 0.5-0.6 (see Fig. 2). This band is cut out of the gel and run on a second gel system following the SDS-PAGE (nonreducing) procedure of Laemmli [1970] with 15% acrylamide (100mV, 19 h). The active fraction in this gel runs as the only diffuse Coomassie blue staining band. This band is cut out and eluted electrophoretically in an apparatus similar to that of Doly and Petek [1977]. The isolate is concentrated in a dialysis bag against a drying reagent (e.g.,

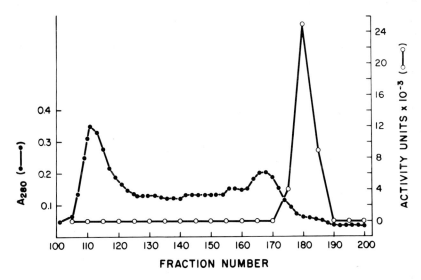

Fig 1.   Elution profile of gel filtration step. From Stenn [1981], with permission.

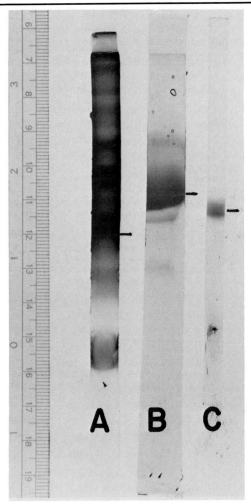

Fig 2. Polyacrylamide gel electrophoresis of preparation after gel filtration (A), after electrophoresis in the Davis [1964] gel system (B), and after electrophoresis in the Laemmli gel system (C). Gel A was electrophoresed by the method of Davis (12.5% acrylamide) and B and C by the method of Laemmli [1970] (15% and 12.5% acrylamide, respectively). Line indicates band of epibolin activity.

Aquacide, Calbiochem, LaJolla, CA), and the residual SDS is removed by chromatography following the procedure of Kapp and Vinogradov [1978] with a 2-cm × 6-cm column of ion retardation resin (AG11A8, Biorad, Richmond, CA).

COMMENT

The procedure is not difficult though it is long. This procedure gives an activity yield at the first gel electrophoresis step of about 1-2%. The final product eluted from the SDS-PAGE gel is not biologically active; however, antibody to this protein produced in a rabbit shows one precipitin arc on immunoelectrophoresis against whole human serum. Moreover, this antibody blocks all the spreading activity in human serum (or plasma) for epidermal cells [Stenn et al., 1983]. Antibody produced to the faster-running, thin band does not block the epidermal cell-spreading activity of whole human serum. That the protein preparation isolated appears to be highly purified is supported by studies of standard electrophoresis, immunoelectrophoresis, and N-terminal amino acid analysis (unpublished).

Early efforts to purify this protein indicated that it was not stable at 4°C in solution. For this reason after the ammonium sulfate fraction step all procedures were conducted at room temperature. Because of stability problems for cell culture studies our laboratory has used the purified preparations within 2 weeks of isolation. The preparation is stored at room temperature in the phosphate buffer plus 0.1 mM sodium chloride plus azide and PMSF (as described above). Alternately, the isoelectric precipitate preparation may be lyophilized, stored frozen, and used weeks to months later.

For some purposes the purest preparation is of value, e.g., in antibody production and in chemical characterization, but for cell physiology studies the isoelectric precipitate product and the gel filtration product are most useful.

REFERENCES

Ali IV, Hynes RO (1978): Effects of LETS glycoprotein on cell motility. Cell 14:439-446.

Barnes D, Wolfe R, Serrero G, McClure D, Sato G (1980): Effects of serum spreading factor on growth and morphology of cells in serum-free medium. J Supramol Struct 14:47-63.

Barnes DW, Darmon M, Orly J (1982): Serum spreading factor: Effect on RFI rat ovary cells and 1003 mouse embryonal carcinoma cells in serum free media. Cold Spring Harbor Symp Cell Prolif 9:155-167.

Coombs VA, Nissen BK, Marks R (1974): The epidermal cell migration promoting activity of serum in guinea pig skin in vitro. Arch Dermatol Forsch 249:367-372.

Davis BJ (1964): Disc electrophoresis. II. Method and application to human serum proteins. Ann NY Acad Sci 121:404-427.

Doly J, Petek F (1977): Polyacrylamide gradient electrophoresis for protein on the milligram scale. J Chromatogr 137:69-81.

Federgreen WR, Stenn KS (1980): Fibronectin does not support epithelial cell spreading. J Invest Dermatol 75:261-263.

Grinnell F, Hays DG (1978): Cell adhesion and spreading factor. Similarity to cold insoluble globulin in human serum. Exp Cell Res 115:221-229.

Hayman EG, Engvall E, A'Hearn E, Barnes D, Pierschbacher M, Ruoslahti E (1982): Cell attachment on replicas of SDS polyacrylamide gel reveals two adhesive plasma proteins. J Cell Biol 95:21-23.

Holmes R (1967): Preparation from human serum of an alpha one protein which induces the immediate growth of unadapted cells in vitro. J Cell Biol 32:297-308.

Kapp OH, Vinogradov SN (1978): Removal of SDS from proteins. Anal Biochem 91:230-235.

Knox P, Griffiths S (1980): The distribution of cell-spreading activities in sera: A quantitative approach. J Cell Sci 46:97-112.

Laemmli UK (1970): Cleavage of structural proteins during the assembly of the head of bacteriophage T4. Nature 227:680-685.

Leonard EJ, Skeel A (1976): A serum protein that stimulates macrophage movement, chemotaxis and spreading. Exp Cell Res 102:434-438.

Levene M (1972): The growth of adult human skin in vitro. Br J Dermatol 86:481-490.

Lipton A, Klinger I, Paul D, Holley RW (1971): Migration of mouse 3T3 fibroblasts in response to a serum factor. Proc Natl Acad Sci USA 68:2799-2801.

Marks R, Bhogal B, Dawber RPR (1972): The migratory property of epidermis in vitro. Arch Dermatol Forsch 243:209-220.

Mitrani E, Marks R (1978): Towards characterization of epidermal cell migration promotion activity in serum. Br J Dermatol 99:513-518.

Ooka H, Yamamoto K, Okuma Y, Suga S, Wakasugi M (1975): The migratory activity of rat epidermal cells in vitro — Age related changes and the effect of serum. Exp Gerontol 10:79-83.

Stenn KS (1978): The role of serum in the epithelial outgrowth of mouse skin explants. Br J Dermatol 98:411-416.

Stenn KS (1981): Epibolin: A protein of human plasma that supports epithelial cell movement. Proc Natl Acad Sci USA 78:6907-6911.

Stenn KS, Blout ER (1972): The activation of bovine prothrombin using an insoluble preparation of activated factor X. Biochemistry 11:4502-4515.

Stenn KS, Dvoretsky I (1979): Human serum and epithelial spread in tissue culture. Arch Dermol Res 264:3-15.

Stenn KS, Madri JA, Tinghitella T, Terranova VP (1983): Multiple mechanisms of dissociated epidermal cell spreading. J Cell Biol 96:63-67.

Tammi R, Jansen CT (1980): Serum epidermal migration stimulating factor — Membrane ultra filtration and physical stability characteristics. Arch Dermol Res 267:319-322.

Whateley, JG, Knox P (1980): Isolation of a serum component that stimulates the spreading of cells in culture. Biochem J 185:349-354.

Wolf L, Lipton A (1973): Studies on serum stimulation of mouse fibroblast migration. Exp Cell Res 80:499-502.

Methods for Preparation of Media, Supplements, and Substrata
for Serum-Free Animal Cell Culture, pages 275–293
© 1984 Alan R. Liss, Inc., 150 Fifth Avenue, New York, NY 10011

17
Preparation of Extracellular Matrices Produced by Cultured Bovine Corneal Endothelial Cells and PF-HR-9 Endodermal Cells: Their Use in Cell Culture

Denis Gospodarowicz

Cell behavior in vitro is known to be strongly affected by the substratum upon which cells will attach, migrate, and proliferate (reviewed by Grinnell [1978]). This reflects the fact that cell proliferation and differentiation in vivo are regulated by the extracellular matrix (ECM) that cells can produce and upon which they migrate. ECMs are mostly composed of collagens, proteoglycans, and glycoproteins [Hay, 1981; Kleinman et al., 1981; Yamada, 1983]. Their compositions differ, depending upon tissue under consideration and its stage of development [Hay, 1981].

It is recognized that in vivo an intact ECM scaffold is required for the maintenance of orderly tissue structure and regeneration. By its presence it defines the spatial relationships among similar and dissimilar types of cells. Such a substrate plays an important role not only in cell attachment and migration [Hay, 1981; Kleinman et al., 1981; Yamada, 1983], but also in the cell response to various growth-promoting factors present in plasma, lymph, or interstitial fluid [Gospodarowicz and Tauber, 1980]. In the early stage of embryonic development, when different tissues composing a given organ are formed as a result of strictly timed and spatially interrelated proliferative and differentiative events, interaction of cells with newly formed ECM has been shown to promote cell proliferation and to stabilize new phenotypic expression

Cancer Research Institute and the Departments of Medicine and Ophthalmology, University of California Medical Center, San Francisco, California 94143

[Wessels, 1977; Hay, 1978; Saxen et al., 1982]. This has been studied best in the case of the formation of the epithelial part of the kidney nephron induced by the invasion of ureter bud into kidney mesenchyme. The metanephric mesenchymal cells are embedded in an ECM composed mostly of isotypes collagen I and III and of fibronectin. Under the directive influence of the invading ureter bud, mesenchymal cells condense and start to synthesize an ECM composed mainly of collagen type IV, heparan sulfate proteoglycan, and laminin. This shift in ECM component production correlates closely with the synchronized development of the epithelial part of the glomeruli and associated nephron tubule [Ekblom, 1981; Ekblom et al., 1980, 1981a,b]. Induction therefore stimulates the production of laminin, type IV collagen and basement membrane proteoglycan, resulting in the formation of a proper substratum for epithelial cell attachment and further differentiation. ECM components have also been implicated as playing a role in inductive tissue interaction in somite chondrogenesis [Kratochwil, 1972], the differentiation of corneal epithelium [Meier and Hay, 1974, 1975; Hay, 1980], salivary gland morphogenesis [Bernfield et al., 1972; Banerjee et al., 1977; Cohn et al., 1977], and tooth germ development [Thesleff and Hurmerinta, 1981; Thesleff et al., 1981]. Likewise, in adult animals the elegant studies of Vracko [1974] have shown that when a given organ is wounded or destroyed, an intact ECM will enable multicellular organisms to reconstitute histological structures of most tissues and organs to what they were prior to loss of cells.

It therefore appears that during the whole of embryogenesis, as well as in neonatal or adult life, an intact ECM scaffold is required for maintenance of orderly tissue structure.

One of the major limitations of tissue culture techniques currently in use is that cells rest on plastic. Unless cultured cells can produce their own ECM, the possibility exists that contact with plastic could inhibit them from proliferating or could cause them to rapidly lose their ability to express their correct phenotype. In the present review, we will outline recent studies of the behavior of cells maintained on dishes coated with ECM of known composition. The use of such a substrate for the long-term culture of normal diploid and tumor cells will be described.

PRODUCTION OF EXTRACELLULAR MATRICES BY CULTURED CORNEAL ENDOTHELIAL AND PF HR-9 ENDODERMAL CELLS AND THEIR PREPARATION FOR IN VITRO USES

The ECM is an organized complex of collagens, proteoglycans, and glycoproteins, all interacting to produce a highly stable structure of whose area only a very small portion is in contact with the cell surface. Elucidation of the components of the ECM involved in controlling cell proliferation either in vivo or in vitro has been made difficult mostly by its intricate nature. Because the correct in vitro reconstruction of the ECM from its isolated native components into the highly ordered structure that it represents in vivo would be a formidable task, advantage was taken of the fact that cultured bovine corneal endothelial (BCE) cells (Figs. 1–3) and PF HR-9 endodermal cells (Fig. 4) have the ability to produce, underneath their basal surface, a thick ECM that adheres strongly to plastic [Gospodarowicz and Tauber, 1980]. The matrix produced by BCE cells is composed of elastin, basement-membrane collagen types IV and V, and interstitial collagen type III [Tseng et al., 1981; Gospodarowicz, 1983]. Also present are proteoglycans, composed of heparan sulfate and dermatan sulfate proteoglycans [Robinson and Gospodarowicz, 1983, 1984], and glycoproteins such as fibronectin and laminin [Gospodarowicz et al., 1981b]. This last glycoprotein has been reported to be a specific marker for basement lamina in vivo [Timpl et al., 1979]. Therefore, the matrix synthesized and secreted by endothelial cells in culture conforms to the chemical criteria, as well as to the biological criterion, of an ECM that is to serve as the substratum upon which cells migrate and divide in vivo.

The same is also true of the HR-9 ECM, which, unlike the ECM of BCE cells, is composed only of basement membrane collagen type IV, heparan sulfate proteoglycans, and glycoproteins such as laminin and entactin [Hogan et al., 1982; Lewo et al., 1982].

Preparation of Primary Cultures of BCE Cells

Eyes from freshly sacrificed cows can be obtained from a local slaughterhouse. The eyes can either be used immediately or stored overnight in a cold room. The corneas are first checked visually for transparency and presence of blood vessels. Eyes with unscratched, thin, and crystal-clear corneas are selected. The corneas are first washed with 95% ethanol. This will sterilize the outside of the corneas and destroy the outer epithelial cell layers. The cornea is then punctured near its edge (at the junction of the cornea and the sclera) with an 18-gauge needle. A dissecting scissors is inserted into the hole and the cornea is dissected out. It is then placed inverted (endothelial side up) in a 10-cm tissue culture dish. The cornea is carefully washed with phosphate-buffered saline (PBS) to remove any traces of iris that could have adhered to the endothelium during its dissection. Such contaminating tissue would appear as small black fragments. Once the corneal endothelium has been extensively washed, it is delicately scraped with a groove director. Too much pressure should not be applied in order that one not go through the Descemet's

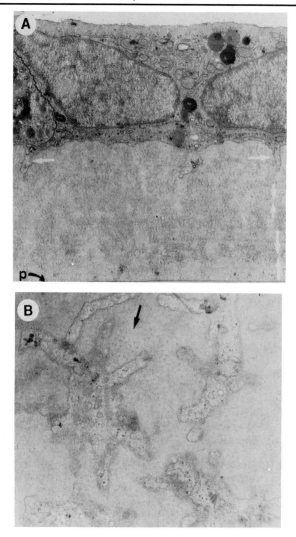

Fig. 1. Electron micrograph of corneal endothelial cells after 3 weeks in culture. A. Cross section showing the cells resting on a membrane 5 μm thick deposited on the plastic surface (p) of the tissue culture dish. In the basal part of the cells, small cytoplasmic processes are entering into the basement membrane and are responsible for the pitted surface observed in Figures 2 and 3. B. Front view of the basement membrane at the level where small processes sent by the cells enter into it (apical surface of the basement membrane). Numerous nodes (arrow) interconnected with 15- to 17-mm-wide filaments can be seen. Similar node distribution can be seen in the upper part of the basement membrane deposited by BCE cells when seen in cross section (A).

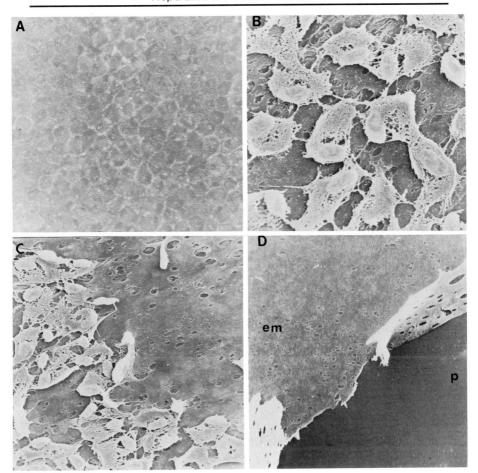

Fig. 2. Scanning electron microscopy of a monolayer of bovine corneal endothelial cells before and after exposure to Triton X-100. A monolayer composed of polygonal, highly flattened, and closely apposed cells can be seen in A (×200). After the monolayer has been treated with Triton X-100 (0.5% in PBS) for 5–10 min at room temperature, it is composed of nuclei and cytoskeletons that no longer attach firmly to the extracellular matrix. In some areas the extracellular matrix has been exposed (B, ×400). Washing the dishes with PBS removed the cytoskeleton and exposed the extracellular matrix present underneath the cells (C, ×200). The plate has been scratched with a needle to expose the plastic (p) to which the extracellular matrix (em) strongly adheres (D, ×200).

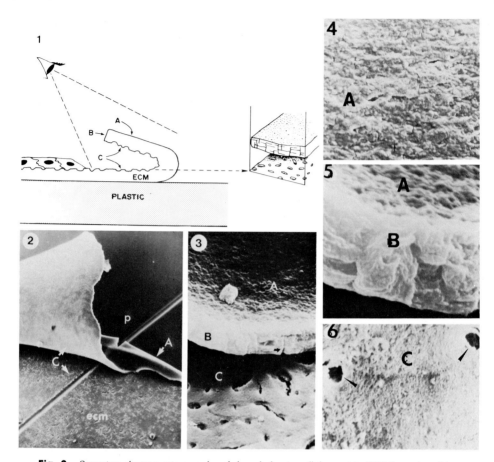

Fig. 3. Scanning electron micrographs of denuded extracellular matrix (ECM) produced by confluent bovine corneal endothelial cells. 1. Schematic illustration of the ECM after removal of the cell monolayer. The cell monolayer was removed following exposure to 20 mM NH₄OH, as described in the text. The plate was then scratched with a needle in order to detach the ECM from the plastic locally. Scanning electron micrographs were taken in the area where the ECM was folded. A is the side that was originally attached to plastic, B is the fracture edge; C is the side on which cells were originally attached to the ECM. 2. Low-power ($\times 200$) scanning electron micrograph of the ECM. The plastic (p) has been exposed, and the trace of the needle running through the ECM (ecm) and the denuded plastic can be seen. A is the ECM side originally attached to the plastic, and C is the side on which cells were originally attached. This side is pitted, the pits corresponding to cell processes entering into the ECM. These pits were probably the sites where the cells were originally anchored to the ECM. 3. The ECM can be observed at a higher magnification ($\times 3,000$). The fracture edge of the ECM (B) can be seen, as can the transverse section (arrow) of a pit in which a cellular process was present before denudation. The ECM side to which the cells were attached is covered with numerous small pits. 4. High magnification ($\times 8,000$) of the side of the ECM previously attached to plastic showing a corrugated pattern. 5. High magnification ($\times 14,000$) of the ECM fracture edge, showing its lamellar structure. 6. High magnification ($\times 16,000$) of the ECM side to which cells were previously attached, showing its fibrillar appearance. Two small empty pits can be seen distinctly.

membrane. The groove director is then dipped into a 6-cm tissue culture dish containing 5 ml of Dulbecco's modified Eagle's medium (DMEM) supplemented with 10% fetal calf serum, 5% calf serum, 50 μg/ml gentamicin, and 0.25 μg/ml Fungizone.

This process can be repeated three times. Visual examination of the plate should reveal small whitish tissue fragments floating into the medium. When examined under phase-contrast microscopy, such tissue fragments appear to be composed of closely apposed hexagonal cells organized into a honeycomb pattern (Fig. 5A). The plates are then incubated at 37°C in a CO_2 incubator with 99% humidity and left untouched for 5 days, except for occasional examination under phase-contrast microscopy. After 2-3 days, the tissue fragments will have attached to the plastic, and cells can be seen migrating out of the explants (Fig. 5B). After 5-6 days the media are changed and fresh media added. Brain or pituitary fibroblast growth factor (FGF) (highly purified, isoelectric focusing step) or partially purified FGF (Sephadex G-75 gel filtration step) is then added to the cultures every other day at a concentration of either 25 or 250 ng/ml, depending on the stage of purity of FGF. In general, for growing primary or maintaining stock cultures the use of highly purified FGF is a waste. The Sephadex G75 fraction (G75-FGF) can therefore be used at a concentration of 250 ng/ml. The less purified FGF fraction eluting with 1 M NaCl, 0.1 M Na phosphates from the carboxymethyl Sephadex C50 column can also be used. In this case, the FGF concentration should be 1 μg/ml. The use of crude brain or pituitary extract is not recommended, since it can induce morphological changes. By day 10, well-developed colonies are clearly visible (Fig. 5C,D). The cell density is then $(5-10) \times 10^5$ cells per 6-cm dish. The primary cultures are trypsinized with STV (0.9% NaCl, 0.01 M Na phosphates, pH 7.4, 0.05% trypsin, and 0.02% versene) at 37°C for 5 min. After the STV solution is removed from the plates, the cells from each primary are resuspended in 6 ml of DMEM supplemented with serum and antibiotics as described above and passaged into 10-cm gelatinized dishes (Table I).

To gelatinize plates, a 0.2% gelatin solution in PBS *free of calcium and magnesium* is prepared. The solution is autoclaved, and then filtered on a 0.2-μm filter after it has cooled. Ten milliters of the 0.2% gelatin solution is added per 10-cm plate, and plates are stored at 4°C for at least 3 h. The gelatin solution is then removed and the plates washed once with 10 ml of PBS. Then 2 ml from the primary cell suspension (1×10^5 to 1.5×10^5 cells) is added to 10-cm gelatinized tissue culture dishes containing 10 ml of DMEM supplemented with 10% fetal calf serum (FCS) and 5% calf serum, antibiotics, and FGF. A total of eighteen 10-cm gelatinized dishes are seeded. FGF is added every other day until the plates are nearly confluent (day 4 after

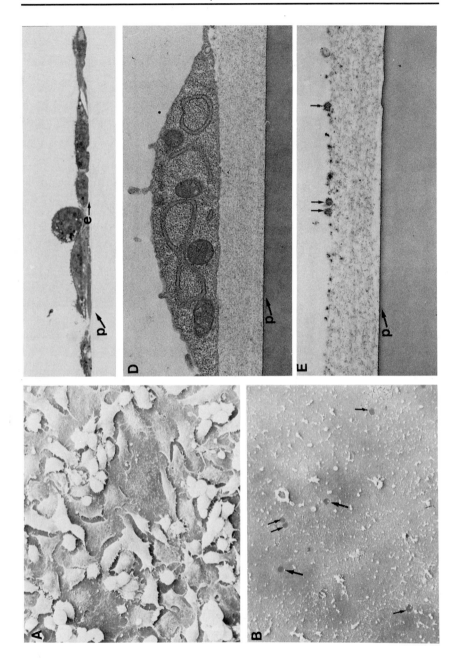

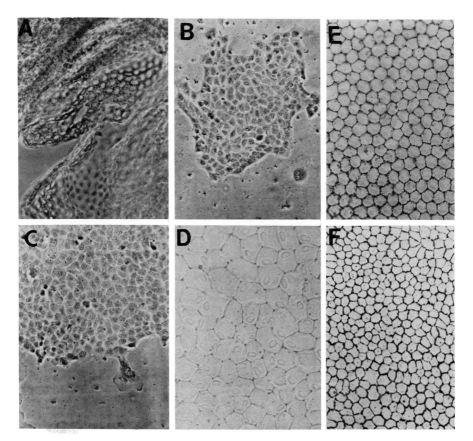

Fig. 5. Bovine corneal endothelial cells in tissue culture. A. Fragments of corneal endothelial tissue immediately after explanation to tissue culture medium (phase-contrast, ×100). B. Colony of endothelial cells arising from a fragment of endothelium. A binucleated cell can be seen at the bottom (phase-contrast, ×100). C. Monolayer outgrowth from endothelium after 7 days in culture (phase-contrast, ×150). D. Endothelial cell monolayer after fixation and silver nitrate staining (phase-contrast, ×300). E, F. Comparison of the morphological appearance of the corneal endothelium in vivo when stained with alizarin red (E) and when similarly stained in vitro (F). Very comparable cellular organization can be seen (bright field, ×100).

Fig. 4. Scanning and transmission micrographs of HR-9 cell cultures and their extracellular matrix. A. Confluent culture of HR-9 cells before treatment with 20 mM NH_4OH. B. After treatment, the cell monolayer has disappeared and the ECM produced by the cells is denuded. In contrast to the corneal epithelial cell extracellular matrix, a bit of cellular debris adhering firmly to the ECM can be seen. The arrows point to a hole in the matrix where the plastic can be seen. C. Bright-field micrograph of the HR-9 cell monolayer showing the ECM (e, arrow) produced by the cells that covered the plastic surface (p). D. Micrograph showing a cross section of confluent HR-9 cells resting on the extracellular matrix they have produced (plastic, p). E. Extracellular matrix after denudation of the cell layer by treatment with 20 mM NH_4OH in water. A bit of cell debris can be seen on top of the matrix (arrows) (plastic, p).

TABLE I. Time Schedule for the Preparation of BCE-ECM-Coated Dishes When One Begins With Primary Cultures

Day 0	A. Primary in one 6-cm dish
Day 10	B. Each primary [(3–4) $\times$ 10^5 cells] is then passaged into three 10-cm gelatinized dishes (total of 18 dishes for 6 primary cultures).
Day 14	C. Subconfluent 10-cm dishes (5 $\times$ 10^6 cells per dish) are trypsinized. The cells harvested from 17 10-cm gelatinized dishes (8.5 $\times$ 10^7) are then distributed into 200 35-mm dishes (4.25 $\times$ 10^5 cells per 35-mm dish). To keep the stock, one 10-cm dish is also passaged into 18 10-cm gelatinized dishes, which become the new stock dishes. These dishes can be processed as described above on day 18 to produced 200 more coated dishes.
Days 17–18	D. The 35-mm plates are confluent. On day 18, subconfluent 10-cm dishes are trypsinized. Cells are then distributed into 200 35-mm dishes and 18 gelatinized 10-cm dishes, as outlined in (C).
Days 22–27	E. Denude the ECM.

seeding, 5 $\times$ 10^6 cells per 10-cm dish). It is best to passage the cells at that time, since after reaching confluence the cells produce so much ECM that they become difficult to trypsinize and can therefore be damaged in the process. This will be reflected in poor plating efficiency and cell growth, as well as aberrant cell morphology (giant multinucleated cells). The cells from one 10-cm dish are then passaged into eighteen 10-cm gelatinized dishes (Table I).

BCE-ECM Preparation

To obtain ECM-coated dishes with a homogeneous matrix across the surface of the plate, it is best that cells be plated at high density in 35-mm dishes. Stated in other words, the strategy is to have the major amount of cell growth occur in the 10-cm stock dishes, with no more than one population doubling occurring in the 35-mm dishes. The surface area in a 10-cm stock dish is approximately equal to that of eight 35-mm dishes. Therefore, to prepare a batch of 200 35-mm dishes (an area equivalent to that of 25 stock dishes), it is advisable to use the cells from approximately 16 stock dishes. A group of 18 stock dishes, upon reaching confluence, can be used as follows. Cells from one 10-cm dish are passaged into 18 10-cm gelatinized stock dishes. The remaining 17 10-cm dishes are used in the production of the 200 ECM-coated 35-mm dishes.

First, 500 ml of DMEM H-16 supplemented with 10% calf serum, 4% Dextran T40, gentamicin (50 μg/ml), Fungizone (0.25 μg/ml), and G75-FGF (250 ng/ml) are prepared. Cells trypsinized from 17 10-cm dishes (8.5 $\times$ 10^7 cells) are then resuspended into the medium, and 2.5-ml aliquots containing 4.25 $\times$ 10^5 cells are distributed into 200 35-mm plastic tissue cultures. FGF is then added every other day until the cultures are confluent (ordinarily within 2–3 days). The cultures are then allowed to incubate for 5–10 more days after confluence without change of media.

BCE-ECM Denudation

To denude the ECM produced by confluent cultured corneal endothelial cells, the medium is first removed. The plates are then exposed to either detergent treatment (0.5% Triton X in PBS, as outlined in Fig. 2) or base treatment (20 mM NH_4OH in water). Both treatments are equally efficient in removing cells from their ECM. However, while detergent treatment works best with cells that have been confluent for a week or so, base treatment is far more efficient than detergent treatment with cultures that have been confluent for 2 or 3 months and have deposited a very thick ECM (Fig. 3). Two milliliters of 20 mM NH_4OH is added to the plates. One of the plates is examined for cell lysis under phase-contrast microscopy. This process will take at most 2–3 min at room temperature. Each plate is then washed with 10 ml of PBS and stored in a refrigerator with 2 ml PBS supplemented with gentamicin (50 μg/ml) and Fungizone (0.25 μg/ml). BCE-ECM-coated dishes can be kept for up to a month without losing their properties. Although the plates are still fully active after a month, fungus contamination resulting from prolonged storage in PBS could start to appear. It is therefore best to use these plates within 2 weeks. Once the primaries have been passaged into 10-cm dishes, the procedure outlined above (Table I) allows preparation of 200 35-mm ECM-coated dishes once a week for at least 10–20 weeks (total of 4,000–8,000 35-mm BCE-ECM-coated dishes from one or two primaries).

Culture of PR-HR-9 Endodermal Cells

PF-HR-9 cells are an established cell line and can be obtained from any of the numerous laboratories currently working with them. Two such laboratories are those of Dr. R. Kramer (University of California Medical Center Dental School, San Francisco, CA) or Dr. E. Ruoslahti (Cancer Research Foundation, La Jolla, CA). Stock plates of PF-HR-9 cells are maintained on plastic tissue culture dishes and grown in the presence of DMEM supplemented with 10% fetal calf serum. Cells are passaged when subconfluent. If increased pleiomorphism of the cultures is observed with time, as reflected by increased appearance of large multinucleated cells, it may be necessary to clone the cells every 3-6 months.

HR-9-ECM Preparation

To make the ECM produced by HR-9 cells adhere firmly to the plastic substratum, it is preferable to coat dishes with fibronectin. Fibronectin is purified from bovine plasma, as described by Engvall et al. [1978], with a Pharmacia 4B Sepharose gelatin affinity column. The fibronectin (10 mg/ml) eluting with 6 M urea is kept at $-20°C$ until it is used. Before use, the stock fibronectin solution is diluted with PBS to a final concentration of 100 μg/ml, and 1 ml is added per 35-mm plate. The plates are left to incubate at room temperature for 1 h. The fibronectin solution is then removed and plates are washed once with PBS (2 ml). Stock plates of PF-HR-9 cells are trypsinized by exposing the cultures to STV for 5 min at 37°C. The trypsinized cells are resuspended into DMEM supplemented with 10% fetal calf serum, and aliquots containing 2×10^5 cells are then added to 35-mm fibronectin-coated dishes containing 2 ml of DMEM supplemented with 10% fetal calf serum, gentamicin (50 μg/ml), and Fungizone (0.25 μg/ml). After 4-5 days in culture, cells become confluent. The medium of the cells is then changed to DMEM supplemented with 5% fetal calf serum. The confluent plates are then incubated for five more days [Gospodarowicz et al., 1984a,b].

HR-9-ECM Denudation

To denude the ECM produced by confluent HR-9 cell cultures, the medium is first removed and, as in the case of BCE cell cultures, 2 ml of 20 mM NH_4OH are then added. After 5 min at room temperature, the cells will have lyzed. In order to remove the cell debris, the plates are washed energetically with 100 ml of PBS. If plates have not been previously coated with fibronectin, the HR-9-ECM will come off the plates during the washing period.

In contrast with the BCE-ECM, which is extremely easy to prepare free of cellular debris, there will always be some debris left adhering to the HR-9 matrix. This is mostly due to the fact that, in contrast to BCE cells, part of the HR-9 cell population dies while proliferating. Cellular debris will therefore be trapped in the ECM continuously produced by the other living cells. Cellular debris adhering firmly to the apical surface of the HR-9-ECM are probably derived from dead cells that are present within the monolayer and that will not fully lyze when exposed to the NH_4OH solution.

USE OF ECM PRODUCED BY BCE CELLS AND HR-9 ENDODERMAL CELLS

ECM-coated dishes are becoming widely used for initiating primaries, for the study of cell migration and metastatic behavior, and for studies on the control of cell proliferation and differentiation.

BCE-ECM

BCE-ECM has been shown to promote the adhesion and migration of various tumor cell types seeded on it [Vlodavsky et al., 1980; Gospodarowicz et al., 1981a]. This includes cell lines, as well as pleural effusion of breast cancer [Hyldahl and Auer, 1982]. It has also been shown to be useful for the development of primary epithelial cell cultures [Wiesel et al., 1983] derived from the upper respiratory tract. In the field of amniocentesis, plating of the cells on ECM promotes their rapid growth, thereby making early diagnosis feasible [Vlodavsky et al., 1982]. ECM-coated dishes have also been useful in restoring the normal phenotypic expression of a number of normal cells that would otherwise have lost it when maintained on plastic [Gospodarowicz and Tauber, 1980; Gospodarowicz et al., 1980; Mason et al., 1982]. It has also been shown that the ECM is useful for maintaining human prolactin-secreting adenoma cells [Bethea and Weiner, 1981; Bethea et al., 1982]. In the field of aging, ECM is of potential use, since cells maintained on it and exposed to serum-supplemented medium have a much longer life span in culture than cells maintained on plastic [Giguere et al., 1982; Gospodarowicz and Massoglia, 1982]. The mechanism(s) by which the ECM prolongs the life span of the cells in vitro is presently unknown. BCE-ECM-coated dishes have also been useful in allowing the maintenance of a number of normal diploid cells and tumor cells under serum-free conditions, so that plasma factors involved in the control of their proliferation could be studied [Gospodarowicz et al., 1982a] (Table II). One should observe in this regard that BCE-ECM best supports the proliferation of mesenchymal cells and cells derived from the neural crest. It does not, as outlined below, support the long-term growth of epithelial cells, even those derived from the mesenchyme, such as normal diploid kidney tubule epithelial cells [Gospodarowicz et al., 1984a]. An exception to that rule are the recent studies of Grove and Pratt who reported that BCE-ECM promotes attachment, spreading, and growth of palate epithelial cells [Grove and Pratt, 1983].

TABLE II. Factors Required for the Proliferation of Normal Diploid or Transformed and Established Cell Lines Maintained in Serum-Free Medium

	HDL (µg protein/ml)	Insulin or Somato C (ng/ml)		FGF or EGF (ng/ml)		Transferrin (µg/ml)	Reference
Normal diploid cells							
Vascular endothelial cells	500	—		—	50	10	[Tauber et al., 1981]
Corneal endothelial cells	250	2,500	100	100	50	10	[Giguere et al., 1982]
Vascular smooth muscle cells	250	2,500	100	100	50	10	[Gospodarowicz et al., 1981c]
Granulosa cells	30	1,000	100	100	50	10	[Savion et al., 1981; Savion and Gospodarowicz, 1981, 1982]
Adrenal cortex cells	30	50	10	100	—	5	[III and Gospodarowicz, 1982]
Lens epithelial cells	250	2,500	100	100	—	10	[Gospodarowicz and Massoglia, 1982]
Kidney tubule cells	750	—	—	—	—	50	[Gospodarowicz et al., 1984b]
Embryo fibroblasts (Rat-1)	1,000	5,000	—	—	25	25	[Giguere and Gospodarowicz, 1983]
Transformed cells (Tumor)							
A-431 carcinoma	500	—	—	—	Toxic	10	[Gospodarowicz et al., 1982b,c]
Colon carcinoma cells	500	—	—	—	—	10	[Gospodarowicz et al., 1982b,c]
Ewing sarcoma cells	500	—	—	—	—	10	[Gospodarowicz et al., 1982b,c]
Rhabdomyosarcoma	500	—	—	—	—	10	[Gospodarowicz et al., 1982b,c]
MDCK (kidney-derived)	500	—	—	—	—	10	[Gospodarowicz et al., 1983]
B-31 cell line	500	—	—	—	—	25	[Giguere and Gospodarowicz, 1983]

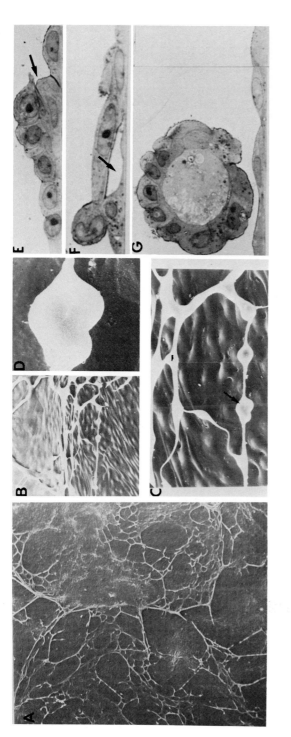

Fig. 6. Scanning electron and cross-section micrograph of kidney tubules forming in vitro. A. Confluent monolayer of kidney tubule cells maintained on HR-9-ECM in the presence of serum-supplemented medium. An extensive network of tubules has formed on the apical surface of the cell layer (× 40). B. An area where tubules are floating over the monolayer is shown. C. A blow-up of a portion of the area shown in B. Tubules joining each other can be seen, and in some regions of the tubules varicosities can be seen (arrow). D. One of the varicosities seen in C is composed of three epithelial cells with distinct borders. E, F. The various steps of tubule formation. In E the first step of tubule formation (arrow) can be seen, with cells starting to overgrow each other. In F a lumen has appeared (arrow) between the cells. G. Cross section of a fully differentiated tubule. The inside of the lumen is filled with amorphous material, and the tubule can be seen floating over the cell monolayer attached to the tissue culture dish.

The BCE-ECM has been used in studies of nerve cell differentiation. The nerve cells can be either tumoral, such as the pheochromocytoma PC-12 cell line, or normal sensory ganglion cells [Fujii et al., 1982; Vlodavsky et al., 1983]. In either case, when cells are maintained on ECM neurite outgrowth is initiated. Nerve growth factor is then required only for the long-term maintenance of the neurites. The neurite outgrowth observed on ECM is also very similar to that observed in vivo, and the ECM factor(s) responsible for neurite initiation and adhesion are presently under investigation. They could either be proteoglycans or glycoproteins intrinsic to the ECM [Lander et al., 1982]. Finally, the metabolic behavior of various cell types seeded on BCE-ECM [Tseng et al., 1983; Kato and Gospodarowicz, 1984] or on ECM depleted of specific proteoglycans [Robinson and Gospodarowicz, 1984] has also been studied.

HR-9-ECM

Because the HR-9-ECM was developed only recently, only a limited number of studies have been performed on it. It will selectively support the proliferation and differentiation of epithelial cells, whereas those of fibroblasts will be repressed. One can therefore use it to select for epithelial cell growth. In the case of kidney tubule epithelial cells, it not only supports their growth but also promotes their differentiation. The cells form tubules that are indistinguishable histologically from those seen in vivo (Fig. 6). HR-9-ECM also allows the development of this cell type under serum-free culture conditions and has led to the identification of the two main plasma factors involved in supporting its growth. These are transferrin and high-density lipoproteins (HDLs) [Gospodarowicz et al., 1984b].

REFERENCES

Banerjee SD, Cohn RH, Bernfield MR (1977): Basal lamina of embryonic salivary epithelia. Production by the epithelium and role in maintaining lobular morphology. J Cell Biol 73:445-463.
Bernfield MR, Banerjee SD, Cohn RH (1972): Dependence of salivary epithelial morphogenesis and branching morphogenesis upon acid mucopolysaccharide-protein (proteoglycan) at the epithelial surface. J Cell Biol 52:674-689.
Bethea CL, Weiner RI (1981): Human prolactin secreting adenoma cells maintained on extracellular matrix. Endocrinology 108:357-360.
Bethea CL, Ramsdell JS, Jaffe RB, Wilson CB, Weiner RI (1982): Characterization of the dopaminergic regulation of human prolactin secreting cells cultured on extracellular matrix. J Clin Endocrinol Metab 54:893-902.
Cohn RH, Banerjee SD, Bernfield MR (1977): Basal lamina of embryonic salivary epithelia. Nature of glycosaminoglycan and organization of extracellular materials. J Cell Biol 73:464-478.

Ekblom P (1981): Formation of basement membranes in the embryonic kidney. An immunohistological study. J Cell Biol 91:1-10.

Ekblom P, Alitalo K, Vaheri A, Timpl R, Saxen L (1980): Induction of a basement membrane glycoprotein in embryonic kidney: Possible role of laminin in morphogenesis. Proc Natl Acad Sci USA 77:485-489.

Ekblom P, Lehtonen E, Saxen L, Timpl R (1981a): Shift in collagen type as an early response to induction of the metanephric mesenchyme. J Cell Biol 89:276-283.

Ekblom P, Miettinen A, Virtanen I, Dawnay A, Wahlstrom T, Saxen L (1981b): In vitro segregation of the metanephric nephron. Dev Biol 84:88-95.

Engvall E, Ruoslahti E, Miller EJ (1978): Affinity of fibronectin to collagens of different genetic types and to fibronectin. J Exp Med 147:1584-1593.

Fujii DK, Massoglia SL, Savin N, Gospodarowicz D (1982): Neurite outgrowth and protein synthesis by PC12 cells as a function of substratum and nerve growth factor. J Neurosci 2:1157-1175.

Giguere L, Gospodarowicz D (1983): Effect of RSV transformation of rat-1 fibroblasts upon their growth factor and anchorage requirements in serum-free medium. Cancer Res 43:2121-2130.

Giguere L, Cheng J, Gospodarowicz D (1982): Factors involved in the control of proliferation of bovine corneal endothelial cells maintained in serum-free medium. J Cell Physiol 110:72-80.

Gospodarowicz D (1983): The control of mammalian cell proliferation by growth factors, extracellular matrix, and lipoproteins. J Inv Dermatology 19:41-50.

Gospodarowicz D, Massoglia SL (1982): Plasma factors involved in the in vitro control of proliferation of bovine lens cells grown in defined medium. Effect of fibroblast growth factor on cell longevity. Exp Eye Res 35:259-270.

Gospodarowicz D, Tauber J-P (1980): Growth factors and extracellular matrix. Endocrine Rev 1:201-227.

Gospodarowicz D, Vlodavsky I, Savion N (1980): The extracellular matrix and the control of proliferation of vascular endothelial and vascular smooth muscle cells. J Supramol Struct 13:339-372.

Gospodarowicz D, Fujii DK, Giguere L, Savion N, Tauber J-P, Vlodavsky I (1981a): The role of the basal lamina in cell attachment, proliferation and differentiation. Tumor cells versus normal cells. In Murphy GP, Sandberg AA, Karr JP (eds): "The Prostatic Cell: Structure and Function. Part A: Morphologic, Secretory, and Biochemical Aspects." Progress in Clinical and Biological Research 75A. New York: Alan R. Liss, pp 95-132.

Gospodarowicz D, Greenburg G, Foidart J-M, Savion N (1981b): The production and localization of laminin in cultured vascular and corneal endothelial cells. J Cell Physiol 107:173-183.

Gospodarowicz D, Hirabayashi K, Giguere L, Tauber J-P (1981c): Factors controlling the proliferative rate, final cell density, and life span of bovine vascular smooth muscle cells in culture. J Cell Biol 89:568-578.

Gospodarowicz D, Cohen DC, Fujii DK (1982a): Regulation of cell growth by the basal lamina and plasma factors: Relevance to embryonic control of cell proliferation and differentiation. In Cold Spring Harbor Conferences on Cell Proliferation, "Hormones and Cell Culture," Vol 9, "Growth of Cells in Hormonally Defined Media." Cold Spring Harbor, New York: Cold Spring Harbor Laboratory, pp 95-124.

Gospodarowicz D, Lui GM, Gonzalez R (1982b): High density lipoproteins and the proliferation of the human tumor cells maintained on extracellular matrix-coated dishes and exposed to defined medium. Cancer Res 42:3704-3713.

Gospodarowicz D, Lepine J, Massoglia S (1984a): Control of cell proliferation and differentiation by extracellular matrices. J Natl Cancer Inst (in press).

Gospodarowicz D, Lepine J, Massoglia S (1984b): Ability of various basement membranes to support differentiation in vitro of normal diploid bovine kidney tubule cells. J Cell Biol (in press).

Gospodarowicz D, Massoglia S, Cohen DC (1983): Effect of high density lipoprotein on the proliferative ability of MDCK cells and HMG CoA reductase activity. J Cell Physiol 117:76-92.

Grinnell F (1978): Cellular adhesiveness and extracellular substrata. Int Rev Cytol 53:65-144.

Grove RI, Pratt RM (1983): Growth and differentiation of embryonic mouse palatal epithelial cells in primary culture. Exp Cell Res 140:195-205.

Hay ED (1978): Embryonic induction and tissue interaction during morphogenesis. In Littlefield JW, de Grouchy J (eds): "Excerpta Medica," Amsterdam-Oxford: Elsevier, pp 126-140.

Hay ED (1980): Development of the vertebrate cornea. Int Rev Cytol 63:263-322.

Hay ED (1981): Extracellular matrix. J Cell Biol 91:205-223.

Hogan BLM, Taylor A, Kurkkinen M, Couchman JR (1982): Synthesis and localization of two sulphated glycoproteins associated with basement membranes and the extracellular matrix. J Cell Biol 95:197-204.

Hyldahl L, Auer G (1982): Permissive effect of the extracellular matrix on attachment and maintenance of human breast carcinoma cells in vitro. Cell Biol Int Rep 6:989-996.

Ill CR, Gospodarowicz D, (1982): Plasma factors involved in supporting the growth and steroidogenic functions of bovine adrenal cortex cells maintained on an extracellular matrix and exposed to a defined medium. J Cell Physiol 113:373-384.

Kato Y, Gospodarowicz D (1984): Effect of extracellular matrix produced by bovine corneal endothelial cells on proteoglycan biosynthesis by rabbit costal chondrocytes. J Cell Physiol (submitted).

Kleinman HK, Klebe FJ, Martin GR (1981): Role of collagenous matrices in the adhesion and growth of cells. J Cell Biol 88:473-485.

Kratochwil K (1972): Tissue interactions during embryonic development. In Tarin D (ed): "Tissue Interactions in Carcinogenesis." New York: Academic, pp 1-47.

Lander AD, Fujii DK, Gospodarowicz D, Reichardt LF (1982): Characterization of a factor that promotes neurite outgrowth: Evidence linking the activity to a heparan sulfate proteoglycan. J Cell Biol 94:574-584.

Lewo I, Alitalo K, Riteli L, Vaheri A, Timpl R, Wartiovaara J (1982): Basal lamina glycoproteins and type IV collagen are assembled into a fine fibered matrix in cultures of a teratocarcinoma-derived endoderm cell line. Exp Cell Res 137:15-23.

Mason RJ, Williams MC, Widdicombe JH, Sanders MJ, Misfeldt DS, Berry LC (1982): Transepithelial transport by pulmonary alveolar type II cells in primary culture. Proc Natl Acad Sci USA 78:6033-6037.

Meier S, Hay ED (1974): Control of corneal differentiation by extracellular materials. Collagen as promoter and stabilizer of epithelial stroma production. Dev Biol 38:249-270.

Meier S, Hay ED (1975): Stimulation of corneal differentiation by interaction between cell surface and extracellular matrix. I. Morphometric analysis of transfilter "induction." J Cell Biol 66:275-291.

Robinson J, Gospodarowicz D (1983): Glycosaminoglycans synthesized by bovine corneal endothelial cells in culture. J Cell Physiol 112:368-376.

Robinson J, Gospodarowicz D (1984): Effect of p-nitrophenyl β-d-xyloside on proteoglycans synthesis and extracellular matrix formation by bovine corneal endothelial cell cultures. J Biol Chem (in press).

Savion N, Gospodarowicz D (1981): Factors controlling the proliferation and phenotypic expression of cultured bovine granulosa cells. In Mahesh VB, Muldoon TG, Saxena BB, Sadler WA (eds): "Developments in Endocrinology, Vol 12. Functional Correlates of Hormone Receptors in Reproduction." New York: Elsevier, North-Holland, pp 437-461.

Savion N, Gospodarowicz D (1982): Role of hormones, growth factors, and lipoproteins in the control of proliferation and differentiation of cultured bovine granulosa cells. In Cold Spring Harbor Conferences on Cell Proliferation, "Hormones and Cell Culture," Vol 9, "Growth of Cells in Hormonally Defined Media." Cold Spring Harbor, New York: Cold Spring Harbor Laboratory, pp 1141-1169.

Savion N, Lui G-M, Laherty R, Gospodarowicz D (1981): Factors controlling proliferation and progesterone production by bovine granulosa cells in serum-free medium. Endocrinology 109:409-421.

Saxen L, Ekblom P, Lehtonen E (1982): The kidney as a model system for determination and differentiation. In Ritzen M (ed): "The Biology of Normal Human Growth." New York: Raven, pp 117-127.

Tauber J-P, Cheng J, Massoglia S, Gospodarowicz D (1981): High density lipoproteins and the growth of vascular endothelial cells in serum-free medium. In Vitro 17:519-530.

Thesleff I, Hurmerinta K (1981): Tissue interactions in tooth development. A review. Differentiation 18:75-88.

Thesleff I, Barrach HJ, Foidart JM, Vaheri A, Pratt RM, Martin GR (1981): Changes in the distribution of type IV collagen, laminin, proteoglycan, and fibronectin during mouse tooth development. Dev Biol 81:182-192.

Timpl R, Rohde H, Robey PG, Rennard SI, Foidart J-M, Martin GR (1979): Laminin—A glycoprotein from basement membrane. J Biol Chem 254:9933-9937.

Tseng SCG, Savion N, Gospodarowicz D, Stern R (1981): Characterization of collagens synthesized by bovine corneal endothelial cell cultures. J Biol Chem 256:3361-3365.

Tseng SCG, Savion N, Gospodarowicz D, Stern R (1983): Modulation of collagen synthesis by a growth factor and by the extracellular matrix: Comparison of cellular response to two different stimuli. J Cell Biol 97:803-809.

Vlodavsky I, Levi A, Lax I, Fuko Z, Schlessinger J (1983): Induction of cell attachment and morphological differentiation in a pheochromocytoma cell line and embryonal sensory cells by the extracellular matrix. Dev Biol 93:285-300.

Vlodavsky I, Lui GM, Gospodarowicz D (1980): Morphological appearance, growth behavior and migratory activity of human tumor cells maintained on extracellular matrix versus plastic. Cell 19:607-616.

Vlodvasky I, Voss R, Yarkoni S, Fuks Z (1982): Stimulation of human amniotic fluid cell proliferation and colony formation by cell plating on a naturally produced extracellular matrix. Prenatal Diagn 2:13-23.

Vracko R (1974): Basal lamina scaffold-anatomy and significance for maintenance of orderly tissue structure. Am J Pathol 77:314-329.

Wessels NK (1977): In "Tissue Interactions and Development." Menlo Park, California: WA Benjamin.

Wiesel JM, Gamiel H, Vlodavsky I, Gay I, Ben Bassat H (1983): Cell attachment, growth characteristics and surface morphology of human upper respiratory tract epithelium cultured on extracellular matrix. Eur J Clin Invest 13 (in press).

Yamada K (1983): Cell surface interaction with extracellular materials. Annu Rev Biochem 52:761-799.

Methods for Preparation of Media, Supplements, and Substrata
for Serum-Free Animal Cell Culture, pages 295–319

18
Analysis of Basement Membrane Synthesis and Turnover in Mouse Embryonal and Human A431 Epidermoid Carcinoma Cells in Serum-Free Medium

David S. Salomon, Lance A. Liotta, Mounanandham Panneerselvam, Victor P. Terranova, Atul Sahai, and Paula Fehnel

Cellular shape is determined by the interaction of cells with each other and the underlying substratum [Gospodarowicz et al., 1978; Folkman and Tucker, 1980; Vlodavsky et al., 1980]. The chemical composition of the substratum can modulate cell shape and in turn control cellular proliferation and differentiation both in vivo and in vitro [Folkman and Tucker, 1980; Gospodarowicz and Tauber, 1980; Kleinman et al., 1981; Cunningham and Fredricksen, 1982]. The extracellular matrix is the natural substrate upon which mesenchymal and epithelial cells rest and proliferate [Vlodavsky et al., 1980; Gospodarowicz and Tauber, 1980; Kleinman et al., 1981]. The basement membrane is a specialized form of the extracellular matrix that is synthesized by epithelial cells and separates epithelial and mesenchymal cells in a variety of tissues [Hay, 1978, 1981; Saxen, 1972; Kefalides et al., 1979]. Components associated with the basement membrane include type IV collagen; glycoproteins such as laminin, entactin, and fibronectin, and proteoglycans; and hyaluronic acid [Kefalides et al., 1979; Bornstein, 1980; Chung et al., 1979; Carlin et al., 1981; Kleinman et al., 1981; Leivo et al., 1982; Tryggvason et al., 1980; Bachinger et al., 1982; Timpl et al., 1982]. The basement functions as an important determinant during embryonic development and differentiation

Laboratory of Tumor Immunology and Biology (D.S.S., M.P., A.S., P.F.) and Laboratory of Pathology (L.A.L.), National Cancer Institute and Laboratory of Developmental Biology (V.P.T.), National Institute of Dental Research, National Institutes of Health, Bethesda, Maryland 20205

[Hay, 1978, 1981; Hogan and Tilly, 1981; Kleinman et al., 1981; Saxen, 1972]. Components of the basement membrane may serve to control morphogenetic movements [Hay, 1981; Gospodarowicz and Tauber, 1980] and cell-cell or tissue-tissue interactions [Saxen, 1972; Kratochwil, 1972] such as occur between the epithelium and adjacent mesenchyme in the developing mammary gland, kidney, lung, pancreas, bone, thyroid, salivary glands, tooth bud, and palate [Salomon and Pratt, 1979; Hay, 1981; Bernfield and Banerjee, 1978; Saxen, 1972; Ekblom, 1981; Foidart and Reddi, 1980; Kratochwil, 1972; Thesleff et al., 1981]. In the adult, the basement membrane provides an important scaffolding for maintaining tissue architecture [Hay, 1978; Vracko, 1978]. This is particularly important for such organs as the mammary gland and cornea, in which an intact basement membrane is crucial for the proliferation and survival of the corneal and mammary epithelial cells [Hay, 1978, 1981; Kidwell et al., 1982; Wicha et al., 1980]. The basement membrane also provides structural stability to the capillaries and functions as a selective permeability barrier in the placenta, capillaries, and glomerular tubules of the kidney [Kefalides et al., 1979; Farquhar, 1981].

Attachment of specific types of cells to different types of collagens is a prerequisite for their subsequent growth and/or differentiation. This attachment is mediated through the interaction of cells with extracellular matrix glycoproteins such as laminin and fibronectin and their subsequent recognition of specific collagen types [Kleinman et al., 1981]. Fibronectin facilitates the selective attachment of mesenchymal cells to the stromal collagens (type I and type III), whereas ectodermal cells (epithelial) and endodermal cells exhibit a preference for attachment to basement membrane collagen (type IV) via laminin [Kleinman et al., 1981; Kidwell et al., 1982; Rizzino et al., 1980; Salomon et al., 1982b; Terranova et al., 1980]. Synthesis and deposition of and attachment to an appropriate extracellular matrix by a particular cell type is therefore required for subsequent proliferation [Vembu et al., 1979; Gospodarowicz and Tauber, 1980; Kidwell et al., 1982; Vlodavsky et al., 1980]. It is also becoming apparent that a variety of growth factors [Golde et al., 1980; Nevo and Laron, 1979] and hormones function as mitogens by regulating the synthesis and/or turnover of various components associated with the extracellular matrix [Chen et al., 1977, 1979; Foidart and Reddi, 1980; Gospodarowicz and Tauber, 1980; Kumegawa et al., 1982; Liotta et al., 1979b; Fredin et al., 1979; Furcht et al., 1979; Marceau et al., 1980; Salomon et al., 1979, 1981, 1982a]. Specifically, the mitogenic response to several different growth factors in various cell types now appears to be related to the cells' ability to interact with and deposit an appropriate extracellular matrix [Vembu et al., 1979; Wicha et al., 1980; Kidwell et al., 1982; Gospodarowicz and Tauber, 1980]. Quantitative or qualitative changes in components of the extracellular matrix may in fact be involved in the pathophysiology of such diverse diseases as cancer, diabetes, atherosclerosis, rheumatoid arthritis, psoriasis, scleroderma, and hepatic cirrhosis [Kefalides et al., 1979; Sporn and Harris, 1981; Liotta et al., 1979a, 1981b].

Identification of novel growth factors that promote cellular proliferation and the determination of the response of normal and neoplastic cells to these mitogens has provided some information with respect to the regulatory processes that may have become deranged following transformation [Kidwell et al., 1982; Rizzino, 1981; Salomon et al., 1982c; Todaro et al., 1981]. Delineating the mechanism(s) by which these tumor-derived growth factors or transforming agents may accentuate or attenuate the synthesis and/or turnover of components associated with the extracellular matrix following transformation [Chen et al., 1979; Hayman et al., 1982; Keski-Oja et al., 1982; Furcht et al., 1979; Sandmeyer et al., 1981] could serve to enhance our understanding of how these factors function as mitogens and possibly transforming agents [Kidwell et al., 1982; Sporn and Harris, 1981]. Toward this end, it is imperative that such studies be conducted on cells maintained in a serum-free, hormone-supplemented growth medium to obviate the problems encountered by the use of serum with its undefined growth components [Kidwell et al., 1982; Salomon et al., 1982b; Barnes and Sato, 1980a,b].

Mouse embryonal carcinoma (EC) cells represent a unique in vitro system in which to screen for these factors and in which to study their effects on cell growth and differentiation as influenced by the synthesis and deposition of a basement membrane [Rizzino et al., 1980; Rizzino, 1981; Salomon, 1980; Salomon et al., 1982a,b,c]. EC cells are the undifferentiated, pluripotential stem cells of teratocarcinomas [Martin, 1980; Mintz and Fleischman, 1981]. Extraembryonic parietal or visceral endoderm cells are one of the first cell types to arise from EC cells during in vivo or in vitro differentiation [Martin, 1980]. Parietal endoderm (END) cells derived from several EC cell lines synthesize and deposit some of the components associated with a basement membrane which resembles Reichert's membrane in the developing embryo. These include type IV collagen, laminin, entactin, fibronectin, and hyaluronic acid [Salomon et al., 1982a,b; Adamson et al., 1979; Carlin et al., 1981; Chung et al., 1979; Hogan, 1980; Bachinger et al., 1982; Cooper et al., 1981; Leivo et al., 1982; Strickland et al., 1980; Wartiovaara et al., 1980]. EC cell lines have now been propagated in serum-free, hormone-supplemented media [Barnes and Sato, 1980b; Rizzino et al., 1980; Salomon, 1980]. The present chapter attempts to provide some methodological information as to the extraction, isolation, characterization, localization, and quantitation of type IV collagen, laminin, and fibronectin from a multipotent EC cell line, OTT-6050, which can be grown in a serum-free medium supplemented with epidermal growth factor (EGF), transferrin, insulin, or multiplication stimulating activity (MSA), Pedersen fetuin, or α_2-macroglobulin (α_2M) [Salomon et al., 1982c] on polylysine or collagen-coated dishes [Salomon, 1980]. The synthesis of type IV collagen and laminin was also investigated in A431 human epidermoid carcinoma cells, as these cells are extremely responsive to EGF [Carpenter et al., 1981] and TPA; they synthesize a basement membrane [Alitalo et al., 1981] and can be propagated in a serum-free, hormone-supplemented medium [Barnes, 1982].

MATERIALS
Growth Factors, Hormones, and Medium Components

Mouse EGF (receptor grade) was obtained from Collaborative Research, Waltham, MA; Bethesda Research Laboratories (BRL), Bethesda, MD; or Seragen, Inc., Boston. MSA was purchased from Collaborative Research. Human transferrin (Tf) was obtained from Collaborative Research or Sigma Chemical Co., St. Louis. Bovine pancreatic insulin (Ins) was obtained from Collaborative Research or Sigma. Pedersen fetuin (Fet, type III) was purchased from Sigma. Poly-D-lysine (high molecular weight) was from Collaborative Research or Sigma. 12-O-Tetradecanoyl-phorbol-13-acetate (TPA) was purchased from P-L Biochemicals Inc., Milwaukee. Human α_2-macroglobulin (α_2M) was obtained from Calbiochem, whereas bovine α_2M was from Boehringer Mannheim, Indianapolis.

Collagen, Laminin, Fibronectin, and Antisera

Commercial sources. Type I (rat tail), type II (bovine sternum), and type III (bovine skin) collagens were purchased from Seragen, Inc. Type IV collagen (mouse EHS sarcoma) and laminin (mouse EHS sarcoma) were from BRL. Bovine or human plasma fibronectin (cold-insoluble globulin) were from BRL, Seragen, or Collaborative Research. Human cellular fibronectin was from BRL. Rabbit antilaminin antiserum was obtained from BRL or E·Y Laboratories, San Mateo, CA. Rabbit anti-bovine and human fibronectin antisera were from BRL, Seragen, or Collaborative Research.

Isolation sources. Type I collagen was prepared from lathyritic rat skin [Liotta et al., 1979b], type II collagen from a rat chondrosarcoma [Smith et al., 1975; Salomon et al., 1979], type III collagen from fetal bovine skin [Liotta et al., 1979a], and type IV collagen and laminin from a mouse EHS basement membrane sarcoma [Liotta et al., 1980; Foidart and Reddi, 1980;

Timpl et al., 1979, 1982]. Mouse and bovine fibronectins were prepared from the respective sera by absorption to and elution from a gelatin-Sepharose 4B column [Kleinman et al., 1981; Furcht et al., 1979]. Rabbit antisera to type IV collagen, laminin, and fibronectin or guinea pig antisera to types I, II, and III collagens were obtained by intradermal injections of 1 mg of the antigens mixed with complete Freund's adjuvant. Two weeks later, a booster injection with the same amount of antigen mixed with incomplete Freund's adjuvant was administered, followed by a third injection 1 month later. After 6 weeks, the animals were exsanguinated and the sera collected. Laminin antibody was purified by affinity chromatography after immunoabsorption on a type IV collagen-Sepharose 4B affinity column and on fibronectin and type V collagen affinity columns. Fibronectin antibody was purified by absorption to a Sepharose 4B column coupled to fibronectin [Timpl et al., 1979, 1982; Timpl, 1982; Foidart and Reddi, 1980; Furcht et al., 1979; Salomon et al., 1982b]. The specificity of all antisera was verified by radioimmunoassay, enzyme-linked immunoassay (ELISA), or immunofluorescence blocking studies [Timpl, 1982; Salomon et al., 1982b; Rennard et al., 1980].

CELLS AND GROWTH IN SERUM-FREE MEDIUM
Cells

Mouse OTT-6050 embryonal carcinoma cells were generously provided by Dr. John Lehman, University of Colorado Medical Center, Denver. Human A431 epidermoid carcinoma cells were obtained from Dr. Joseph DeLarco, Laboratory of Viral Carcinogenesis, NCI, Frederick, MD. Stock cultures of EC cells were maintained in Eagle's minimal essential medium (MEM) and A431 cells in Dulbecco's modified Eagle's medium (DMEM). Both media were supplemented with 4 mM glutamine, 20 mM HEPES (pH 7.4), 100 units/ml penicillin, 100 μg/ml streptomycin, and 5% fetal calf serum (FCS; GIBCO) for EC cells or 10% FCS for A431 cells. Cells were maintained at 37°C in an atmosphere of 5% CO_2 and 95% air. Cells were subcultured every 3–4 days with 0.1% trypsin (vol./vol.) containing 0.05% EDTA in Dulbecco's modified phosphate-buffered saline, pH 7.4 (PBS), without Ca^{++} and Mg^{++} and seeded at approximately 2×10^4 cells per 1 cm^2 in 75-cm^2 tissue culture flasks.

Serum-free (FEIT) medium. A431 and EC cells were propagated in improved MEM-zinc option [Salomon, 1980; Salomon et al., 1981] supplemented with 4 mM glutamine, 20 mM HEPES (pH 7.4), antibiotics, Pedersen fetuin (1 mg/ml), insulin (5 μg/ml), and transferrin (5 μg/ml), and where indicated with EGF (10 ng/ml for EC or 250 ng/ml for A431 cells) on 75-

cm^2 tissue culture flasks or 35-mm bacteriologic petri dishes previously coated with poly-D-lysine or various collagen types in the absence or presence of laminin or fibronectin. Cells were removed from the flasks or dishes with trypsin-EDTA and counted in a ZBI Coulter particle counter [Salomon, 1980].

Substrate Coating

Poly-D-lysine was dissolved in triple-distilled water at a concentration of 1 mg/ml, and dishes and flasks were coated as described by Salomon [1980].

Collagens were dissolved in 0.1 M acetic acid at a concentration of 1 mg/ml and diluted to various concentrations in 0.1 M acetic acid. Bacteriologic petri dishes (35 mm) were coated with 1 ml of each solution and allowed to air-dry under ultraviolet light (UV) as previously described [Terranova et al., 1980; Rao et al., 1982; Salomon et al., 1982b].

Laminin and fibronectin were dissolved in cold 10 mM phosphate buffer (pH 7.4) containing 0.15 M NaCl at a concentration of 1 mg/ml and were diluted in serum-free DMEM at the indicated concentrations before treatment of collagen-coated dishes for 1 h at 27°C prior to the cell attachment assays.

ATTACHMENT OF CELLS TO COLLAGEN VIA GLYCOPROTEINS
Cell Attachment Assay

The attachment of EC or A431 cells to various collagen substrates on 35-mm bacteriologic petri dishes in the presence or absence of laminin or fibronectin was assayed as described previously [Terranova et al., 1980; Rao et al., 1982]. Briefly, 10^5 cells were preincubated with or without cycloheximide (CYX, 25 μg/ml) for 4 h in order to inhibit protein synthesis. Cells were then trypsinized with 0.01% trypsin-EDTA, washed, and resuspended in 1 ml of serum-free MEM in the absence or presence of CYX (25 μg/ml) containing 200 μg/ml bovine serum albumin (BSA). The cells were seeded in the absence or presence of laminin or fibronectin at various concentrations on dishes coated with 10 μg per plate of type I or type IV collagen. The percentage cells attached after 2 h at 37°C was determined after the plates were washed with HEPES-buffered PBS. Cells were removed with trypsin and counted in a ZB1 Coulter counter (Table I).

LOCALIZATION OF COLLAGEN, LAMININ, AND FIBRONECTIN
Immunofluorescence

EC or A431 cells were grown for 2 days on Lab-Tek eight-chamber tissue culture slides (Miles Laboratories) previously coated with poly-D-lysine. The

TABLE I. Preferential Attachment of EC and A431 Cells to Type IV Collagen Mediated by Laminin[a,b]

| Cell type | Attachment factor | Substrate | | |
		Type IV collagen	Type I collagen	Plastic
A431	Laminin (5 µg/ml)	73	34	48
	Fibronectin (5 µg/ml)	42	40	39
	Media alone	39	24	24
EC	Laminin (10 µg/ml)	78	—	—
	Fibronectin (10 µg/ml)	—	10	10
	Media alone	20	10	5
	Laminin + antilaminin antiserum (1:1000 dilution)	20[c]	—	—

[a]All values represent the average from duplicate or triplicate determinations.
[b]Values represent the percentage of attached cells after 120 min at 37°C following pretreatment of the cells with cycloheximide (25 µg/ml) for 4 h at 37°C. The concentrations of laminin and fibronectin represent the optimal concentrations as determined from dose-response curves.
[c]EC cells were plated with laminin (10 µg/ml) on type IV collagen-coated dishes in medium containing antilaminin antiserum, which inhibited the selective attachment of these cells to type IV collagen. Data reproduced from Rao et al. [1982] and Salomon et al. [1982b].

slides were washed three times with PBS and allowed to air-dry. Purified collagen, laminin, and fibronectin antibodies (20-50 µg/ml) diluted in PBS were added to the slides and incubated for 30 min in a moist chamber at 22°C as previously described [Foidart and Reddi, 1980; Furcht et al., 1979; Salomon et al., 1982b; Liotta et al., 1979b]. Normal preimmune rabbit and guinea pig IgG were used as controls. The cells were rinsed five times, each time for 5 min, with PBS to remove excess unbound antiserum, were air-dried and were incubated at 22°C for 30 min with either fluorescein-conjugated goat anti-rabbit or anti-guinea pig antibodies (Cappel Laboratories, Cochranville, PA) at a dilution of 1:50 in PBS. The slides were then washed with three changes of PBS and overlaid with a solution of 90% glycerol-10% PBS and mounted with coverslips (Fig. 1). Fluorescence was examined with an epifluorescence-illuminated Leitz Ortholux II fluorescence microscope equipped with an ultraviolet mercury arc lamp.

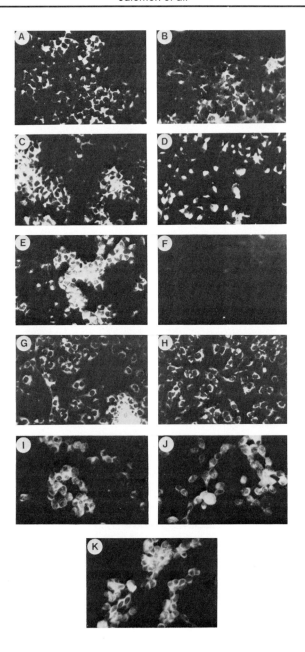

QUANTITATION OF COLLAGEN, LAMININ, AND FIBRONECTIN
Enzyme-Linked Immunoassay (ELISA)

Laminin and fibronectin were extracted and solubilized from the matrix of EC or A431 cells with 6 M urea, 4% sodium dodecylsulfate (SDS), 20 mM dithiothrietol (DTT) in 50 mM Tris-HCl (pH 8.6) for 24 h at 4°C. The solubilized extracts were centrifuged at 20,000g for 30 min. The supernates were dialyzed against 50mM Tris-HCl (pH 8.6) for 18 h at 4°C and lyophilized. Type IV collagen associated with the medium was precipitated by the addition of solid ammonium sulphate (175 mg/ml) to the medium in the presence of 10 mM EDTA, 10 mM N-ethylmaleimide (NEM), and 1 mM phenylmethylsulfonyl fluoride (pMSF). The precipitation was carried out for 18 h at 4°C. Media and matrix samples were subsequently dialyzed for 24 h against PBS (pH 7.4), and aliquots were analyzed by ELISA as previously described [Rennard et al., 1980] (Table II).

Immuno-Dot-Blot Analysis (IDBA)

Laminin, fibronectin, and type IV collagen were extracted from the cells or media as described above for ELISA. Aliquots of the samples or standards were transferred to nitrocellulose membrane disks and processed by an Immun-blot-assay kit, Bio-Rad Laboratories, Rockville Center, NY. After binding of the antigen, membranes were soaked in PBS containing BSA (20 mg/ml) for 1 h at 22°C, washed, and reacted with rabbit antibodies (diluted 1:200) for 3 h at 22°C. The membranes were then washed 5 times and incubated for 1 h at 22°C with goat anti-rabbit IgG coupled to horseradish peroxidase (diluted 1:200). The filters were washed 5 times, immersed in the development solution (3,3'-diaminobenzidine-tetrahydrochloride, 5 mg; 20 μl H_2O_2 in 25 ml PBS) for 15 min, and rinsed in distilled water (Fig. 2). Alternatively, ^{125}I-Staph A protein (New England Nuclear) can be utilized to quantitate the amount of antigen-antibody complexes that are bound to the membranes (see Bio-Radiations, 42, 1982, Bio-Rad Labs).

Immunoprecipitation

Aliquots (approximately 200 μl containing 5 × 10^5 cpm) of labeled media or cell extract samples were added to 100 μl of a BSA solution (10 mg/ml) in

Fig. 1. Indirect immunofluorescence of EC and A431 cells reacted with antibodies to type I and type IV collagen, fibronectin, and laminin. EC cells were cultured in serum-free, FEIT (FGF, EGF, insulin, transferrin, and Pedersen fetuin) medium in the absence (A-F) or presence of retinoic acid (G,H) for 7 days. Cultures were stained with type I antiserum (A), type IV antiserum (B,G), fibronectin antiserum (C), laminin antiserum (D,E,H), or normal preimmune rabbit serum (F). A431 cells were stained with type I antiserum (I), type IV antiserum (J), and laminin antiserum (K).

TABLE II. ELISA Analysis of EC Cells and Medium for Fibronectin and Type IV Collagen

Matrix component	FCS		FEIT	
	Medium (ng/ml)	Cells (ng/10^6)	Medium (ng/ml)	Cells (ng/10^6)
Fibronectin	739	95	1,180 (100)	93 (100)
			508[a] (43)	50 (54)
			760[b] (64)	74 (80)
			783[c] (66)	55 (59)
Type IV collagen	130	ND	455 (100)	ND
			140[a] (30)	
			155[b] (34)	
			270[c] (59)	

Cells (5 × 10^4) were grown for 6 days in medium containing 5% fetal calf serum (FCS) or serum-free medium (FEIT). Medium and cells were then collected and analyzed separately by ELISA. Cells were also grown in serum-free medium devoid of [a]EGF, [b]insulin, or [c]transferrin. Values represent the mean from quadruplicate determinations. Values in parentheses represent percentage ± 10% of control (100%) values for cells in complete FEIT medium. ND, not detected.

NP-40-NET buffer (150 mM NaCl, 5mM EDTA, 50 mM Tris-HCl, pH 7.4, and 0.5% NP-40) with 100 μl of the appropriate rabbit antibody (diluted 1:100 with preimmune rabbit serum) in Eppendorf microfuge tubes. The incubations were carried out at 4°C for 18 h. Staph A protein coupled to Sepharose (Pharmacia Fine Chemicals, Piscataway, NJ), 50 μl of a 50% suspension in NP-40-NET buffer, was added to each tube and incubated at 37°C for 2 h. NP-40-NET buffer (200 μl) was then added and the mixtures were centrifuged in an Eppendorf 5412 microfuge for 3 min. The pellets were washed 5 times with 300 μl of NP-40-NET buffer, subsequently boiled for 3 min in 200 μl of SDS-urea-PAGE buffer (see below), and recentrifuged, and aliquots were counted and/or used for electrophoretic analysis (Figs. 3,4).

Bacterial Collagenase Assay [Peterkofsky and Diegelmann, 1971]

Labeled media or cell extract samples from ^{3}H-proline-labeled cultures (see Table III) were lyophilized and dissolved in 2 ml of collagenase buffer (50 mM Tris-HCl, pH 7.6, 20 mM NaCl, 5 mM CaCl$_2$) to give at least 10^5 cpm per assay. To duplicate tubes, 400 μl of samples was added in the absence or presence of 100 μg (50 units) protease-free, bacterial collagenase (Advanced Biofractures, form III, Lynbrook, NY) that was previously dissolved in collagenase buffer containing 2 mM NEM. Samples were incubated at 37°C for 10 h and chilled to 4°C for 10 min. All assays then received 20 μl BSA (5 mg/ml) and 100 μl 10% trichloroacetic acid (TCA) containing 0.5% tannic acid

A

9   8   7   6   5   4   3   2   1

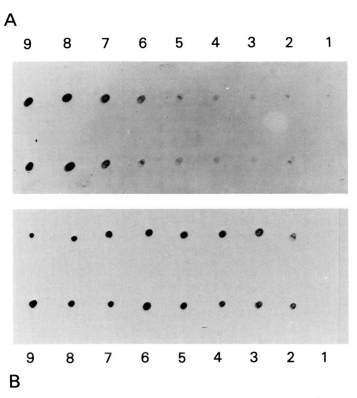

9   8   7   6   5   4   3   2   1

B

Fig. 2. Immuno-dot blot analysis of type IV collagen (A) and laminin (B). Duplicate standards of 5 ng BSA (1), or of 1 ng (2), 5 ng (3), 10 ng (4), 25 ng (5), 50 ng (6), 100 ng (7), 250 ng (8), and 500 ng (9) of type IV collagen or laminin were spotted on nitrocellulose and processed as described in the text.

and 2 mM proline. Tubes were incubated at 4°C for 90 min and centrifuged at 2,000 rpm for 10 min at 4°C. Aliquots (200 μl) of the supernates (collagenase-sensitive material) were counted and the pellets were washed with collagenase buffer 2 times and counted (collagenase-insensitive material) (Table III). Collagen percentage of total cell or medium proteins = cpm in collagenase-sensitive material divided by the sum of a) total cpm incorporated into TCA-precipitated material and b) 5.4 × cpm in collagenase-insensitive material.

LABELING, EXTRACTION, ISOLATION, AND ELECTROPHORETIC CHARACTERIZATION OF COLLAGEN, LAMININ, AND FIBRONECTIN
Labeling

EC or A431 cells were preincubated in serum-free improved Eagle's minimal essential medium (IMEM) supplemented with 50 μg/ml ascorbate and 10

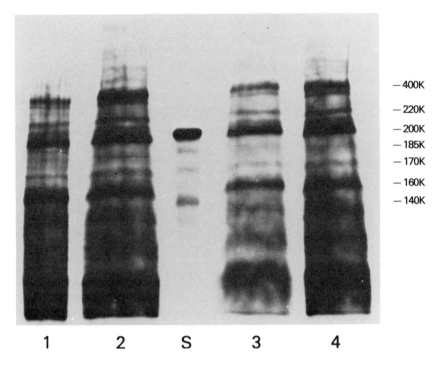

Fig. 3. Fluorograph of ^{14}C-proline- and ^{14}C-lysine-labeled cell-associated polypeptides synthesized by A431 cells. A431 cells were labeled for 8 h in serum-free DMEM containing ascorbic acid (50 μg/ml) and βAPN (10 μg/ml) with 5 μCi each of ^{14}C-proline and ^{14}C-lysine. During the last 2 h the cells were treated in the absence (track 1) or presence of TPA (10^{-7} M, track 2), EGF (250 ng/ml, track 3), or EGF (250 ng/ml) plus TPA (10^{-7} M, track 4). Track S, ^{14}C-myosin standard (200 kdaltons).

μg/ml β-aminoproprionitrile (βAPN, to inhibit cross-linking of the collagen chains) containing Fet, Ins, and Tf in the absence or presence of EGF for 30–60 min at 37°C in 5% CO_2-95% air. L-^{14}C-proline (sp. act. 285 mCi/mM, New England Nuclear [NEN] Corp.) or 2,3,4,5-^{3}H-proline (sp. act. 115 Ci/mM, NEN) was added to the EC cell medium at 5 μCi/ml. For A431 cells, L-^{14}C-proline and L-^{14}C-lysine (sp. act. 290 mCi/mM, ICN Pharmaceuticals, Irvine, CA) were added, each at a concentration of 5 μCi/ml.

    EC and A431 cells were labeled for 6–8 h prior to removal of the cells and media. Cells were labeled in proline- (EC cells) or proline- and lysine-free (A431 cells) IMEM. A short labeling period was selected [Salomon et al., 1982b], since type IV collagen in these cells is turning over extremely rapidly, possibly owing to the elaboration by both of these cell types of a type IV-

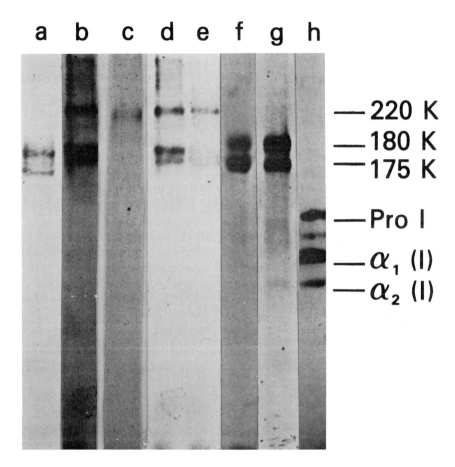

Fig. 4. Fluorograph of ^{14}C-proline-labeled medium-associated polypeptides synthesized by EC cells. EC cells were cultured for 7 days in serum-free FEIT medium in the absence or presence of 10^{-8} M retinoic acid (RA). EC cells were labeled with ^{14}C-proline, 5 μCi/ml, for 6 h in medium containing ascorbic acid, 50 μg/ml, and β-aminoproprionitrile (βAPN), 10 μg/ml. Migration was from the top toward the anode. Track a: Pro-type IV collagen standard; tracks b and d: labeled polypeptides from medium of EC cells in FEIT medium; tracks f and g: labeled collagenous EC polypeptides purified by salt precipitation from medium of cells grown in the presence (f) or absence (g) of RA; track c: same as track b except treated with bacterial collagenase; track e: same as track d except medium treated with EHS sarcoma-derived type IV collagenase; track h: type 1 (α1, α2) collagen and procollagen standards.

TABLE III. Collagen Production by EC Cells

Treatment	Percent collagen/mg cellular protein
FEIT	3.6, 4.5 (4.0)
FEIT + ascorbic acid (50 μg/ml)	2.7, 3.9 (3.3)
FEIT + retinoic acid (10^{-8} M)	10.2, 15.1 (12.7)
FCS	3.6

Cells (5×10^4) were seeded in serum-free FEIT medium on polylysine-coated (1 mg/ml) 75-cm^2 flasks, in the absence or presence of ascorbic acid or retinoic acid, or in medium containing 5% FCS. Cells were cultured for 7 days and labeled for the last 6 h with ^{3}H-proline (5 μCi/ml). After labeling, the medium was removed and extracted with 0.5 M acetic acid containing 4 mM NEM for 48 h at 4°C, was dialyzed against acetic acid, and was precipitated with 10% sodium chloride. The precipitates were resuspended in the bacterial collagenase assay buffer, and duplicate aliquots were analyzed for the incorporation of ^{3}H-proline into collagenase-sensitive proteins. All values in parentheses represent the average of two separate experiments. Data reproduced from Salomon et al. [1982b].

specific collagenase (see below) [Salomon et al., 1982a,b; Liotta, unpublished observations].

Extraction and Isolation

Collagen was extracted from cells by scrapping the cultures into 0.5 M acetic acid containing 8 mM EDTA and 4 mM NEM. Extraction was carried out at 4°C for 24 h. The suspension was then clarified by centrifugation at 27,000g for 30 min and dialyzed against the acetic acid extraction solution for 12 h at 4°C. Collagen associated with the medium was prepared by dialyzing the medium against 0.5 M acetic acid containing the same protease inhibitors for 48 h at 4°C. Collagenous proteins in the cell and medium acid extracts were isolated by precipitation with 1.71 M NaCl (10%, final concentration). The precipitates were collected by centrifugation at 40,000g for 30 min, dissolved in 0.5 M acetic acid, and lyophilized.

Cell-associated laminin and fibronectin were extracted with 6 M urea, 4% SDS, 20 mM DTT in 50 mM Tris-HCl (pH 8.6) as previously described under the ELISA section.

Electrophoretic Characterization

Polyacrylamide gel electrophoresis (PAGE) in 5% SDS-slab gels was performed according to the method of Laemmli [1970] in a Tris-HCl (pH 8.3)-

glycine (0.2 M) buffer containing 0.1% SDS with the following modifications. 3.5% Stacking and 5% separating gels contained 2% SDS in 0.75 M Tris-HCl (pH 8.8). Separating gels also contained 1 M urea to improve the separation of the type IV procollagen ($\alpha_1 = 185$ KD; $\alpha_2 = 170$ KD), type IV collagen ($\alpha_1 = 160$ KD; $\alpha_2 = 140$ KD), and laminin ($\beta = 400$ KD; $\alpha = 200$ KD) polypeptide chains. Samples were dissolved in 62.5 mM Tris-HCl (pH 6.8) containing 0.5 M urea, 10% sucrose, 2% SDS, and 5 mM DTT and were denatured at 110°C for 5 min [Crouch and Bornstein, 1978]. Electrophoresis was carried out in a water-cooled system for 4 h at 22°C, 25 mA per gel. After electrophoresis, the gels were stained with 0.5% R-250 Coomassie brilliant blue (Bio. Rad Labs) in 10% acetic acid, 30% methanol, and 2.5% TCA. The gels were subsequently destained, impregnated with Autofluor (National Diagnostics), and dried onto filter paper and exposed to Kodak X-Omat autoradiographic film at −70°C [Bonner and Laskey, 1974]. Chemically ^{14}C-alkylated type I and type IV collagen standards were electrophoresed in parallel (Fig. 4).

ASSAY FOR PLASMINOGEN ACTIVATOR, TYPE IV COLLAGENASE, AND LAMININ-DEGRADING ACTIVITIES
Plasminogen Activator (PA)

[3]H-Fibrin was prepared according to the method of Barret et al. [1977]. Costar 24-well, 16-mm tissue culture dishes were coated with ^{3}H-fibrin. Conditioned medium from cells were collected at various intervals and placed in the fibrin-coated wells in the absence or presence of plasminogen (25 μg/ml, Sigma) and incubated at 37° for 4 h. Release of soluble radioactive fibrinopeptides was used to assess PA activity [Liotta et al., 1981b; Salomon et al., 1981, 1982b]. Trypsin (0.125%) was added to some wells to determine the total amount of ^{3}H-fibrin coated onto the wells (Table IV).

Type IV Collagenase

Type IV-specific collagenase was assayed as described by Liotta et al. [1979a, 1980]. Costar 16-mm cluster tissue culture dishes were coated with ^{14}C-proline-labeled type IV collagen obtained from a mouse EHS basement membrane sarcoma. ^{14}C-Labeled type IV collagen (approximately 2.5×10^3 cpm) in 200 μl of 10 mM acetic acid was added to each well and air-dried under UV light. Cells were grown in the absence or presence of plasminogen (25 μg/ml) on the labeled type IV collagen-coated dishes for 2 days. The soluble radioactivity released into the medium following centrifugation at 2,000 rpm at 4°C for 10 min was measured and used to assess activity

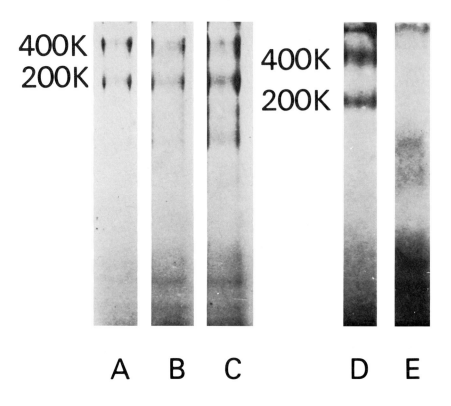

Fig. 5. Fluorograph of the biosynthesis and degradation of laminin by EC cells cultured in serum-free, FEIT medium. A: Immunoprecipitation of EHS ^{14}C-laminin with antilaminin antiserum; B: immunoprecipitation of ^{14}C-proline-labeled EC culture medium with antilaminin antiserum; C: immunoprecipitation of ^{14}C-proline-labeled EC culture medium from retinoic acid-treated (RA, 10^{-8} M) cells (7 days); D,E: ^{14}C-laminin (D) and ^{14}C-laminin + EC (E) culture medium obtained from cells treated with RA (10^{-8} M) for 7 days + 10 μg plasminogen and digested for 8 h at 25°C. Note: Both the α-chain (200-kdalton) and β-chain (400-kdalton) of laminin in (E) are degraded by EC-conditioned medium. Data obtained from Salomon et al. [1982b].

TABLE IV. Plasminogen Activator, Type IV Collagenase, and Laminase Production by EC Cells[a]

Treatment	dpm substrate solubilized/10^6 cells	
Plasminogen activator (PA)		
FEIT	1,200	(1.8)
FEIT + plasminogen (25 μg/ml)	1,475	(2.2)
FEIT + RA	1,500	(2.3)
FEIT + RA + plasminogen (25 μg/ml)	16,520	(22)
Trypsin (1.25 mg/ml)	66,000	(100)
Type IV collagenase		
FEIT	169	(6)
FEIT + plasminogen (25 μg/ml)	442	(16)
FEIT + RA	1,274	(47)
FEIT + RA + plasminogen (25 μg/ml)	2,197	(80)
Background	292	(10.6)
Bacterial collagenase (10 μg/ml)	2,730	(100)
Laminase		
FEIT	5,884	(28)
FEIT + plasminogen (10 ng/ml)	6,326	(31)
FEIT + RA	7,579	(37)
FEIT + RA + plasminogen (10 ng/ml)	20,363	(98)
Background	282	(1.3)
Plasmin (10 ng/ml)	20,676	(100)
Trypsin (10 ng/ml)	21,293	(102)
Plasminogen (10 ng/ml)	389	(1.9)

[a]EC cells were cultured in medium containing Fet, EGF, Ins, and Tf (FEIT) in the absence or presence of retinoic acid (RA, 10^{-8} M) on polylysine-coated dishes. For PA and type IV collagenase determinations, the medium was removed after 2 days and assayed in the absence or presence of plasminogen on ^{3}H-fibrin and ^{14}C-type IV collagen-coated dishes. For laminase, cells were cultured for an additional 5 days, and the medium was removed and assayed on ^{14}C-laminin-coated wells. All values represent the average of duplicate or triplicate determinations. Values in parentheses represent the percentage of substrate solubilized with conditioned medium as expressed against the total amount of fibrin, collagen, or laminin digested with trypsin, bacterial collagenase, and plasmin, respectively. Data reproduced from Salomon et al. [1982b].

associated with or on the cells. "Blank" values (approximately 300 cpm) obtained from parallel wells that contained no cells, but to which medium with porcine plasmin (5 μg/ml, Sigma) had been added, were subtracted from the experimental values. The total amount of labeled collagen that was coated onto the wells was determined by incubation of parallel wells with protease-free, bacterial collagenase (10 μg/ml) for 4 h at 37°C. Collagenase activity associated with 1-ml aliquots of conditioned medium was assayed by addition to ^{14}C-type IV collagen-coated wells and incubated in the absence or presence of plasminogen (25 μg/ml) and soybean trypsin inhibitor (50 μg/ml, Sigma) for 20 h at 37°C. Release of soluble radioactivity was used to assess activity in the medium (Table IV). The inclusion of plasminogen in the culture medium of cells propagated on ^{14}C-labeled type IV collagen or to medium aliquots was necessary to measure any latent enzyme [Liotta et al., 1979a, 1980; Salomon et al., 1981, 1982a,b].

Laminin-Degrading Activity (Laminase)

[14]C-Proline-labeled laminin was prepared by biosynthetically labeling organ cultures of the mouse EHS basement membrane sarcoma [Liotta et al., 1981a,b]. ^{14}C-Labeled laminin was extracted from the tumor explants with 0.5 M NaCl [Timpl et al., 1979, 1982]. This material consisted of the β-polypeptide (400 KD)and α-polypeptide (200 KD) chains as demonstrated by SDS-urea gel electrophoresis (Fig. 5). Laminin degradation was assessed by incubation of medium aliquots (500 μl) with 30 μg of authentic laminin in 50 μl of PBS (pH 7.4) containing 10 μg of plasminogen for 8 h at 25°C. Laminin (30 μg) was also incubated with 10 CU of pure porcine plasmin or urokinase (Sigma) (Fig. 5). Digestion products were identified after electrophoresis on 5% SDS-urea slab gels [Rao et al., 1982]. Quantitation of laminase production was accomplished by the addition of 1-ml conditioned medium aliquots to ^{14}C-laminin-coated wells (Costar, 16-mm tissue culture cluster plates) [Salomon et al., 1982a,b]. ^{14}C-Labeled laminin was dissolved in 0.1 M acetic acid (500 μl) and air-dried onto the wells for 12 h under UV light. Incubations were performed in the absence or presence of plasminogen (10 ng/ml) for 3 h at 37°C. Parallel wells were digested with porcine plasmin (10 ng/ml) or trypsin (10 ng/ml) to determine the total amount of ^{14}C-laminin coated onto the wells (Table IV).

CONCLUSIONS

It was the goal of the present chapter to provide a brief methodological synopsis of some of the more routine procedures for the isolation, characterization, localization, and quantitation of type IV collagen, laminin, and fibronectin and for assaying those enzymes (plasminogen activator, type IV collagenase, and laminase) that are involved in regulating the turnover of these basement membrane components. Moreover, the attachment of specific cells via laminin or fibronectin to type IV or type I collagen, respectively, may provide a useful screen for selecting an appropriate in vitro matrix on which to propagate and select [Liotta et al., 1982] for various cell types in serum-free, hormone-defined media [Kleinman et al., 1981; Terranova et al., 1980], since the attachment and spreading of cells to a compatible, defined substratum is an absolute prerequisite for their subsequent growth and/or differentiation [Gospodarowicz and Tauber, 1980; Kidwell et al., 1982; Vlodavsky et al., 1980].

It should be emphasized that alternative protocols exist for quantitating collagen, laminin, and fibronectin. Various radioimmunoassays (both liquid- and solid-phase) and rocket immunoelectrophoresis techniques have been utilized for measuring these matrix components [Timpl et al., 1982; Timpl, 1982]. Likewise, determination of the amount of 3- and 4-hydroxyproline in acid-hydrolyzed samples by amino acid analysis or high-performance liquid chromatography has proven to be a reliable method for quantifying the total amount of cellular or medium-associated collagen [Liotta et al., 1979; Salomon et al., 1981; Berg, 1982]. Molecular sieve and ion-exchange chromatography in conjunction with SDS-PAGE have also been successfully used to identify various collagen species [Liotta et al., 1979b; Miller and Rhodes, 1982]. Therefore, depending upon the system under study and the experimental questions being addressed, a wide array of assays are available to the investigator.

Studies utilizing rat mammary epithelial cells [Liotta et al., 1979b; Salomon et al., 1981; Kidwell et al., 1982], mouse embryonal carcinoma cells [Salomon et al., 1982a,b,c], and human A431 epidermoid carcinoma cells have demonstrated that a variety of hormones (glucocorticoids, insulin), vitamins (ascorbic acid, retinoids), growth factors (EGF, mammary tumor factor), serum components (α^2M), and tumor promoters (TPA) can affect the synthesis, turnover, and secretion of several components associated with the basement membrane such as type IV collagen, laminin, and fibronectin. The scheme presented in Figure 6 summarizes the presently envisioned pathway for type IV collagen and laminin degradation as regulated by plasminogen activator, plasmin, thrombin, or an endogenous protease (laminase) and type IV collagenase [Liotta et al., 1979a, 1980, 1981a,b, 1982]. It is possible that particular hormones and growth factors alone or in combination [Salomon et al., 1981] may in some manner modulate the level and/or activity of these enzymes

Cascade for Basement Membrane Degradation

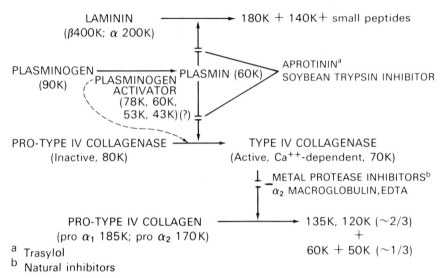

Trasylol
Natural inhibitors

Fig. 6. Degradation of laminin and type IV collagen. The level or activity of plasminogen activator (PA) can be regulated by glucocorticoids, retinoids, TPA, and fetuin. PA converts plasminogen to plasmin, which can in turn degrade both the α-chain (200-kdalton) and β-chain (400-kdalton) of laminin to specific cleavage products (180-kdalton and 140-kdalton). Thrombin (not depicted) selectively degrades only the β-chain to a 180-kdalton product. Neither plasmin nor thrombin degrade type IV collagen. Endogenous cellular proteases that are plasminogen-independent also can degrade native laminin. Plasmin can also activate latent (inactive) pro-type IV collagenases, which when active degrade pro-type IV collagen to a mixture of 135-kdalton/ 120-kdalton products and 60- and 50-kdalton products in a ratio of 2:1. Glucocorticoids, retinoids, and TPA can also modulate the level of type IV collagenase, whereas α_2M can directly inhibit the activity of this neutral, Ca^{++}-dependent protease. This latter property of α_2M may be significant, since α_2M has been shown to be associated with Pedersen fetuin and is probably one of the growth-promoting principles associated with this preparation of fetuin [Salomon et al., 1982c]. In contrast, growth factors such as EGF, MSA, and mammary tumor factor [Bano et al., 1983], and hormones such as insulin, can stimulate type IV collagen synthesis without affecting turnover by these enzymes [Liotta et al., 1979b; Salomon et al., 1982b; Kidwell et al., 1982; Salomon et al., 1981].

and thereby accentuate or attenuate the deposition of a basement membrane, which is necessary for the subsequent proliferation of specific cells. Elaboration of an appropriate extracellular matrix by a cell may be necessary for the cell's subsequent ability to mitogenically respond to a particular hormone or growth factor [Gospodarowicz and Tauber, 1980]. Analysis of these enzymes and the synthesis of various components associated with the extracellular matrix in cell culture systems maintained in serum-free, hormone-defined media will aid in the identification of other novel growth factors (e.g., transforming growth factors: Todaro et al. [1981]) and in the delineation of the mechanism(s) by which these factors may function as mitogens and possibly transforming agents.

ACKNOWLEDGMENTS

The authors wish to express their gratitude to Drs. Jean-Michel Foidart, Pietro Gullino, William R. Kidwell, George Martin, and Stephen I. Rennard for their advice and support of/or contributions to this work.

REFERENCES

Adamson ED, Gaunt SJ, Graham CF (1979): The differentiation of teratocarcinoma stem cells is marked by the types of collagen which are synthesized. Cell 17:469–476.

Alitalo K, Keski-Oja J, Vaheri A (1981): Extracellular matrix proteins characterize human tumor cell lines. Int J Cancer 27:733–741.

Bachinger HA, Fessler LI,, Fessler JH (1982): Mouse procollagen IV: Characterization and supramolecular association. J Biol Chem 257:9796–9803.

Bano M, Zwiebel JA, Salomon DS, Kidwell WR (1983): Detection and partial characterization of collagen synthesis stimulating activities in rat mammary adenocarcinomas. J Biol Chem 258:2729–2735.

Barnes DW (1982): Epidermal growth factor inhibits growth of A431 human epidermoid carcinoma cells in serum-free cell culture. J Cell Biol 93:1–4.

Barnes DW, Sato G (1980a): Serum-free cell culture: A unifying approach. Cell 22:649–655.

Barnes DW, Sato G (1980b): Methods for growth of cultured cells in serum-free medium. Anal Biochem 102:255–270.

Barret JC, Crawford BD, Tso PO (1977): Quantitation of fibrinolytic activity of Syrian hamster fibroblasts using ^{3}H-labeled fibrinogen produced by reductive alkylation. Can Res 37:1182–1185.

Berg RA (1982): Determination of 3- and 4-hydroxproline. In Cunningham LW, Fredricksen DW (eds): "Methods in Enzymology," Vol 82: "Extracellular Matrix." New York: Academic, pp 372–398.

Bernfield MR, Banerjee SD (1978): The basal laminin in epthelial-mesenchymal interaction. In Kefalides NA (ed): "Biology and Chemistry of Basement Membranes." New York: Academic, pp 137–148.

Bonner WM, Laskey RA (1974): A film detection method for tritium-labeled proteins and nucleic acids on polyacrylamide agels. Eur J Biochem 46:83–88.

Bornstein P (1980): Structurally distinct collagen types. Ann Rev Biochem 49:957–1003.

Carlin B, Jaffe R, Bender B, Chung AE (1981): Entactin, a novel basal laminin-associated sulfated glycoprotein. J Biol Chem 256:5209–5214.

Carpenter G, King L, Cohen S (1981): Coupling of protein phosphorylation to the epidermal growth factor: Receptor complexes. In Middlebrook JL, Kohn LD (eds): "Receptor-Mediated Binding and Internalization of Toxins and Hormones." New York: Academic, pp 163–177.

Chen LB, Gudor RC, Sun T-T, Chen AB, Mosesson MW (1977): Control of cell surface major glycoprotein by epidermal growth factor. Science 197:776-778.

Chen LB, Summerhoyes I, Hsieh P, Gallimore PH (1979): Possible role of fibronectin in malignancy. J Supramol Struct 12:139-150.

Chung AE, Jaffe R, Freeman IL, Vergnes J-P, Braginski JE, Carlin B (1979): Properties of a basement membrane-related glycoprotein synthesized in culture by a mouse embryonal carcinoma-derived cell line. Cell 16:277-287.

Cooper AR, Kurkinen M, Taylor A, Hogan BLM, (1981): Studies on the biosynthesis of laminin by murine parietal endoderm cells. Eur J Biochem 119:189-197.

Crouch E, Bornstein P (1978): Collagen synthesis by human amniotic fluid cells in culture: Characterization of a procollagen with three identical pro αl (I) chains. Biochemistry 17:5499-5510.

Cunningham LW, Fredricksen DW (eds) (1982): "Methods in Enzymology," Vol. 82: "Extracellular Matrix." New York: Academic.

Ekblom P (1981): Formation of basement membranes in the embryonic kidney: An immunohistological study. J Cell Biol 91:1-10.

Farquhar MG (1981): The glomerular basement membrane: A selective macromolecular filter. In Hay ED (ed): "Cell Biology of the Extracellular Matrix." New York: Plenum, pp 335-378.

Foidart J-M, Reddi AH (1980): Immunofluorescent localization of type IV collagen and laminin during endochronal bone differentiation and regulation by pituitary growth hormone. Dev Biol 75:13-136.

Folkman J, Tucker RW (1980): Cell configuration, substratum and growth control. In Subtelny S, Wessells NK (eds): "The Cell Surface: Mediator of Developmental Processes." New York: Academic, pp 259-275.

Fredin BL, Seifert SC, Gelehrter TD (1979): Dexamethasone-induced adhesion in hepatoma cells: The role of plasminogen activator. Nature 277:312-313.

Furcht LT, Mosher DF, Wendelschafer-Crabb G, Foidart J-M (1979): Reversal by glucocorticoid hormones of a fibronectin and procollagen matrix around transformed human cells. Can Res 39:2077-2083.

Golde DW, Herschmann HR, Lusis A, Groopman JE (1980): Growth factors. Ann Int Med 92:650-662.

Gospodarowicz D, Tauber J-P (1980): Growth factors and the extracellular matrix. Endocrine Rev 1:201-227.

Gospodarowicz D, Greenburg G, Birdwell CR (1978): Determination of cellular shape by the extracellular matrix and its correlation with the control of cellular growth. Can Res 38:4155-4171.

Hay ED (1978): Role of basement membranes in development and differentiation. In Kefalides NA (ed): "Biology and Chemistry of Basement Membranes." New York: Academic, pp 119-136.

Hay ED (1981): Collagen and embryonic development. In Hay ED (ed): "Cell Biology of the Extracellular Matrix." New York: Plenum, pp 379-405.

Hayman EG, Oldberg A, Martin GR, Ruoslahti E (1982): Codistribution of heparan sulfate proteoglycan, laminin and fibronectin in the extracellular matrix of normal rat kidney cells and their coordinate absence in transformed cells. J Cell Biol 94:28-35.

Hogan BL (1980): High molecular weight extracellular proteins synthesized by endoderm cells derived from mouse teratocarcinoma cells and normal extraembryonic membranes. Dev Biol 76:275-285.

Hogan BL, Tilly R (1981): Cell interactions and endoderm differentiation in cultured mouse embryos. J Embryol Exp Morphol 62:379-394.

Kefalides NA, Alper R, Clark CC (1979): Biochemistry and metabolism of basement membranes. Int Rev Cytol 61:167-228.

Keski-Oja J, Gahmberg CG, Alitalo K (1982): Pericellular matrix and cell surface glycoproteins of virus-transformed mouse epithelial cells. Can Res 42:1147-1153.

Kidwell WR, Salomon DS, Liotta LA, Zweibel JA, Bano M (1982): Effects of growth factors on mammary epithelial cell proliferation and basement membrane synthesis. In Sato G, Sirbasku D (eds): "Cold Spring Harbor Conferences on Cell Proliferation," Vol 9: "Growth of Cells in Hormonally Defined Media." Cold Spring Harbor, NY: Cold Spring Harbor Press, pp 807-818.

Kleinman HK, Klebe RJ, Martin GR (1981): Role of collagenous matrices in the adhesion and growth of cells. J Cell Biol 88:473-485.

Kratochwil K (1972): Tissue interaction during embryonic development: General properties. In Tarin D (ed): "Tissue Interactions in Carcinogenesis." New York: Academic, pp 1-47.

Kumegawa M, Hiramatsu M, Vajima T, Hatukeyama K, Hosoda S, Namba M (1982): Effect of epidermal growth factor on collagen formation in liver-derived epithelial clone cells. Endocrinology 110:607-612.

Laemmli UK (1970): Cleavage of structural proteins during the assembly of the head of bacteriophage T4. Nature 227:680-685.

Leivo I, Alitalo K, Risteli L, Vaheri A, Timpl R, Wartiovaara J (1982): Basal lamina glycoproteins, laminin and type IV collagen are assembled into a fine-fibered matrix in cultures of a teratocarcinoma derived endodermal cell line. Exp Cell Res 137:15-23.

Liotta LA, Abe S, Robey PG, Martin GR (1979a): Preferential digestion of basement membrane collagen by an enzyme derived from a metatastic murine tumor. Proc Natl Acad Sci USA 76:2268-2272.

Liotta LA, Wicha MS, Foidart J-M, Rennard SI, Garbisa S, Kidwell WR (1979b): Hormonal requirements for basement membrane collagen deposition by cultured rat mammary epithelium. Lab Invest 41:511-518.

Liotta LA, Tryggvason K, Garbisa S, Hart I, Foltz CM, Shafie S (1980): Metastatic potential correlates with enzymatic degradation of basement membrane collagen. Nature 284:67-68.

Liotta LA, Goldfarb RH, Terranova VP (1981a): Cleavage of laminin by thrombin and plasmin: Alpha thrombin selectively cleaves the beta chain of laminin. Thrombosis Res 21:663-673.

Liotta LA, Goldfarb RH, Brundage R, Siegal GP, Terranova V, Garbisa S (1981b): Effect of plasminogen activator (urokinase) plasmin and thrombin on glycoprotein and collagenous components of basement membrane. Can Res 41:4629-4636.

Liotta LA, Terranova VP, Lanzer WL, Russo R, Siegel GP, Garbisa S (1982): Basement membrane attachment and degradation by metastatic tumor cells. In Kuehn K, Schoen H, Timpl R (eds): "New Trends in Basement Membrane Research." New York: Raven, pp 277-286.

Marceau N, Goyette R, Valet JP, Deschenes J (1980): The effect of dexamethasone on formation of a fibronectin extracellular matrix by rat hepatocytes in vitro. Exp Cell Res 125:497-502.

Martin GR (1980): Teratocarcinomas and mammalian embryogenesis. Science 209:768-776.

Miller EJ, Rhodes RK (1982): Preparation and characterization of different types of collagen. In Cunningham LW, Fredricksen DW (eds): "Methods in Enzymology," Vol. 82: "Extracellular Matrix." New York: Academic, pp 33-64.

Mintz B. Fleischman RA (1981): Teratocarcinoma and other neoplasms as developmental defects in gene expression. Adv Can Res 34:211-278.

Nevo Z, Laron Z (1979): Growth factors. Am J Dis Child 133:419-428.

Peterkofsky B, Diegelmann R (1971): Use of a mixture of protinease-free collagenase for the specific assay of radioactive collagen in the presence of other proteins. Biochemistry 10:988-994.

Rao CN, Margolies IMK, Tralka TS, Terranova VP, Madei JA, Liotta LA (1982): Isolation of a subunit of laminin and its role in molecular structure and tumor cell attachment. J Biol Chem 257:9740-9744.

Rennard SI, Berg RA, Martin GR, Robey PG, Foidart J-M (1980): Enzyme-linked immunoassay (ELISA) for connective tissue components. Anal Biochem 104:205-214.

Rizzino A (1981): Growth factors from virally transformed cells stimulate the growth of differentiated cells derived from embryonal carcinoma cells. Abstract #11009. J Cell Biol 91:195a.

Rizzino A, Terranova V, Rohrbach D, Crowley C, Rizzino H (1980): The effects of laminin on the growth and differentiation of embryonal carcinoma cells in defined media. J Supramol Struct. 13:243-253.

Salomon DS (1980): Correlation of receptors for growth factors on mouse embryonal carcinoma cells with growth in serum-free, hormone-supplemented medium. Exp Cell Res 128:311-321.

Salomon DS, Pratt RM (1979): Involvement of glucocorticoids in the development of the secondary palate. Differentiation 13:141-154.

Salomon DS, Paglia LM, Verbruggen L (1979): Hormone-dependent growth of a rat chrondrosarcoma in vivo. Can Res 39:4387-4395.

Salomon DS, Liotta LA, Kidwell WR (1981): Differential response to growth factor by rat mammary epithelium plated on different collagen substrate in serum-free medium. Proc Natl Acad Sci USA 78:382-386.

Salomon DS, Liotta LA, Foidart J-M, Yaar M (1982a): Synthesis and turnover of basement-membrane components by mouse embryonal carcinoma cells in serum-free, hormone-supplemented medium. In Sato G, Sirbasku D (eds): "Cold Spring Harbor Conferences on Cell Proliferation," Vol. 9: "Growth of Cells in Hormonally Defined Media." Cold Spring Harbor, NY: Cold Spring Harbor Press, pp 203-207.

Salomon DS, Liotta LA, Rennard SI, Foidart J-M, Terranova V, Yaar M (1982b): Stimulation by retinoic acid of synthesis and turnover of basement membrane in mouse embryonal carcinoma-derived endoderm cells. Collagen Rel Res 2:93-110.

Salomon DS, Bano M, Smith KB, Kidwell WR (1982c): Isolation and characterization of a growth factor (embryonin) from bovine fetuin which resembles α^2-macroglobulin. J Biol Chem 257:14093-14101.

Sandmeyer S, Smith R, Kiehn D, Bornstein P (1981): Correlation of collagen synthesis and procollagen messenger RNA levels with transformation in rat embryo fibroblasts. Can Res 41:830-838.

Saxen L (1972): Interactive mechanisms in morphogenesis. In Tarin D (ed): "Tissue Interaction in Carcinogenesis." New York: Academic, pp 49-80.

Smith BD, Martin GR, Miller EJ, Dorfman A, Swarm R (1975): Nature of collagen synthesized by a transplantable chondrosarcoma. Arch Biochem Biophys 166:181-186.

Sporn MB, Harris ED (1981): Proliferative diseases. Am J Med 70:1231-1236.

Strickland S, Smith KK, Marotti KR (1980): Hormonal induction of differentiation in teratocarcinoma stem cells: Generation of parietal endoderm by retinoic acid and dibutyryl cAMP. Cell 21:347-355.

Terranova VP, Rohrbach DH, Martin GR (1980): Role of laminin in the attachment of PAM 212 (epithelial) cells to basement membrane collagen. Cell 22:719-726.

Thesleff I, Barrach HJ, Foidart J-M, Vaheri A, Pratt RM, Martin GR (1981): Changes in the distribution of type IV collagen, laminin, proteoglycan and fibronectin during mouse tooth development. Dev Biol 81:182-192.

Timpl R (1982): Antibodies to collagens and procollagens. In Cunningham LW, Fredricksen DW (eds): "Methods in Enzymology," Vol 82: "Extracellular Matrix." New York: Academic, pp 472-498.

Timpl R, Rohde H, Robey PG, Rennard SI, Foidart J-M, Martin GR (1979): Laminin—a glycoprotein from basement membranes. J Biol Chem 254:9933-9937.

Timpl R, Rohde H, Risteli L, Ott U, Robey PG, Martin GR (1982): Laminin. In Cunningham LW, Fredricksen DW (eds): "Methods in Enzymology," Vol 82: "Extracellular Matrix." New York: Academic, pp 831-838.

Todaro GJ, DeLarco JE, Fryling, C, Johnson PA, Sporn MB (1981): Transforming growth factors (TGFs): Properties and possible mechanisms of action. J Supramol Struct 15:287-301.

Tryggvason K, Robey PG, Martin GR (1980): Biosynthesis of type IV procollagens. Biochemistry 19:1284-1289.

Vembu D, Liotta LA, Paranjpe M, Boone CW (1979): Correlation of tumorigenicity with resistance to growth inhibition by cis-hydroxyproline. Exp Cell Res 124:247-252.

Vlodavsky I, Lui GM, Gospodarowicz D (1980): Morphological appearance, growth behavior and mitogenic activity of human tumor cells maintained on extracellular matrix versus plastic. Cell 19:607-616.

Vracko R (1978): Anatomy of basal lamina scaffold and its role in maintenance of tissue structure. In Kefalides NA (ed): "Biology and Chemistry of Basement Membranes." New York: Academic, pp 43-56.

Wartiovaara J, Leivo I, Vaheri A (1980): Matrix glycoproteins in early mouse development and in differentiation of teratocarcinoma cells. In Subtelny S, Wessells NK (eds): "The Cell Surface: Mediator of Developmental Processes." New York: Academic, pp 305-324.

Wicha MS, Liotta LA, Vonderhaar BK, Kidwell WR (1980): Effects of inhibition of basement membrane collagen deposition on rat mammary gland development. Dev Biol 80: 253-266.

Methods for Preparation of Media, Supplements, and Substrata
for Serum-Free Animal Cell Culture, pages 321–337
© 1984 Alan R. Liss, Inc., 150 Fifth Avenue, New York, NY 10011

19
Cell Attachment and Spreading on Extracellular Matrix-Coated Beads

Shing Mai and Albert E. Chung

The basement membrane represents a component of the extracellular matrix and consists of an array of complex macromolecules whose organization is being slowly unraveled. The extracellular matrix and the basement membrane have been implicated in the processes of tissue induction [Grobstein, 1953], cell differentiation [Hay, 1978; Thesleff et al., 1978; Wicha et al., 1982], morphogenesis [Bernfield et al., 1972], and cell attachment [Kleinman et al., 1981]. The vital functions of the extracellular matrix have resulted in intensive efforts to characterize individual macromolecular components. The identification of type IV collagen [Bornstein and Sage, 1980], laminin [Chung et al., 1977b, 1981; Foidart et al., 1980; Hogan et al., 1980; Madri et al., 1980; Timpl et al., 1979], heparan sulfate [Hassall et al., 1980; Kanwar and Farquhar, 1979], and entactin [Bender et al., 1981; Carlin et al., 1981; Hay, 1978; Hogan et al., 1982] has increased our understanding of the organization and function of the basement membrane. Laminin mediates cell adhesion [Carlsson et al., 1981; Couchman et al., 1983; Johansson et al., 1981; Terranova et al., 1980; Vlodavsky and Gospodarowicz, 1981], heparan sulfate provides a filtration barrier for negatively charged macromolecules [Farquhar, 1981], and type IV collagen serves a structural role [Engel et al., 1981]. The elucidation of the detailed interactions between cells and the basement membrane or extracellular matrix has been hampered by the unavailability of good model systems. This problem has been partially overcome by several recently described preparations. Gospodarowicz et al. [1980] have obtained an extracellular matrix preparation from cultured corneal endothelial

Department of Biological Sciences, University of Pittsburgh, Pittsburgh, Pennsylvania 15260

cells that promotes cell proliferation; similarly, Rojkind et al. [1980] have described a connective tissue biomatrix that supports the long-term culture of normal rat hepatocytes. Liotta et al. [1980] have utilized intact basement membranes from human palcenta to explore the metastatic potential of tumor cells, and Cammarata and Spiro [1982] have isolated bovine lens capsule basement membranes to study the attachment and interaction of lens epithelial cells. The general utility of these matrices is somewhat compromised because it is difficult to prepare them. In an effort to provide a useful and readily available extracellular matrix system, we have developed a matrix-coated bead that supports cell adhesion and spreading. The characterization of these beads and examples of their application are presented in this chapter.

MATERIALS AND METHODS
Cells and Cell Culture

African green monkey kidney BSC-40 cells were obtained from Dr. James Pipas, Department of Biological Sciences, University of Pittsburgh. Human mammary tumor MCF-7 cells were obtained from Dr. David Barnes, Department of Biological Sciences, University of Pittsburgh. Rat hepatoma H-4-II-E, ATCC CRL 1548, and normal rat liver clone 9, ATCC CRL 1439, were donated by Dr. Peter Ove, Department of Anatomy and Cell Biology, University of Pittsburgh Medical School. Mouse fibroblasts L929 were obtained from the Institute for Medical Research, Camden, NJ. The mouse parietal endoderm M1536-B3 was isolated by Chung et al. as previously described [Chung et al., 1977a]. BSC-40 cells were grown on Eagle's minimal essential medium (MEM); MCF-7, L929, rat hepatoma, and M1536-B3 cells were grown on Dulbecco's modification of Eagle's medium (DME); and normal rat liver cells were grown on Ham's F10 medium. All media were supplemented with 10% fetal calf serum for routine culture. Cell cultures were grown in humidified 10% CO_2/air at 37°C in Lux tissue culture dishes. Culture medium and fetal calf serum were obtained from Gibco, Grand Island, NY. Cells were routinely transferred on 2- to 3-day schedule after trypsinization. Serum-free growth medium for attachment and growth experiments on matrix-coated beads was prepared according to Murakami et al. [1982].

Preparation of Extracellular Matrix-Coated Beads

Cytodex 2 beads (Pharmacia Fine Chemicals, Piscataway, NJ), $(0.5-1) \times 10^5$, were placed in a 15-mm $\times$ 100-mm plastic petri dish with 2×10^6 M1536-B3 cells at a cell density of 5×10^4 cells/1 ml of DME containing 10% fetal calf serum. The suspension was incubated for 5 days, during which time the beads became covered with a monolayer of cells. The cell-coated beads were harvested by centrifugation at 600 rpm in a Beckman J-6B centrifuge for 2 min. The supernatant medium was removed and the beads were suspended in 2 ml of phosphate-buffered saline (PBS) supplemented with Ca^{2+} and Mg^{2+} at 100 μg/ml and containing 10 μg of cytochalasin B/ 1 ml. After 2 h of incubation at 37°C with gentle shaking, the cells were completely dislodged by agitation for 30 sec on a vortex mixer. The matrix-coated beads were separated from the dislodged cells with repeated washings by sedimentation under gravity. The progress of cell removal was monitored by phase contrast microscopy. The matrix coating of the bead was solubilized by boiling the coated beads for 2 min in sample buffer containing 10% vol./ vol. glycerol; 2% wt./vol. sodium dodecyl sulfate; 5% vol./vol. 2-mercap-toethanol in 0.0625 M Tris-HCl, pH 6.8. Aliquots of the supernatant solution obtained after centrifugation for 2 min in an Eppendorf microfuge were applied to a 5–20% gradient polyacrylamide gel slab and subjected to electro-phoresis [Laemmli, 1970].

Scanning Electron Microscopy

Matrix-coated and control uncoated cytodex 2 beads were placed on squares of nitrocellulose paper (Millipore Corp., Bedford, MA) and air-dried. The samples were then coated with gold with a Polaron E5100 sputter coater and examined in an AMR Model 1000 scanning electron microscope.

The morphology of cells attached to matrix-coated and uncoated beads was also examined by scanning electron microscopy. Squares of nitrocellulose paper were placed into 35-mm-diameter Lux tissue culture dishes together with either coated or uncoated beads. To each dish 2.5×10^5 cells in 2 ml of growth medium were added. After the appropriate incubation times the nitrocellulose paper with attached beads and cells was removed. The cells were fixed with 3% glutaraldehyde in 0.05 M sodium cacodylate buffer, pH 7.4, for 1–2 h. The specimens were dehydrated in a graded series of ethanol. Finally, samples were subjected to critical-point drying in a Polaron drying apparatus, coated with gold, and examined by scanning electron microscopy.

Growth of Cells on Matrix-Coated and Uncoated Beads

Petri dishes, 35 mm in diameter, were coated with a 1% wt./vol. solution of bovine serum albumin in PBS and rinsed once with PBS. The dishes were seeded with $(1.2–2) \times 10^5$ cells and 1.5×10^5 coated or uncoated beads in 2 ml of serum-free medium that contained 1.0 μCi of ^{3}H-thymidine, 20 Ci/ mmole (New England Nuclear, Boston). The dishes were incubated at 37°C and duplicate dishes removed at 24-h intervals. The medium containing the beads and cells was poured into a small glass liquid scintillation vial. The beads and cells were harvested and washed five times with PBS by centrifugation. To each vial 0.2 ml of Protosol and 5 ml of Formula 963 liquid scintillation fluid (New England Nuclear) were added and the incorporation of radioactivity was determined.

Cell Attachment Assay

Cells were labeled with ^{3}H-thymidine, 1 μCi/ml medium, for 16 h. The labeled medium was removed and the cells were incubated for an additional 2 h in unlabeled medium. In those experiments in which protein synthesis was blocked, the unlabeled medium contained 25–50 μg of cycloheximide. The labeled cells were detached by trypsinization; and 3×10^5 cells together with 1.5×10^5 coated or uncoated beads in 2 ml serum-free medium were added to a 35-mm albumin-coated petri dish. The dishes were incubated at 37°C, and duplicate or triplicate samples were removed at 1-h intervals. The culture medium containing beads and unattached cells was poured into a 6-ml glass scintillation vial, the dish was rinsed three times with PBS, and the rinses were combined. The glass vial was agitated vigorously for 30 sec on a vortex mixer to dislodge loosely attached cells. Upon standing for 2–3 min to allow the beads to sediment, the supernatant solution containing free cells was transferred to a second vial by careful pipetting. The beads were resuspended in PBS and the procedure was repeated. In order to prevent the transfer of beads during removal of the supernatant solution, a small cone of lens paper was attached to the moistened tip of the pipet. The free cells passed through the cone but the beads remained on the outer surface of the cone. The cone with attached beads was left in the scintillation vial for counting. The cells in the supernatant fluid were collected by centrifugation. The radioactivity associated with the beads and the free cells was determined as described above. Cell attachment, converted to percentage, was calculated from the radioactivity associated with the beads divided by the total radioactivity in the free and bead-associated cells.

Inhibition of Cell Attachment by Monoclonal Antibodies

Hybridoma cells Lam-I, Lam-II, and Lam-IV that produce monoclonal antibodies against the GP-2 subunit of laminin [Chung et al., 1983] were grown in serum-free medium for 2 days at an initial density of 5×10^6 cells per 15 ml medium. The growth media were obtained by centrifugation and the antibodies precipitated at 50% saturation with ammonium sulfate. The precipitated antibody was redissolved in PBS and dialyzed against three 1,000-ml changes of of PBS. Petri dishes, 35 mm in diameter, were seeded with 2×10^5 ^{3}H-thymidine-labeled cells and 1.5×10^5 matrix-coated beads in 2 ml of serum-free medium. Monoclonal antibodies, 1.4–1.6 mg protein, were added and the dishes were incubated for 4 h. The beads with attached cells were separated from the free cells as described previously. The radioactivity in each fraction was determined, and the extent of the inhibition of cell attachment was calculated from the data.

RESULTS

Characterization of Matrix-Coated Beads

The extracellular matrix associated with the cytodex beads after removal of M1536-B3 cells consisted predominantly of laminin (GP-1 and GP-2 subunits), entactin A and entactin B, and variable quantities of a protein of approximately 100 kilodaltons. The polypeptide components of the matrix-coated beads and extracellular matrix isolated from spherical aggregates of M1536-B3 cells are shown in Figure 1. As can also be seen in this figure the

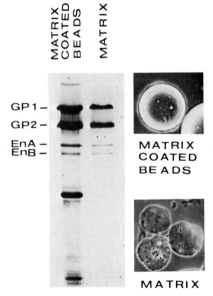

Fig. 1. A comparison of matrix-coated beads and the extracellular matrix sac elaborated by M1536-B3 cells. Phase contrast micrographs of matrix-coated bead and matrix sacs are shown on the right. The bead and sacs are approximately 100 μ in diameter. The electrophoretic patterns of the matrix on the beads and matrix isolated from M1536-B3 cell aggregates on polyacrylamide gels are shown on the left. EnA and EnB represent entactin A and entactin B.

coated beads are devoid of cells. In addition to the major polypeptides smaller quantities of polypeptides of lower molecular weights may also be seen. Metabolic labeling of M1536-B3 cells on cytodex beads with ^{35}S-methionine revealed that the label coincided with the laminin and entactin bands. No label was present in the 100-kilodalton band, which suggested that it was derived from the serum in the growth medium. Matrix-coated beads were compared with uncoated beads by scanning electron microscopy and, as displayed in Figure 2, an organized cell-free matrix covered the beads derived from M1536-B3 cells.

Attachment of Cells to Matrix-Coated Beads

Several cell lines were used to examine the influence of the extracellular matrix on their attachment. The cell lines exhibited different responses, as shown in Figure 3. Eighty percent of added African green monkey kidney BSC-40 cells and normal rat liver clone 9 (CL9) cells attached within 6 h of incubation in serum-free medium to matrix-coated beads (panels A and B). In contrast to the matrix-coated beads only 20% of BSC-40 cells attached to the uncoated beads. The CL9 cells, however, did not discriminate between coated and uncoated beads. The initial rate of attachment of L929, MCF-7, and H-4-II-E was greater on uncoated than on coated beads (panels C, D, E). For each of these cell lines, after 2 h of incubation approximately twice as many cells had attached to the uncoated beads than to the coated beads. It might be noted that these three cell lines do not display contact inhibition when grown on plastic in contrast to the BSC-40 and CL9 lines. The mechanisms of attachment differed for the cells as demonstrated by studies in which cycloheximide was added to inhibit protein synthesis. In all cases the attachment of the cells to uncoated beads was markedly inhibited (Fig. 3A-D); the most striking example was CL9, where cycloheximide was approximately 70% inhibitory at 5 h of incubation. The effect of cycloheximide on the attachment of BSC-40 cells was to depress even further the relatively low number of cells attached to uncoated beads. The attachment of cells to uncoated beads therefore appears to be dependent on the synthesis of new protein molecules. Control experiments had indicated that protein synthesis as determined by incorporation of ^3H-leucine was inhibited more than 90% at the concentrations of cycloheximide employed. Examination of the response of the cells to cycloheximide when attachment to coated beads was studied revealed that the cells could be divided into two groups. The first group, which consisted of BSC-40 and CL9 cells, were only moderately affected by cycloheximide, as shown in Figure 3A,B. The attachment of the second group, L929, MCF-7, and H-4-II-E, was more severely affected; e.g.,

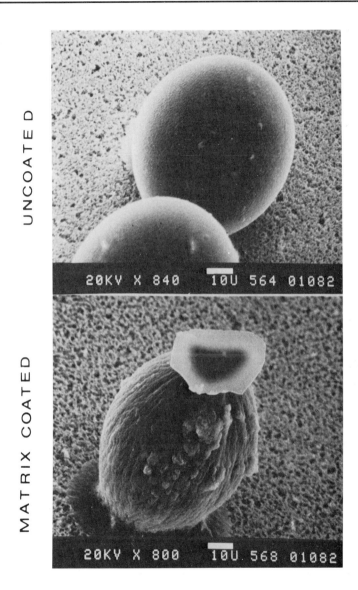

Fig. 2. Fine structure of matrix-coated and uncoated beads by scanning electron microscopy. The upper panel shows the smooth surface of the uncoated bead and the lower panel the rough surface of an extracellular matrix-coated bead. BSC-40 (A), CL9 (B), L929 (C), MCF-7 (D), and H-4-II-E (E).

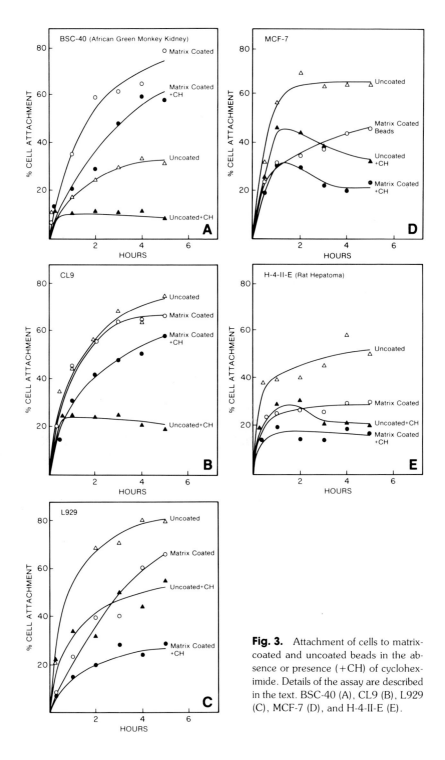

Fig. 3. Attachment of cells to matrix-coated and uncoated beads in the absence or presence (+CH) of cycloheximide. Details of the assay are described in the text. BSC-40 (A), CL9 (B), L929 (C), MCF-7 (D), and H-4-II-E (E).

L929 attachment decreased from 65% to approximately 25% at 5 h. These observations indicate that the mechanism of attachment of cells to the matrix is complex and that for BSC-40 and CL9 cells new protein synthesis was not obligatory.

Morphology of Cells Attached to Coated and Uncoated Beads

The extracellular matrix on the beads had marked effects on the morphology of most of the lines examined. The cells that attached to the matrix were flattened and tightly adherent. In Figure 4 it may be seen that mouse fibroblast L929 cells, when examined by phase contrast microscopy, had spread on the matrix-coated beads and assumed a spindle-like shape (panel B), whereas cells attached to uncoated beads remained rounded after 24 h (panel A). In contrast to the L929 cells, CL9 attached and spread equally well on uncoated (panel C) or matrix-coated (panel D) beads. The attachment of BSC-40 and

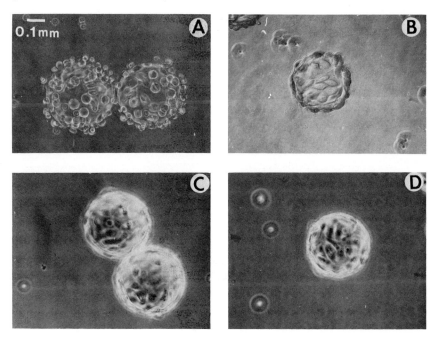

Fig. 4. Morphology of cells grown on matrix-coated and uncoated beads examined by phase contrast microscopy. Cells were examined after 24 h incubation. L929 cells remain rounded on uncoated beads (A), whereas on the matrix-coated beads the cells are flattened and spindle-shaped (B). On the other hand, clone 9 cells attach equally well to uncoated beads (C) and coated beads (D).

H-4-II-E hepatoma cells was examined by scanning electron microscopy. These results are shown in Figure 5. In panel A, BSC-40 cells remain rounded after 24 h of incubation with uncoated beads; this was in sharp contrast to the flattened, tightly adherent cells on the matrix-coated beads (panel B), where the cells appeared to be embedded in the matrix. As noted previously, rat hepatoma cells attached more rapidly to uncoated beads and, as shown in panel C, the cells are flattened with one or more processes extending from

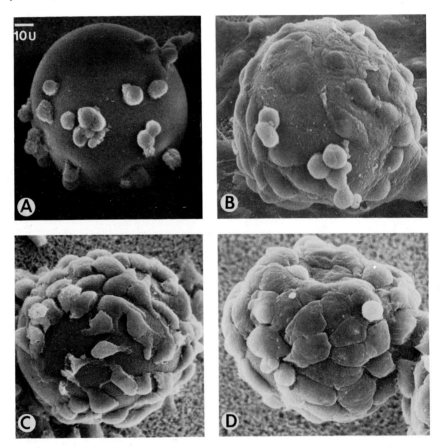

Fig. 5.   Morphology of cells grown on matrix-coated or uncoated beads examined by scanning electron microscopy. Cells were incubated for 24 h. BSC-40 cells attach to the uncoated bead but remain rounded (A); on the matrix-coated beads the cells attach, spread, and become imbedded in the matrix (B). H-4-II-E cells attach and spread on uncoated (C) and matrix-coated beads (D). The H-4-II-E cells appear to be flatter and to have fewer processes on the matrix-coated bead.

the cell bodies. The cells, however, do not appear to exhibit the pavement-like morphology with intimate cell contact typical of epithelium. On the coated beads (panel D) a more normal epithelial morphology is apparent. The cells are in close contact with fewer processes and appear to be more flattened. Examination of MCF-7 cells also revealed that the matrix promoted spreading of these cells. The influence of the extracellular matrix on the behavior of MCF-7 cells over longer periods of cell culture was interesting. On matrix-coated beads, MCF-7 cells formed a monolayer of closely packed cells that persisted up to 12 days in culture. A typical cell-coated bead is shown in Figure 6B. The results obtained with uncoated beads were in sharp contrast. After initial attachment and growth on uncoated beads, the cells retracted and formed tight cellular aggregates as shown in panel A. The matrix thus influenced the morphology and behavior of these cells in a manner that was independent of the initial rate of attachment. The MCF-7 cells elaborate a matrix themselves on uncoated beads, as shown in panel C. The nature and elationship of this matrix to cell attachment are not known at present.

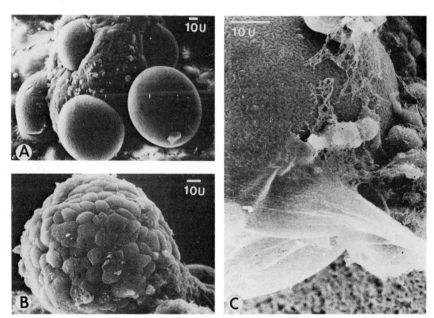

Fig. 6. Scanning electron micrographs of MCF-7 cells grown on matrix-coated and uncoated beads. Cells were incubated for 5 days and then processed for microscopy. The cells in panel A have retracted from the uncoated beads to form a large mass. On the matrix-coated beads, in panel B, the cells form a monolayer. In panel C, it can be seen that the MCF-7 cells deposit a matrix of unknown composition on the uncoated bead and around the cells themselves.

Influence of Matrix on Cell Proliferation

The rate of incorporation of ^{3}H-thymidine into cells on matrix-coated and uncoated beads was determined. It is apparent from Figure 7 that the incorporation of label was markedly stimulated by the matrix for the four cell lines tested. The results were most dramatic for BSC-40 (panel A) and CL9 (panel C) cells. It is noteworthy that the initial attachment of CL9 on uncoated and coated beads was similar, yet the matrix stimulated cell proliferation as determined by ^{3}H-thymidine incorporation. The stimulation of proliferation of the rat hepatoma and L929 cells was less dramatic.

Inhibition of Cell Attachment by Antilaminin Monoclonal Antibodies

One of the major components of the extracellular matrix was shown to be laminin (Fig. 1), and the cell line most dependent on the matrix for attachment, spreading, and proliferation was BSC-40. Experiments were carried out to determine if rat monoclonal antibodies directed against the GP-2 component of laminin influenced cell attachment. The data obtained are summarized in Table I. It is apparent that all three monoclonal antibodies inhibited cell attachment. The attachment of the cells is probably quite complex; however, the results suggest that laminin is partially responsible for the interaction of the matrix with BSC-40 cells.

DISCUSSION

One of the objectives of this study was to develop a reproducible, readily available, extracellular matrix system that would be of utility in exploring cell-extracellular matrix interactions. Furthermore, the biochemical characterization of the matrix components would provide a rational basis for exploring the molecular interactions between the cell surface and the matrix. A simple technique has been described that results in the isolation of large quantities of matrix-coated cytodex 2 beads. The technique involves the growth of mouse endodermal M1536-B3 cells in suspension culture on cytodex 2 beads and removal of the cells by treatment with cytochalasin B. The resulting beads are coated with a network of laminin, entactin, and lesser quantities of other components. The advantage of this matrix is that its native organization as elaborated by the M1536-B3 cells is unperturbed by harsh extraction proce-

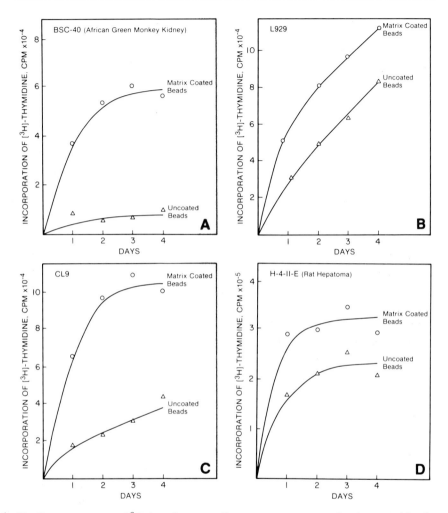

Fig. 7. Incorporation of ^{3}H-thymidine into cells grown on matrix-coated and uncoated beads. Details of the experiment are described in the text. BSC-40 (A), L929 (B), CL9 (C), and H-4-II-E (D).

dures. It is therefore expected that cell-matrix interactions will more closely mimic physiologic conditions. A series of experiments have been carried out that illustrate the utility of the system in studies on cell attachment and cell growth.

TABLE I. Effect of Antilaminin Monoclonal Antibodies on the Attachment of BSC-40 Cells to Matrix-Coated Beads

Antibody	Attachment (%)
None	100
Lam-I	46
Lam-II	57
Lam-IV	35

Several cell lines were selected to explore the function of the matrix on cell attachment and morphology. The results, not surprisingly, reconfirm the complexity of cell surface-matrix interactions. In examining the kinetics of cell attachment, it became apparent that some cell lines (for example, BSC-40 cells) were strongly dependent on the matrix for attachment, but other cells such as rat hepatoma did not show a requirement for initial attachment. The other cells spanned the spectrum of dependence on the matrix. A second interesting feature of the cell-matrix interaction was that in the attachment process, BSC-40 and normal rat liver CL9 cells did not require new protein synthesis; however, L929 mouse fibroblasts were strongly dependent on new protein synthesis. These observations indicate that the former cells contain stable receptors or attachment factors that interact with the matrix, whereas in L929 cells the opposite was true. It was of interest that both BSC-40 and CL9, which were strongly contact-inhibited when grown on plastic, exhibited the greatest affinity for matrix-coated beads. The attachment of the five cell lines to uncoated cytodex beads revealed a strong dependence on new synthesis of proteins, since the presence of cycloheximide markedly decreased the rate of cell attachment. These results clearly indicate that attachment to matrix-coated beads proceeds by mechanisms that are different from attachment to uncoated beads.

In each of the five cell lines examined, attachment to the matrix-coated beads resulted in cell spreading. The most dramatic example of this was observed with BSC-40 cells, where the attachment and spreading could be observed within an hour after incubation with the coated beads, whereas the cells remained rounded even after overnight incubation on uncoated beads. It was curious that normal rat liver CL9 cells attached and flattened equally well on coated and uncoated beads. Since these cells are not in reality normal, this dual affinity probably reflects an intermediate level of surface modification between the truly normal state and a transformed state. The matrix affected not only the behavior of cells during relatively short periods of incubation, but appeared to influence the long-term behavior of some cells, as illustrated by the mammary tumor MCF-7 cell line. These cells attached more rapidly to uncoated beads than to coated beads; however, after long-term incubation on uncoated beads the cells exhibited a greater affinity for adjacent cells than for the cytodex 2 bead surface. On the other hand, these cells attached to matrix-coated beads, proliferated, and remained attached to these beads as monolayers even after 12 days of culture. It is unclear at present how the matrix effects this dramatic alteration in cell behavior.

In addition to modifying the morphology of several of the cell lines studied, the matrix also stimulated the proliferation of BSC-40, CL9, and rat hepatoma when compared with uncoated beads. This proliferation is apparently not readily correlated with the initial attachment and spreading of the cells, since CL9 and hepatoma cells attached rapidly and spread on uncoated beads. It could be speculated that the proper cell surface-matrix contacts are required for efficient cell division and growth, although the nature of these contacts needs to be explored.

The role of laminin in cell attachment has been previously explored [Couchman et al., 1983; Terranova et al., 1980], and experiments were carried out to determine if laminin could be involved in the attachment of BSC-40 cells. By using indirect immunofluorescent antibody techniques it was determined that BSC-40 cells did not synthesize laminin or entactin. Treatment of the matrix-coated beads with three independent monoclonal antibodies revealed that attachment of BSC-40 cells was markedly inhibited. These observations suggest that laminin was at least partially involved in the attachment process. The laminin antibodies have been shown to be directed against the GP-2 or 220,000-dalton subunit of laminin.

In summary, the isolation of an extracellular matrix-coated bead has been described. Experiments on the influence of the matrix on cell attachment, spreading, and proliferation reveal heterogeneity or specificity of responses with different cell lines. It is hoped that the matrix-coated bead will aid a) in elucidating the mechanisms of cell attachment, b) in the isolation and growth of primary cell cultures, c) in maintaining differentiated cell functions, and d) in understanding problems of cell transformation and metastasis. The isolation procedure for the matrix-coated bead from M1536-B3 cells may also be applicable to other cells that elaborate extracellular matrices. This would permit the examination of the specificities of cell and matrix interactions not previously possible.

ACKNOWLEDGMENTS

This research was supported by grants CA21246 and GM25690 from the National Institutes of Health.

REFERENCES

Bender BL, Jaffe R, Carlin B, Chung AE (1981): Immunolocalization of entactin, a sulfated basement membrane component, in rodent tissues: A comparison with GP-2. Am J Pathol 103:419-426.

Bernfield MR, Banerjee SD, Cohn RH (1972): Dependence of salivary epithelial morphology and branching morphogenesis upon acid mucopolysaccharide-protein (proteoglycan) at the epithelial surface. J Cell Biol 52:674-689.

Bornstein P, Sage H (1980): Structurally distinct collagen types. Annu Rev Biochem 49:957-1003.

Cammarata PR, Spiro RG (1982): Lens epithelial cell adhesion to lens capsule: A model system for cell-basement membrane interaction. J Cell Physiol 113:273-280.

Carlin B, Jaffe R, Bender B, Chung AE (1981): Entactin, a novel basal lamina-associated sulfated glycoprotein. J Biol Chem 256:5209-5214.

Carlsson R, Engvall E, Freeman A, Ruoslahti E (1981): Laminin and fibronectin in cell adhesion: Enhanced adhesion of cells from regenerating liver to laminin. Proc Natl Acad Sci USA 78:2403-2406.

Chung AE, Estes LE, Shinozuka H, Braginski J, Lorz C, Chung CA (1977a): Morphological and biochemical observations on cells derived from the in vitro differentiation of the embryonal carcinoma cell line PCC4-F. Cancer Res 37:2072-2081.

Chung AE, Freeman IL, Braginski JE (1977b): A novel extracellular membrane elaborated by a mouse embryonal carcinoma-derived cell line. Biochem Biophys Res Commun 79:859-868.

Chung AE, Jaffe R, Freeman IL, Vergnes JP, Braginski JE, Carlin B (1981): Properties of a basement membrane-related glycoprotein synthesized in culture by a mouse embryonal carcinoma-derived cell line. Cell 16:227-287.

Chung AE, Jaffe R, Bender B, Lewis M, Durkin M (1983): Monoclonal antibodies against the GP-2 subunit of laminin. Lab Invest 49:576-581.

Couchman JR, Hook M, Rees DA, Timpl R (1983): Adhesion, growth, and matrix production by fibroblasts on laminin substrates. J Cell Biol 96:177-183.

Engel J, Odermatt E, Engel A, Madri JA, Furthmayr H, Rohde H, Timpl R (1981): Shapes, domain organizations and flexibility of laminin and fibronectin, two multifunctional proteins of the extracellular matrix. J Mol Biol 150:97-120.

Farquhar MG (1981): The glomerular basement membrane: A selective macromolecular filter. In Hay ED (ed): "Cell Biology of Extracellular Matrix." New York/London: Plenum, pp 335-378.

Foidart J-M, Bere EW, Yaar M, Rennard SI, Guillino M, Martin GR, Katz SI (1980): Distribution and immunoelectron microscope localization of laminin, a noncollagenous basement membrane glycoprotein. Lab Invest 42:336-343.

Gospodarowicz D, Delgado D, Vlodavsky I (1980): Permissive effect of the extracellular matrix on cell proliferation in vitro. Proc Natl Acad Sci USA 77:4094-4098.

Grobstein C (1953): Morphogenetic interaction between embryonic mouse tissues separated by a membrane filter. Nature 172:869-871.

Hassall JR, Robey PG, Barrach HJ, Wilchek J, Rennard SI, Martin GR (1980): Isolation of heparin sulfate-containing proteoglycan from basement membrane. Proc Natl Acad Sci 77:4494-4498.

Hay ED (1978): Role of basement membranes in development and differentiation. In Kefalides NA (ed): "Biology and Chemistry of Basement Membranes." New York/London: Academic, pp 119-136.

Hogan BLM, Cooper AR, Kurkinen M (1980): Incorporation into Reichert's membrane of laminin-like extracellular proteins synthesized by parietal endoderm cells of the mouse embryo. Dev Biol 80:289-300.

Hogan BLM, Taylor A, Kurkinen M, Couchman JR (1982): Synthesis and localization of two sulfated glycoproteins associated with basement membranes and the extracellular matrix. J Cell Biol 95:197.

Johansson S, Kjellen L, Hook M, Timpl R (1981): Substrate adhesion of rat hepatocytes: A comparison of laminin and fibronectin as attachment proteins. J Cell Biol 90:260-264.

Kanwar YS, Farquhar MG (1979): Presence of heparin sulfate in the glomerular basement membrane. Proc Natl Acad Sci 76:1303-1307.

Kleinman HK, Klebe RJ, Martin, GR (1981): Role of collageneous matrices in adhesion and growth of cells. J Cell Biol 88:473-485.

Laemmli UK (1970): Cleavage of structural proteins during assembly of the head of bacteriophage T4. Nature (Lond) 227:680-685.

Liotta LA, Lee CW, Morakis DJ (1980): New method for preparing large surfaces of intact basement membrane for tumor invasion studies. Cancer Lett 11:141-152.

Madri JA, Roll, FJ, Furthmayr M, Foidart J-M (1980): Ultrastructural localization of fibronectin and laminin in the basement membranes of the murine kidney. J Cell Biol 86:682-687.

Murakami H, Samui H, Sato GH, Sueoka N, Chow TP, Kano-Sueko T (1982): Growth of hybridoma cells in serum free medium: Ethanolamine is an essential component. Proc Natl Acad Sci USA 79:1158-1162.

Rojking M, Gatmaitan Z, Mackensen S, Giambrone MA, Ponce P, Reid LM (1980): Connective tissue biomatrix: Its isolation and utilization for long-term cultures of normal rat hepatocytes. J Cell Biol 87:255-263.

Terranova VP, Rohrbach DH, Martin GR (1980): Role of laminin in the attachment of PAM 212 (epithelial) cells to basement membrane collagen. Cell 22:719-726.

Thesleff I, Lehtonen E, Saxen L (1978): Basement membrane formation in transfilter tooth culture and its relation to odontoblast differentiation. Differentiation 10:71-79.

Timpl R, Rohde H, Robey PG, Rennard SI, Foidart J-M, Martin GR (1979): Laminin—a glycoprotein from basement proteins. J Biol Chem 254:9933-9937.

Vlodavsky I, Gospodarowicz D (1981): Respective roles of laminin and fibronectin in adhesion of human carcinoma and sarcoma cells. Nature 289:304-306.

Wicha MS, Lowrie G, Kohn E, Bagavandoss P, Mahn T (1982): Extracellular matrix promotes mammary epithelial growth and differentiation in vitro. Proc Natl Acad Sci USA 79:3213-3217.

Index

Acid-ethanol extraction, transforming growth factors (β type), 185-187
Acid-urea PAGE, 136
Affinity chromatography
  EGF, purification from human urine, 153-157
    preparation, 153, 155
    rabbit antibody, 155-156
  gelatin-Sepharose, fibronectin, 224-225
  see also Chromatography
A431 cells. See under Basement membrane synthesis and turnover in serum-free medium
Agar, soft. See Soft agar
Amino acids and media preparation, 29-30, 44-45, 49
AM77/B autoclavable medium 41, 47-54; see also under Media preparation procedures, laboratory
Anchorage
  dependence, substrata, 209
  independence, transforming growth factors (β type), 182
Antibiotics in media, 38-39; see also specific antibiotics
Anti-epibolin antibody, rabbit, 273
Antisera, BM synthesis and turnover in serum-free medium, 298-299
Arginase in FBS, 25

Arginine, carboxy terminal, lack of, EGF purification from guinea pig prostate, 144
Attachment, cell
  BM synthesis and turnover in serum-free medium, 296, 300
  and ECM coated bead. See ECM-coated beads, cell attachment and spreading
Autocrine, cf. endocrine and paracrine, 34

Basal nutrient media, formulation methods, 3-18
  definition and importance, 3-5
  optimized, 5-9
    advantages, 6-9
    defined and undefined supplements, 6
    fibroblast cf. epithelial cells, 8-9
    history, 5-6
  optimized, procedures for developing, 7-17
    first-limiting-factor theory, 13-15
    growth factors, 13, 17
    hormones, 13
    initiating growth of desired cell type, 10
    lipids, 16

339